AF361700

Soil Nitrogen Supply: Linking Plant Available N to Ecosystems Functions and Productivity

Soil Nitrogen Supply: Linking Plant Available N to Ecosystems Functions and Productivity

Editor

Jacynthe Dessureault-Rompré

MDPI • Basel • Beijing • Wuhan • Barcelona • Belgrade • Manchester • Tokyo • Cluj • Tianjin

Editor
Jacynthe Dessureault-Rompré
Laval University
Canada

Editorial Office
MDPI
St. Alban-Anlage 66
4052 Basel, Switzerland

This is a reprint of articles from the Special Issue published online in the open access journal *Nitrogen* (ISSN 2504-3129) (available at: https://www.mdpi.com/journal/nitrogen/special_issues/n_supply).

For citation purposes, cite each article independently as indicated on the article page online and as indicated below:

LastName, A.A.; LastName, B.B.; LastName, C.C. Article Title. *Journal Name* **Year**, *Volume Number*, Page Range.

ISBN 978-3-0365-5081-7 (Hbk)
ISBN 978-3-0365-5082-4 (PDF)

Cover image courtesy of Jacynthe Dessureault-Rompré

Contents

About the Editor

Jacynthe Dessureault-Rompré

Jacynthe Dessureault-Rompré is currently a professor in soil conservation and health at Laval University. She completed her bachelor's degree in agronomy and her master's degree in soil chemistry at Laval University (Canada), and obtained a PhD degree from the ETH Zurich in Switzerland, working specifically on the soil–plant relationship. She has completed several years of post-doc, studying nitrogen cycling in soil in collaboration with Agriculture Canada and Dalhousie University (Canada). She specialized in the conservation and restoration of cultivated peatland from 2014, her work being focused on carbon equilibrium through biomass production and soil amendment. Her scientific interests include soil conservation in general and that of cultivated peatland in particular, the soil–plant relationship and climate change, the optimization of the N and C cycles in soils for sustainable agriculture, and the regeneration of degraded soils using phytotechnology.

 *Nitrogen*

MDPI

Editorial

Soil Nitrogen Supply: Linking Plant Available N to Ecosystem Functions and Productivity

Jacynthe Dessureault-Rompré

Soil and Agri-Food Engineering Department, Laval University, Quebec, QC G1V 0A6, Canada; Jacynthe.dessureault-rompre@fsaa.ulaval.ca

Citation: Dessureault-Rompré, J. Soil Nitrogen Supply: Linking Plant Available N to Ecosystem Functions and Productivity. *Nitrogen* **2022**, *3*, 455–457. https://doi.org/10.3390/nitrogen3030030

Received: 22 July 2022
Accepted: 6 August 2022
Published: 10 August 2022

Nitrogen (N), a common chemical element in the atmosphere (78% of our atmosphere) yet less common within the Earth's crust (less than 2%), is a crucial nutrient for life. It is an essential constituent of many cells, such as amino acids, proteins, chlorophyll, and even DNA, and is involved in various processes, such as photosynthesis, energy transfer, growth and reproduction. Up to 90% of the N in surface soils is organic in nature and is contained in soil organic matter (SOM). N cycling is complex, with numerous interacting controlling factors. Scientists around the world have been working for many decades now to understand, model and predict the soil's capacity to supply N to plants in different ecosystems, whether agricultural, forest, grassland or urban. Plant available nitrogen (PAN) is a pillar of the functioning and productivity of any ecosystem.

There is no exact definition for PAN. Sometimes considered as a pool of inorganic N, sometimes referred to as an amount related to plant uptake during a growing season, PAN is now generally viewed as a continuum of soluble inorganic and organic N, with the latter composed of molecules of varied structural complexity, which cycles continuously in a dynamic way. The transient nature of available forms of N has led them to be defined in many models as pools and rates, and they continue to be the focus of numerous research studies.

Efforts to gain a better understanding of the soil N supply and PAN are oriented towards optimizing N cycling. In this context, optimization is related to improving the synchronicity between the soil N supply and plant demand and reducing N losses to the atmosphere or water, which can have environmental and economic consequences. Improving the productivity of ecosystems necessarily involves optimizing the N cycle. Higher productivity implies that the many ecosystem-related functions will also improve, notably biomass production and carbon sequestration.

Developing models that can more accurately predict N mineralization will contribute to optimizing N cycling. In this regard, Morvan et al. [1] present an original experimental design aimed at quantifying and modeling the net mineralization of organic nitrogen in the field using generalized additive models. The innovative aspect of their model is the inclusion of a cropping system indicator that increases the accuracy of the mineralization predictions. Although the final accuracy of the model remained relatively moderate, the results evidenced that N mineralization cannot be predicted from a single test but instead requires a combination of site-specific information, including land use, basic soil properties, and chemical and biological indicators. New knowledge based on machine learning algorithms might be the next step towards improving soil N supply predictions and optimizing N cycling.

N cycling, supply and availability to plants are all influenced by management practices, such as tillage, cover crops, crop residues and soil amendment. In a laboratory lysimeter experiment comparing the effect of autumn inversion tillage (AuT) against no-till (NT) on the reduction of N loss through leaching, with volunteer winter rye as a cover crop, Miranda-Vélez and Vogeler [2] observed that the presence of the cover crop significantly reduced leaching. They thus propose that fall tillage be avoided in order to take advantage of the cover crop effect.

In addition to protecting against soil erosion and nutrient leaching, cover crops have the potential to improve soil health. In the particular context of a barley cash crop for malt production, Siller et al. [3] investigated how the use of summer cover crops, namely sunn hemp and crimson clover, alone or in a mixture, might improve barley crop production the following winter. Their results showed an increase in nitrate content under sunn hemp as compared to no cover crop. Although no improvements in the quality parameters of the barley crop were observed, the authors concluded that barley can be successfully integrated into crop rotations with leguminous plants without negative impacts on barley growth, yield or grain quality.

The type of crop residue and its management can also have a significant impact on N mineralization processes and therefore on PAN. In an interesting incubation experiment, Alghamdi et al. [4] investigated changes in N mineralization under successive surface applications of crop residues (simulating the no-till system). They observed that in the long term, corn, wheat and soybean crop residues reduced available N as compared to pea or forage radish residues. Although their study did not take into account the contribution of the root system to PAN, the results showed that the long-term accumulation of certain types of crop residues might negatively impact the soil's capacity to supply N to plants. For their part, Tanjila et al. [5] observed in a laboratory incubation experiment that soil respiration was higher under decomposing high-N corn residue and that the effect was exacerbated by fertilization. N and C cycling are intimately linked, and excessive N fertilization could potentially lead to soil C depletion, even in no-till systems.

Organic soil amendments are known to have a positive impact on many soil properties, including N pools and cycling. In a two-year field experiment, Omara et al. [6] showed that biochar applied in combination with inorganic N can improve N availability and potentially increase crop N uptake in coarse-textured soils. Furthermore, Dessureault-Rompré et al. [7] showed that cultivated peatland amended with miscanthus straw or willow wood chips contained less available N, offering opportunities to improve N sequestration in these highly mineralizing soils that are prone to N loss through leaching and denitrification. Phillips et al. [8] investigated N loss by runoff and N-use efficiency using a combination of inorganic N and organic soil amendment. They observed that the inorganic-organic amendment combination doubled N-use efficiency as compared to mineral fertilizer alone and significantly reduced N runoff, allowing for better synchronization between N availability and plant uptake.

Numerous studies have focused on the use of urease or nitrification inhibitors to reduce N losses through volatilization, denitrification or leaching. Lasisi and Akinremi [9] showed that urease and nitrification inhibitors can be combined without impairing their individual effects and that soil pH is an important factor for the persistence and efficacy of urease inhibitors. Furthermore, Guo et al. [10] showed that the efficacy of nitrification inhibitors is highly dependent on soil texture, but when the right combination is used, these inhibitors can reduce N_2O loss by up to 88% and CO_2 loss by up to 73%.

N supply and availability are a concern in urban landscapes as well. Bukomba and Lusk [11] investigated the small-scale variability of soil nitrogen (N) properties in a single urban landscape featuring distinctly different patches or types of cover. They found that N mineralization varied widely over just a few meters. Future studies should focus on the mechanisms that act on the soil N supply and PAN in such landscapes, which are commonly fertilized and irrigated, and are at risk of N loss.

Besides soil, climate and management, plant root traits and functions have a straightforward, often neglected, impact on soil N cycling. Because root traits and functions vary between crops, and between cultivars of a same crop, their influence on N cycling, including N capture, also varies. Kupcsik et al. [12] evaluated 55 modern winter oilseed rape cultivars for their ability to use nitrogen. They observed that root biomass production and morphological traits could be positive indicators of above-ground biomass production and that N uptake capacity was, to some extent, root morphology-dependent. These results have major implications with regard to improving oilseed rape production and optimizing

N cycling under this crop. While N-fixing bacteria are directly involved in N cycling and plant nutrition, the N-contributing efficacy of rhizobia varies widely. *Sandhu* et al. [13] evaluated how *Bradyrhizobium* growth is affected by a soil nutrient environment as compared to more traditional growth media, such as mineral salts or arabinose. They observed marked differences in nodulation efficiency in the soil nutrient environment and propose that the nodulation efficiency of *Bradyrhizobium* be evaluated in soils from specific sites prior to planting soybeans. Better knowledge of leguminous crop symbiosis is another key component of strategies aimed at optimizing N cycling.

In conclusion, this Special Issue, Soil Nitrogen Supply: Linking Plant Available N to Ecosystem Functions and Productivity, presents insightful new research papers that advance our understanding of this relevant topic. Modeling, cover cropping, crop residues, soil amendment, root traits and functions, and the symbiotic N-fixation capacity of leguminous crops and rhizobium bacteria are discussed. Further research on soil nitrogen supply should go beyond investigating mineral forms of N and focus more on the rhizosphere environment and on the interconnection between C and N cycling.

Funding: This research received no external funding.

Conflicts of Interest: The author declares no conflict of interest.

References

1. Morvan, T.; Beff, L.; Lambert, Y.; Mary, B.; Germain, P.; Louis, B.; Beaudoin, N. An Original Experimental Design to Quantify and Model Net Mineralization of Organic Nitrogen in the Field. *Nitrogen* **2022**, *3*, 197–212. [CrossRef]
2. Miranda-Vélez, J.F.; Vogeler, I. Autumn Tillage Reduces the Effect of Plant Cover on Topsoil Nitrogen Leaching. *Nitrogen* **2022**, *3*, 186–196. [CrossRef]
3. Siller, A.; Darby, H.; Smychkovich, A.; Hashemi, M. Winter Malting Barley Growth, Yield, and Quality following Leguminous Cover Crops in the Northeast United States. *Nitrogen* **2021**, *2*, 415–427. [CrossRef]
4. Alghamdi, R.S.; Cihacek, L.; Wen, Q. Simulated Cropping Season Effects on N Mineralization from Accumulated No-Till Crop Residues. *Nitrogen* **2022**, *3*, 149–160. [CrossRef]
5. Jesmin, T.; Mitchell, D.T.; Mulvaney, R.L. Short-Term Effect of Nitrogen Fertilization on Carbon Mineralization during Corn Residue Decomposition in Soil. *Nitrogen* **2021**, *2*, 444–460. [CrossRef]
6. Omara, P.; Aula, L.; Otim, F.; Obia, A.; Souza, J.L.B.; Arnall, D.B. Biochar Applied with Inorganic Nitrogen Improves Soil Carbon, Nitrate and Ammonium Content of a Sandy Loam Temperate Soil. *Nitrogen* **2022**, *3*, 90–100. [CrossRef]
7. Dessureault-Rompré, J.; Gloutney, A.; Caron, J. Nutrient Availability for *Lactuca sativa* Cultivated in an Amended Peatland: An Ionic Exchange Study. *Nitrogen* **2022**, *3*, 26–42. [CrossRef]
8. Phillips, I.; Paungfoo-Lonhienne, C.; Tahmasbian, I.; Hunter, B.; Smith, B.; Mayer, D.; Redding, M. Combination of Inorganic Nitrogen and Organic Soil Amendment Improves Nitrogen Use Efficiency While Reducing Nitrogen Runoff. *Nitrogen* **2022**, *3*, 58–73. [CrossRef]
9. Lasisi, A.A.; Akinremi, O.O. Degradation of *N*-(*n*-butyl) Thiophosphoric Triamide (NBPT) with and without Nitrification Inhibitor in Soils. *Nitrogen* **2022**, *3*, 161–169. [CrossRef]
10. Guo, Y.; Naeem, A.; Mühling, K.H. Comparative Effectiveness of Four Nitrification Inhibitors for Mitigating Carbon Dioxide and Nitrous Oxide Emissions from Three Different Textured Soils. *Nitrogen* **2021**, *2*, 155–166. [CrossRef]
11. Bukomba, J.; Lusk, M.G. Spatial Variability in Inorganic Soil Nitrogen Production in a Mixed-Vegetation Urban Landscape. *Nitrogen* **2022**, *3*, 118–127. [CrossRef]
12. Kupcsik, L.; Chiodi, C.; Moturu, T.R.; De Gernier, H.; Haelterman, L.; Louvieaux, J.; Tillard, P.; Sturrock, C.J.; Bennett, M.; Nacry, P.; et al. Oilseed Rape Cultivars Show Diversity of Root Morphologies with the Potential for Better Capture of Nitrogen. *Nitrogen* **2021**, *2*, 491–505. [CrossRef]
13. Sandhu, A.K.; Subramanian, S.; Brözel, V.S. Surface Properties and Adherence of *Bradyrhizobium diazoefficiens* to *Glycine max* Roots Are Altered When Grown in Soil Extracted Nutrients. *Nitrogen* **2021**, *2*, 461–473. [CrossRef]

 *Nitrogen*

Article

An Original Experimental Design to Quantify and Model Net Mineralization of Organic Nitrogen in the Field

Thierry Morvan [1,*], Laure Beff [2], Yvon Lambert [2], Bruno Mary [3], Philippe Germain [1], Benjamin Louis [4] and Nicolas Beaudoin [3]

[1] INRAE, Institut Agro, SAS, F-35000 Rennes, France; philippe.germain@inrae.fr
[2] Chambre d'Agriculture de Bretagne, F-35000 Rennes, France; laure.beff@bretagne.chambagri.fr (L.B.); yvon.lambert@bretagne.chambagri.fr (Y.L.)
[3] INRAE, UR1158 Agro-Impact, F-02000 Laon, France; bruno.mary@inrae.fr (B.M.); nicolas.beaudoin@inrae.fr (N.B.)
[4] Benjamin U Consultance Formation, F-63110 Beaumont, France; contact@benjaminlouis-stat.fr
* Correspondence: thierry.morvan@inrae.fr

Abstract: Improving the assessment and prediction of soil organic nitrogen (N) mineralization is essential: it contributes significantly to the N nutrition of crops and remains a major economic and environmental challenge. Consequently, a network of 137 fields was established in Brittany, France, to represent the wide diversity of soils and cultivation practices in this region. The experimental design was developed to measure net N mineralization for three consecutive years, in order to improve the accuracy of measuring it. Net N mineralization was quantified by the mineral N mass balance, which was estimated from March to October for a maize crop with no N fertilization. The effect of climate on mineralization was considered by calculating normalized time (ndays) and, then, calculating the N mineralization rate (Vn) as the ratio of the mineral N mass balance to normalized time. Strict screening of the experimental data, using agronomic and statistical criteria, resulted in the selection of a subset of 67 fields for data analysis. Mean Vn was relatively high (0.99 kg N ha^{-1} nday^{-1}) over the period and varied greatly, from 0.62 to 1.46 kg N ha^{-1} nday^{-1} for the 10th and 90th percentiles, respectively. The upper soil layer (0–30 cm) was sampled to estimate its physical and chemical properties, particulate organic matter carbon and N fractions (POM-C and POM-N, respectively), soil microbial biomass (SMB), and extractable organic N (EON) determined in a phosphate borate extractant. The strongest correlations between Vn and these variables were observed with EON (r = 0.47), SMB (r = 0.45), POM-N (r = 0.43), and, to a lesser extent, the soil N stock (r = 0.31). Vn was also strongly correlated with a cropping system indicator (r = 0.39). A modeling approach, using generalized additive models, was used to identify and rank the variables with the greatest ability to predict net N mineralization.

Keywords: field experiment network; soil nitrogen mineralization; soil properties; cropping system; modeling; STICS model

Citation: Morvan, T.; Beff, L.; Lambert, Y.; Mary, B.; Germain, P.; Louis, B.; Beaudoin, N. An Original Experimental Design to Quantify and Model Net Mineralization of Organic Nitrogen in the Field. *Nitrogen* **2022**, *3*, 197–212. https://doi.org/10.3390/nitrogen3020015

Academic Editor: Jacynthe Dessureault-Rompré

Received: 7 March 2022
Accepted: 11 April 2022
Published: 15 April 2022

Publisher's Note: MDPI stays neutral with regard to jurisdictional claims in published maps and institutional affiliations.

1. Introduction

The balance sheet method is widely used to predict nitrogen (N) fertilization of crops [1–5]. With this method, a balance sheet is drawn up, in which fertilizer requirements are calculated as crop N requirements minus soil N availability. The accuracy of this method, thus, depends on that of estimating N mineralization, which if overestimated can lead to yield losses, or if underestimated can lead to N losses through leaching.

Certain soil properties, soil conditions (especially water content and temperature), and cropping practices are known to determine mineralization of organic N in the soil [6]. Laboratory incubations are widely used to identify and rank the soil physical and chemical parameters that strongly influence mineralization, especially the organic N content, texture [6–11], calcium carbonate content [12], and pH [5,12,13]. More recently, researchers have focused on the particle-size fractions of organic matter, particularly particulate organic matter (POM). POM is sensitive to management [14,15], and its turnover is significantly higher than that of the heavy fraction of organic matter [16]. POM is known to be an organic matter compartment that is rapidly biodegradable [17–20], but which has, however, contrasting results for the mineralization of its N (POM-N). Some studies observed a positive correlation between N supply and POM-N [14,21], while others tended to observe that mineralization resulted more from the biodegradation of the heavy fraction of organic matter [17]. Nonetheless, there is a consensus on the utility of considering POM when studying mineralization [22].

Laboratory experiments can also assess effects of cropping practices studied in long-term field experiments, revealing the significant influence of crop rotations [23–26], intercrops [27], introduction of legume crops, the type of soil tillage, and mineral and organic fertilization on mineralization [28,29]. Laboratory experiments are ultimately useful for evaluating the many extractable organic N (EON) indicators of mineralization, based on chemical extraction of a fraction of the total N in a soil sample [30–32]. The meta-analysis of Ros et al. [31] identified indicators with a greater ability to predict mineralization than the organic N content of the soil.

Laboratory experiments are, ultimately, useful for evaluating the many extractable organic N (EON) indicators of mineralization based on the chemical extraction of a fraction of the total N in a soil sample [30–32]. The meta-analysis of [31] identified indicators with a greater ability to predict mineralization than the organic N content of the soil.

However, these laboratory data are poor predictors of mineralization under field conditions, due to the lack of considering (i) interactions between microorganisms and mesofauna, which are active in decomposition [33]; (ii) mineralization in deep soil layers; and (iii) plant effects on N mineralization–organization processes stimulated by rhizodeposition, which strongly influences net mineralization under crops. In addition, fluctuations in the environmental conditions that drive these processes also help to understand why laboratory experiments can only partially explain mineralization under field conditions.

These factors justify studying N mineralization under field conditions and quantifying it, which requires a modeling approach to estimate losses from nitrate leaching and assess the influence of weather conditions. An initial approach, developed by Mary et al. [34], was based on frequently measuring the water and mineral N contents of the soil (divided into several layers), calibrating the LIXIM model with these data, and predicting net mineralization and leaching for each time step. This approach was applied to many experimental sites in France, to create reference values for mineralization in French soils under contrasting soil and cropping conditions [12,35]. However, it has the disadvantage of being labor intensive and limited in the number of fields to which it can be applied. A second, simpler approach consists of estimating net mineralization using the N mass balance of a crop, which is based on measuring the N taken up by the crop and the difference between the initial and final contents of soil mineral N [4,5,36,37]. This method has the advantage of being based on the functioning of the soil–plant system under field conditions.

Mineralization estimated from these field experiments depends on the dynamics of soil water content and temperature, which are influenced by the weather conditions during each experiment. Consequently, it is necessary to control for the influence of weather to be able to assess the effects of the cropping system and soil properties [34], thus, converted "true time" into normalized time, which is calculated as the product of a temperature function and a water-content function, using the parameters developed by Rodrigo et al. [38]. Mineralization during a measurement period is, thus, estimated as a daily normalized mineralization rate (Vn) multiplied by the normalized time calculated during the period.

Applying the N mass balance method in France led to a compartment approach [39], in which one estimates, separately, the mineralization of residues of the previous crop, recent applications of organic waste, recent plowing of grassland, and the "baseline mineralization" of soil organic N (SON). Experiments to measure baseline mineralization based on the N mass balance in the field are rendered more complicated by this coexistence of flows from other compartments. Consequently, some studies have used models to subtract mineralization of residues of the previous crop [4,5], thus adding uncertainty to estimates of basal mineralization.

To create the best conditions possible for quantifying mineralization of organic N in the field, we developed an original experimental design, supported by five years of monitoring, of a network of 137 fields in Brittany, in western France. Net N mineralization (Mn) was quantified by field measurements of the mineral N mass balance of a maize crop that remained unfertilized for all five years, and whose aboveground biomass was completely removed from the field at harvest, in order to minimize the amount of crop residues returned to the soil. Only data from the last three years were analyzed, in order to limit biases resulting from inputs of fertilizers and crop residues incorporated into the soil before the experiment began. The innovations of this experimental approach were, thus, (i) to create the best possible conditions for estimating these N flows and (ii) to measure these flows frequently over a long period to obtain more accurate estimates of mineralization.

2. Materials and Methods

For more details on the methods, see Morvan et al. [40].

2.1. Network Presentation

Experiments were performed in a network of 137 cultivated fields located throughout Brittany (Figure 1). The soil was sampled in each field to determine its depth, layers, and textural class. In the upper layer (0–30 cm), most soils had a silty loam (n = 81) or loamy (n = 33) texture. The other soils were sandy loam (n = 15), clay loam (n = 4), silty clay loam (n = 3), and silty clay (n = 1).

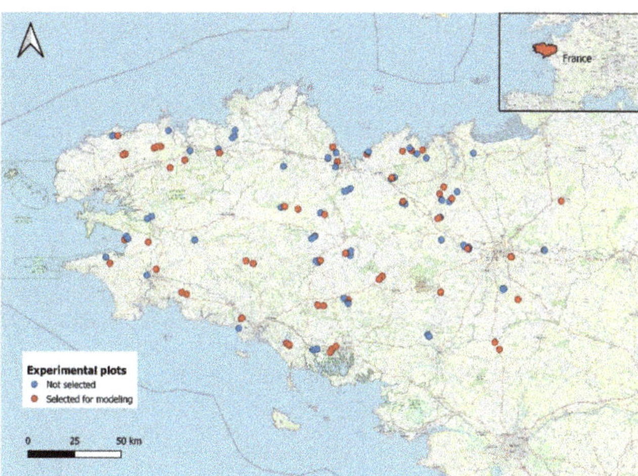

Figure 1. Locations of experimental fields in the network in Brittany, France (red points indicate the 67 fields selected).

Before the experiments, 82 fields had annual crop rotations, 30 fields had grassland in their rotations, and the remaining 25 fields were summer fallow or cultivated with vegetables. Animal waste was regularly applied to half (n = 65) of the fields, especially on maize crops; of these fields, 26 received manure every year, with one or two applications per year. Fifty-seven fields received at least one application every four years of cattle manure (n = 36), pig slurry (n = 14), cattle slurry (n = 8), or poultry manure (n = 5).

2.2. Climate

The climate in Brittany is mild oceanic temperate, with a pronounced east-west rainfall gradient. The weather differed among the three experimental years analyzed. The year 2012 was variable, with dry periods in winter (January and February), rainy periods in April and June, and rainfall close to the mean observed from 1994–2014 in July (Figure 2). In contrast, 2013 was dry, especially in summer, with monthly rainfall that was much lower than the mean, while 2014 was rainy, especially in winter (January and February) and in summer (July and August). In addition to this inter-annual variability, the weather varied greatly among fields within a given year.

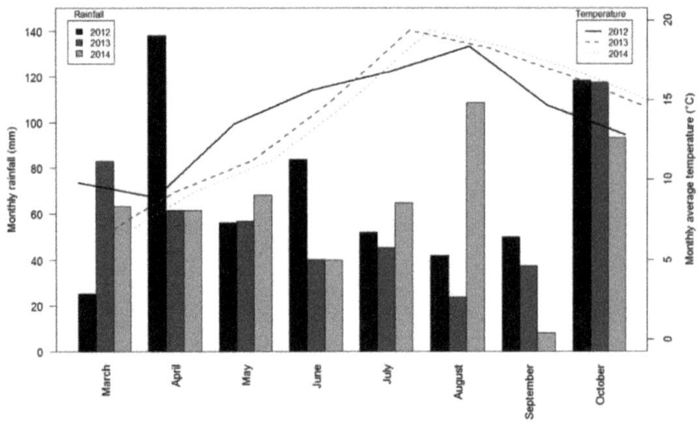

Figure 2. Monthly rainfall and mean air temperature from March-October for the 3 years of the experiment.

2.3. Experimental Design

The objective of the experiment was to quantify the "baseline" mineralization of soil organic N, since the N mass balance used in this experiment included N mineralization not only from this compartment, but also from other compartments, including animal waste and annual crop or grassland residues recently incorporated into the soil. To this end, (i) the 137 experimental fields were cropped with silage maize for four (since 2011) or five (since 2010) consecutive years without any mineral or organic fertilization, and (ii) only mineralization data for 2012, 2013 and 2014 were considered, to exclude N flows resulting from inputs of fertilizers and crop residues incorporated into the soil before the experiment began. The experimental design was, thus, based on estimating N mineralization for three consecutive years. Experimental monitoring was performed on an area of 1485 m^2 (33 m × 45 m), divided into three subplots of 45 m^2 (6.0 m × 7.5 m) in the middle for replicate measurements.

2.4. Calculating Net Soil N Mineralization

Mn was calculated from the end of winter to the beginning of autumn from the mineral N mass balance of a maize crop not fertilized with N, as follows:

$$Mn = Nf - Ni + N_{uptake} + N_{leached} \qquad (1)$$

with Ni and Nf corresponding to the soil mineral N content in the 0–90 cm soil profile in March and October, respectively; N_{uptake} corresponding to N uptake by the plant (kg N ha^{-1}); and $N_{leached}$ corresponding to nitrate leaching that may occur in spring, after measurement of Ni (kg N ha^{-1}).

Ni, Nf, and N_{uptake} were measured in triplicate. $N_{leached}$ was estimated using the STICS model [41], which was parameterized with the soil properties of each field, and initialized at the measurement date of Ni. Equation (1) is a simplified approach for estimating the mineral N mass balance, but it is valid in situations without N fertilization. Gaseous N losses can be assumed to be very low and compensated by atmospheric deposition and symbiotic fixation of N.

Since N mineralization depends strongly on weather conditions, we controlled for the influence of weather on mineralization by converting true time into normalized time (nday), using functions integrated into STICS [41], in order to model the effects of temperature and soil water content on N mineralization. The effect of soil temperature on mineralization is described by a logistic function, which is roughly exponential from 0–25 °C. This function is similar to an Arrhenius function, with an activation energy of 78 kJ mol^{-1} K^{-1} from 0–35 °C, and also equivalent to a Van 't Hoff function, with a Q_{10} coefficient of 3.15 from 0–35 °C from 0–25 °C. The effect of soil water content on mineralization is described by a linear function. Mineralization in temperate soils peaks when soil water content equals field capacity and stops when the ratio of soil water content to field capacity is less than 0.3 [38]. The Mn estimated from the N mass balance for each year and field was then divided by the normalized time for each year, which gives a daily "normalized" rate of mineralization Vn (kg N ha^{-1} nday^{-1}). Mean Vn (Vn$_{mean}$) was calculated for each field by averaging Vn for all three years of measurements.

2.5. Data Screening

Data were screened to exclude the fields in which agronomic or measurement problems had occurred. They were represented in a decision tree with three nodes (Figure 3): (i) the presence of weeds, which can compete with maize and bias estimates of Mn; (ii) non-homogeneous maize cover at harvest, which can bias measurements of Nf and N uptake and, thus, estimates of Mn; and (iii) excessively high variability in Mn among the three subplots. We considered, as outliers, fields whose coefficient of variation of Mn was greater than or equal to 25% (i.e., the 95th percentile) in 2012, 2013, or 2014. To retain the most consistent data and, thus, decrease uncertainty in estimates of Mn, we excluded these fields from further analysis and modeling.

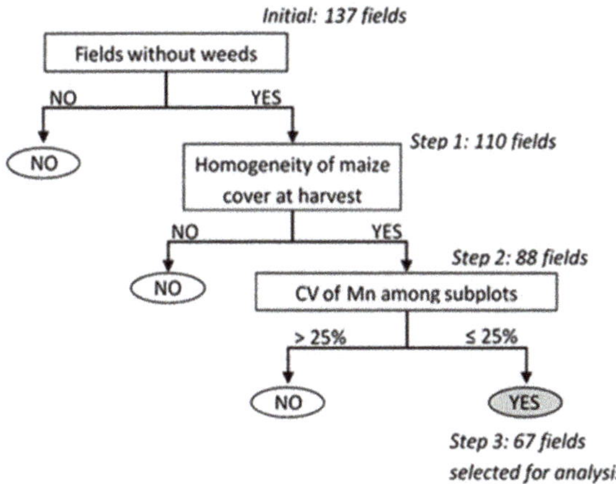

Figure 3. Decision tree used to filter the data from each field. Conditions had to be true for 2012, 2013, and 2014 to pass to the next step. CV: coefficient of variation, Mn: soil net nitrogen mineralization.

2.6. Soil and Plant Analysis

Initial soil mineral N content (Ni) was measured at the end of winter (March), and final soil mineral N content (Nf) was measured at the beginning of autumn (October), before resumption of nitrate leaching. A composite sample of 10 soil cores was created for each subplot, for three layers (0–30, 30–60 and 60–90 cm). Soil mineral N was extracted in a 1M

KCl solution using a soil/KCl ratio of 1:2, and the NH_4^+ and NO_3^- contents of the soil extracts were then determined by continuous flow colorimetry by the methods developed by [42] and [43], respectively.

The soil of the upper layer was sampled in March 2013 to estimate soil properties. Samples consisted of 10 cores, which were pooled. Total carbon (C) and N were determined by the Dumas dry-combustion method. Cation exchange capacity was established using the Metson method [44], and pH was obtained in water [45]. Soil texture was based on measuring the particle size of five fractions: clay (<2 µm), fine silt (2–20 µm), coarse silt (20–50 µm), fine sand (50–200 µm), and coarse sand (200–2000 µm) [46]. POM was determined using a simple fractionation method, by wet sieving under water with a 50 mm sieve [16]. POM was recovered on the sieve, dried, weighed, and finely ground before analyzing the C and N contents by the Dumas dry-combustion method using a Thermo Finnigan Flash EA 1112 Series analyzer.

The fumigation-extraction method was used to estimate soil microbial biomass (SMB) [47], using 40 g of an oven-dried equivalent of soil, shaken in 200 mL of 0.025M K_2SO_4 for 45 min. Non-fumigated soils were also extracted in the same way. Oxidizable C in the fumigated and non-fumigated K_2SO_4 was determined using a V-WS SHIMADZU TOC analyzer. A k_{EC} conversion factor of 0.38 was applied to convert the flush of oxidizable C into SMB. EON was determined using the method of [48]: 4 g of soil were steam distilled in 40 mL of phosphate borate extractant (buffered at pH 11.2) for 8 min, and ammonium in the distillate was back titrated using 0.0025 M H_2SO_4. The amount of organic N hydrolyzed, which corresponded to EON, was obtained by subtracting the total NH_4-N extracted from native NH_4-N. Mean soil properties, POM-N, SMB, and EON are summarized in Table 1.

Table 1. Measured mean (±1 standard deviation) soil variables for the 137 fields of the network and the 67 fields selected. Physico-chemical properties of the 0–30 cm soil layer include texture, C and N contents, pH, Metson cation exchange capacity (CEC), soil microbial biomass (SMB), extractable organic N (EON), particulate organic matter (POM-N), and soil organic nitrogen (SON) stocks.

Variable	Unit	137 Fields	67 Fields
Clay	g kg^{-1}	194 (±52.0)	183 (±36.0)
Fine Silt	g kg^{-1}	247.8 (±87.5)	237.4 (±73.7)
Coarse Silt	g kg^{-1}	268.3 (±118.8)	264.9 (±112.4)
Silt	g kg^{-1}	516 (±135.0)	502 (±132.0)
Fine Sand	g kg^{-1}	125.6 (±53.6)	129.8 (±52.0)
Coarse Sand	g kg^{-1}	164.5 (±129.3)	185.0 (±136.4)
Sand	g kg^{-1}	290 (±145.0)	315 (±145.0)
C content	g C kg^{-1}	19.8 (±6.5)	20.2 (±6.1)
N content	g N kg^{-1}	1.8 (±0.6)	1.8 (±0.5)
pH		6.1 (±0.5)	6.0 (±0.5)
CEC Metson	meq 100g^{-1}	9.8 (±2.5)	9.9 (±2.2)
SMB	mg C kg^{-1}	171.3 (±45.3)	168.4 (±38.2)
EON	mg N kg^{-1}	27.3 (±8.4)	27.7 (±7.4)
POM-N	t N ha^{-1}	0.67 (±0.24)	0.68 (±0.24)
SON	t N ha^{-1}	6.5 (±1.5)	6.5 (±1.5)

Bulk density was measured once, in triplicate, in 2011, for each experimental field and each of the three soil layers (0–30, 30–60 and 60–90 cm), using an 8 cm diameter root auger, which cored undisturbed samples of a known volume. The bulk density of the fine-earth fraction calculated from the dry mass and the core volume was used to convert the mineral N content of the samples to kg N ha^{-1} and to calculate the stocks of SON and POM-N (t N ha^{-1}), as well as those of EON and SMB (kg N ha^{-1}).

Aboveground biomass and N content of the maize crop were quantified at harvest, when maize plants were harvested and weighed in all subplots. Total N uptake of maize was calculated by multiplying N in the aboveground biomass by 1.15 to estimate its belowground N at harvest [49].

2.7. Calculating an Indicator of the Cropping System

An indicator of the cropping system (I_Sys) was calculated to integrate the diversity of field management (i.e., crop rotation and organic waste application) among fields, considering a period of 15 years before the year of interest. I_Sys was calculated by summing an indicator of the effect of the N returned to the soil in crop residues and an indicator of the effect of repeated applications of organic waste on soil mineralization. See [40] for details on calculation of the indicator. To assess the influence of I_Sys on the N mass balance and soil N mineralization, we classified its values into three levels using k-means clustering: low ($\leq$63 kg N ha^{-1}), moderate (63–98 kg N ha^{-1}), and high (>98 kg N ha^{-1}). These three classes contained 32%, 53%, and 15% of the fields, respectively.

2.8. Statistical Analysis

All statistical analyses were performed with R, v 4.1.0 [50]. Pearson correlations were used to assess relations between variables. Analysis of variance (ANOVA) was used to compare means when data were normally distributed (according to the Shapiro–Wilk test) and had homogenous variances (Levene's test); if not, the Kruskal–Wallis test was used.

Generalized additive models (GAMs) were used to predict Vn [51]. This innovative method consists of selecting soil properties for a given GAM, which can best explain the variation in N dynamics [51,52]. GAMs can distinguish the relative effects of covariates by allowing for nonlinear relations between them and the variable studied (i.e., Vn$_{mean}$). Variables were chosen to obtain models with the smallest mean square error of prediction (MSEP) [53]. This criterion was calculated by applying a "leave-one-out" strategy, which represented an internal validation of the model and avoided over-fitting the model to the data. The agreement between predictions and observed Vn$_{mean}$ was evaluated statistically by calculating the coefficient of determination (R^2) and the root mean squared error (RMSE) [53]. Additionally, the ratio of performance to inter-quartile distance (RPIQ) was calculated as the ratio of the inter-quartile range to the square root of the MSEP [51,54]. RPIQ represents the degree to which the dispersion of the response variable exceeds the model's prediction error. We also calculated the indicator dMSEP, which represents the proportional increase in prediction error (MSEP) when a given variable is removed from the model, as a metric of the relative importance of the variable.

3. Results

3.1. Accuracy of the Dataset

The criteria applied using the decision tree (Figure 3) led to the final selection of 67 fields (26, 22, and 19 fields were excluded at the 1st, 2nd, and 3rd nodes of the tree, respectively). The selection was severe because a given field had to avoid problems all three years (2012, 2013, and 2014). The 67 fields selected remained representative of the diversity of soil characteristics, cropping systems (Table 1 and Figure 4) and geography of the initial network (Figure 1). To ensure that screening the data had not introduced bias, we verified that the distribution and mean of Mn had remained similar. Mn varied little among subplots over the three years (mean standard deviation of 15 kg N ha^{-1}, with a mean CV of 10% for the 67 fields), which illustrates the high precision of the N mass balance in the selected dataset.

3.2. Soil N Mineralization: Mass Balance Components and N Rates

Mean, minimum or maximum Mn from March to October were similar for the 67 fields in all three years (e.g. mean Mn was 162, 146 and 154 kg N ha^{-1} in 2012, 2013 and 2014, respectively) (Table 2). Mean N uptake by maize was the main component of the N mass balance, particularly in 2014, when it repesented 95% of Mn vs. 88% in 2012 and 73% in 2013 (Table 2). Lower N uptake in 2013 was explained by the weather conditions, with low rainfall from June to September that induced hydric stress. Mean predicted nitrate leaching in early spring was low in 2013 and 2014 but was a significant component of the N mass balance in 2012 (21 kg N ha^{-1}) (Table 2); this difference can be explained by higher Ni due to less leaching during the preceding winter and high rainfall in April (138 $\pm$ 32 mm).

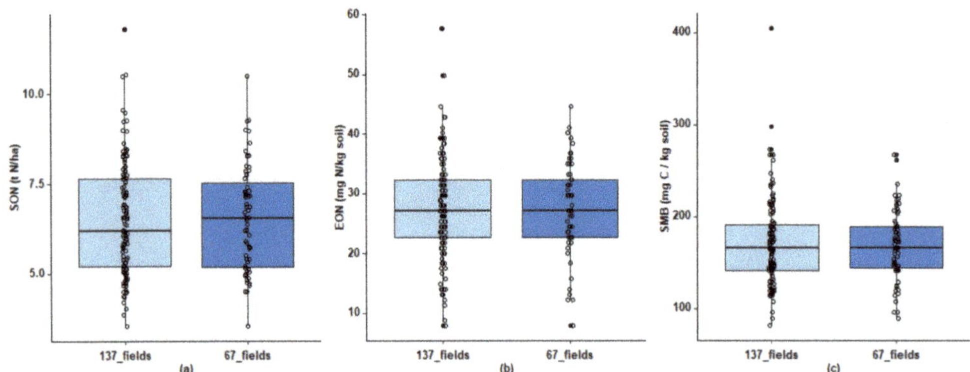

Figure 4. Boxplots of (**a**) soil organic nitrogen (SON), (**b**) extractable organic nitrogen (EON), and (**c**) soil microbial biomass (SMB) for all 137 fields and for the 67 fields selected from the network. Whiskers extend to 1.5 times the interquartile range.

Table 2. Mean, minimum, maximum, and standard deviation (SD) of the mass-balance components and nitrogen (N) mineralization rates for the 67 fields selected. N uptake is N taken up by maize plants, Ni and Nf are, respectively, the initial and final amount of mineral N in the soil profile (0–90 cm), N leached is N leached out of the soil profile predicted by the STICS crop model, Mn is net N mineralization, ndays is normalized time between initial and final N measurements, and Vn is normalized N mineralization rate.

Year	Metric	N Uptake kg N ha^{-1}	Ni kg N ha^{-1}	Nf kg N ha^{-1}	N Leached kg N ha^{-1}	Mn kg N ha^{-1}	tn nday	Vn kg N ha^{-1} nday^{-1}
2012	mean	144	47	46	22	162	165	0.99
	min	50	6	7	1	53	112	0.28
	max	292	145	155	63	302	202	1.68
	SD	47	29	30	14	52	16	0.31
2013	mean	106	36	71	9	146	139	1.06
	min	39	12	17	1	66	102	0.41
	max	257	82	167	30	326	181	2.3
	SD	41	13	30	6	60	16	0.45
2014	mean	147	34	38	6	154	172	0.92
	min	73	10	8	0	75	122	0.39
	max	334	116	168	27	454	213	2.39
	SD	56	17	28	5	71	20	0.44

Although the fields were unfertilized and soil N mineralization was the main source of N for plants, the mean N Nutrition Index (NNI) [55] was high for unfertilized crops, particularly in 2012 (0.88) and 2014 (0.92) (vs. 0.73 in 2013), reflecting high availability of N during the crop cycle. Mean Vn was similar in 2012, 2013, and 2014 (0.99, 1.06, and 0.92 kg N ha^{-1} nday^{-1}, respectively), in agreement with the hypotheses on which the experimental design was developed (Table 2).

3.3. Correlations between Vn, Soil Properties, EON, and I_Sys

Vn$_{mean}$ correlated most strongly with EON (r = 0.47), SMB (r =0.45), POM-N (r = 0.43), and, to a lesser extent, SON (r = 0.31) (Table 3). Texture, particularly clay content, can influence N mineralization strongly, and Vn$_{mean}$ had a significant but weak negative correlation with clay content (r = −0.19), yet a stronger positive correlation with the coarse sand content (r = 0.32). Vn_{mean} also correlated strongly with I_Sys (r = 0.39), which was highlighted by a significant effect of I_Sys class on Vn_{mean} ($p < 0.05$) (Figure 5b): mean Vn_{mean} of the high I_Sys class was 29% higher than that of the low I_Sys class (1.11 vs. 0.86 kg N ha^{-1} nday^{-1}, respectively). This difference was due to much lower Vn measured in fields with low I_Sys, which corresponded to fields without grassland in their rotation and without organic manure application. SON was strongly correlated with EON (r = 0.64) and POM-N (r = 0.75), while POM-N was also correlated with EON (r = 0.50).

Table 3. Correlation coefficients (r) between Vn for the years 2012, 2013, 2014, Vn_{mean}, the cropping system indicator (I_Sys), and soil properties for the 67 fields selected. ($p < 0.05$ for r > 0.24, $p < 0.01$ for r > 0.31, and $p < 0.001$ for r > 0.39). F: fine, C: Coarse.

	Vn12	Vn13	Vn14	Vn_{mean}	I_Sys	SON	EON	SMB	POM-N	Clay	F Silt	C Silt	Silt	F Sand	C Sand	Sand	pH	CEC
Vn12	1.00	0.63	0.53	0.81	0.37	0.38	0.34	0.45	0.45	−0.12	−0.08	−0.10	−0.13	−0.21	0.24	0.15	0.06	0.26
Vn13		1.00	0.56	0.88	0.38	0.27	0.40	0.36	0.45	−0.07	−0.17	−0.16	−0.23	−0.10	0.28	0.23	−0.03	0.16
Vn14			1.00	0.84	0.25	0.17	0.44	0.35	0.21	−0.25	−0.20	−0.04	−0.15	−0.18	0.28	0.20	−0.14	−0.03
Vn_{mean}				1.00	0.39	0.31	0.47	0.45	0.43	−0.17	−0.19	−0.12	−0.21	−0.19	0.32	0.23	−0.05	0.14
I_Sys					1.00	0.14	0.12	0.20	0.17	0.07	−0.09	0.11	0.04	−0.05	−0.04	−0.06	0.04	0.01
SON						1.00	0.64	0.07	0.75	0.16	0.15	−0.48	−0.33	−0.16	0.33	0.26	−0.18	0.78
EON							1.00	0.14	0.50	0.07	0.02	−0.23	−0.19	−0.15	0.22	0.15	−0.40	0.42
SMB								1.00	0.25	0.25	0.18	−0.11	0.01	−0.08	−0.04	−0.07	0.10	−0.04
POM-N									1.00	0.05	0.18	−0.45	−0.29	−0.15	0.32	0.25	−0.02	0.57
Clay										1.00	0.54	−0.04	0.27	−0.15	−0.46	−0.49	0.04	0.25
F Silt											1.00	−0.05	0.52	−0.48	−0.46	−0.60	0.06	0.09
C Silt												1.00	0.83	−0.09	−0.76	−0.74	0.32	−0.47
Silt													1.00	−0.34	−0.90	−0.97	0.31	−0.35
F Sand														1.00	−0.01	0.35	−0.08	0.05
C Sand															1.00	0.93	−0.28	0.25
Sand																1.00	−0.29	0.25
pH																	1.00	−0.11
CEC																		1.00

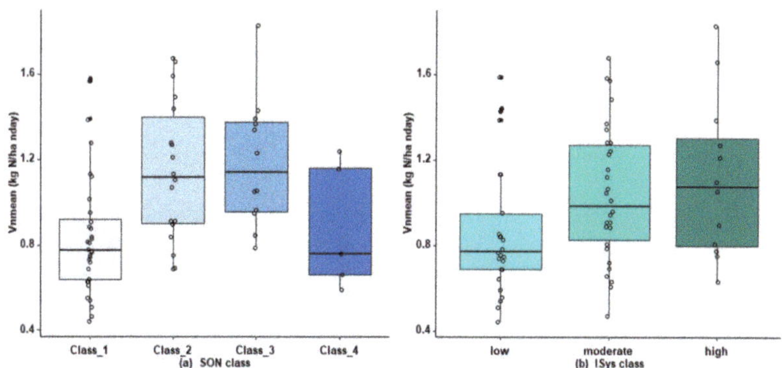

Figure 5. Boxplots of Vn_{mean} as a function of (**a**) soil organic nitrogen (SON) class (class_1: SON < 5.7 t N ha^{-1}; class_2: [5.7; 7.15]; class_3: [7.15; 8.2]; class_4: >8.2 t N ha^{-1}) and (**b**) I_Sys class. Whiskers extend to 1.5 times the interquartile range.

3.4. Modeling Normalized N Mineralization Rate

A model that predicted Vn_{mean} using only soil properties and the I_Sys indicator (model 1) explained only 47% of the variance in mineralization (Table 4, Figure 6a). The soil properties selected were SON, clay, coarse sand and coarse silt. A quadratic relation was observed with coarse sand and clay, with a negative effect of clay contents that exceeded 22 g kg^{-1}. A positive linear relation was observed with I_Sys, which was the most influential variable after coarse sand (Table 4). The model's RPIQ value of 1.8 was moderate.

Table 4. Assessment of the models, covariates selected, and dMSEP values. Relation indicates the type of relation selected: L, linear; P2, 2nd-degree polynomial.

Model	R^2	MSEP	RPIQ	Covariates	Relation	dMSEP
1	0.47	0.075	1.82	I_Sys	L	0.214
				Coarse Sand	P2	0.244
				Clay	P2	0.100
				SON	L	0.053
				Coarse Silt	L	0.024
2	0.67	0.05	2.25	SMB	L	0.320
				EON	L	0.245
				Coarse Sand	P2	0.250
				Clay	P2	0.200
				I_Sys	L	0.159
				Coarse Silt	L	0.090

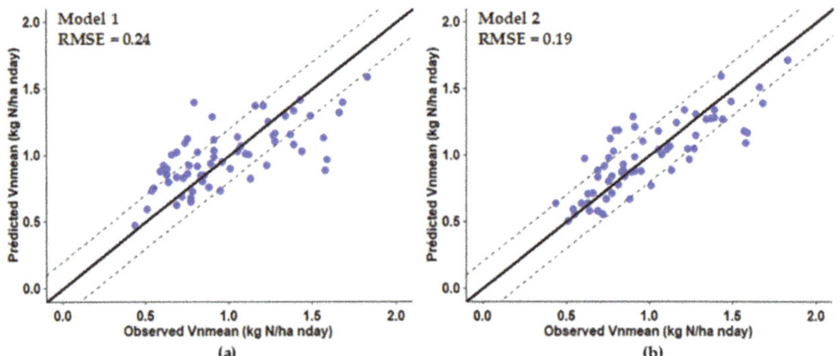

Figure 6. Comparison of observed Vn_{mean} to that predicted by (**a**) model 1, whose covariates were the basic soil parameters and I_Sys, and (**b**) model 2, with SMB and EON as additional variables. Solid lines are 1:1 lines, while dashed lines indicate $\pm$ 0.2 kg N ha^{-1} nday^{-1} around each 1:1 line. (RMSE: Root Mean Square Error).

We then developed a model with EON, SMB and POM-N as additional input variables (model 2). The same soil physical properties as in model 1 were selected, as was the I_Sys indicator, and they had similar relationships (Table 4). The additional variables selected were EON and SMB, which significantly increased the proportion of variance explained (R^2 = 0.67) (Figure 6b) but increased the RPIQ value only moderately (2.2). EON and SMB were selected because they were the two variables most correlated with Vn_{mean} but not correlated with each other (r = 0.14). Because they provided complementary information, it was useful to include them both in the model. SON was not selected due to its strong correlation with EON.

4. Discussion

4.1. Soil N Mineralization and N Rates

The approaches used to estimate Mn in crop fields agree that crop N uptake is the main component of the N mass balance [5,9,37,56,57]. Nonetheless, we observed relatively high variability in the contribution of N uptake to the mineral N mass balance among the three years: 95% in 2014, but only 73% in 2013 [40]. This can be explained by differing weather conditions between years, with high hydric stress in summer 2013, which may have decreased both soil N mineralization and plant growth. In addition, this drought period was followed by strong rainfall events at the end of summer, which could have created favorable conditions for N mineralization just before Nf was measured [58,59]. This can explain why the mean difference between Nf and Ni equaled 24% of Mn in 2013, and why Nf was higher that year. Calculating the N mass balance from March to October, thus including mineralization at the beginning of autumn, explains why mean Mn was similar among the three years. Mn likely would have differed even more if Nf had been measured immediately after harvest, as illustrated by the large inter-annual variability observed by Delin and Linden [9].

Comparing the Mn measured in the field to data from the literature is difficult because measurement periods can vary from 5–8 months depending on the crop (e.g., wheat, spring barley, maize, sugar beet), which has a huge influence on mineralization [36,37]. Comparisons must, thus, be made with daily Vn, either provided directly by studies (which is rare) or estimated later from their data. Delin and Linden [9] reported mean daily rates of 0.34 ± 0.12, 0.50 ± 0.17, and 0.69 ± 0.16 kg N ha^{-1} day^{-1} in a field experiment with 34 cereal fields studied for three consecutive years. Vn_{mean} estimated from data of Engels and Kuhlmann [36] equals 0.37 ± 0.17 under wheat and 0.67 ± 0.23 under sugar beet, while those estimated from data of [37] and [60] under maize equal 0.71 and 0.68 ± 0.19 kg N ha^{-1} day^{-1}, respectively, which lie in the same range as those we measured in our network (Table 2). These reference values show the high variability in Mn among fields.

Expressing Vn in normalized time is the best basis for comparison, since it controls for the influence of weather, but it has rarely been used in the literature. From a database of 65 soils, Clivot et al. [12] report a range of 0.17–1.67 kg N ha^{-1} nday^{-1} (mean = 0.72 $\pm$ 0.32 kg N ha^{-1} nday^{-1}). Oorts et al. [35] calculated normalized Vn of 0.57 and 0.62 kg N ha^{-1} nday^{-1} at two experimental sites with field crops,. The normalized Vn that we calculated from the network (mean = 0.99 kg N ha^{-1} nday^{-1}), thus, lie near the top of the range reported in the literature.

4.2. Effect of Cropping System on Net N Mineralization

Many studies of data from long-term experiments have demonstrated the influence of crop rotation and management practices on soil N availability [20,26,61–65]. The wide variety of cropping systems studied in the network confirmed this influence (Figure 5b) via the I_Sys indicator, which integrated the influence of the crop rotation and organic waste application only over the medium term, since the experiment was designed so that no organic waste had been applied or grassland had been plowed within the previous three years. I_Sys values varied widely, ranging from 14–185 kg N ha^{-1} and integrated the effects of crop rotation well, especially the presence of grassland in the rotation and the frequency of organic waste application [40]. I_Sys values were low for forage maize monocultures and even lower in fields in which no organic waste had been applied. In contrast, I_Sys values were highest for rotations with grazed grassland and increased as the age of grassland increased; it is well known that grassland age influences mineralization, since soil N mineralization increases as the age of temporary grassland increases and decreases for the annual crops planted for one–three years after a grassland [66]. I_Sys values were intermediate for common rotations, such as grain maize/wheat/rapeseed/barley (with maize and rapeseed fertilized with pig slurry) or grain maize/wheat/three years of grazed grassland (with maize fertilized with cattle manure).

Regular application of organic waste increases stocks of C and N over the medium term, which influence POM [26], SMB, and mineralization activity over the long term [67]. However, this influence depends greatly on the type of waste, the amount applied, and the frequency of application; frequent application of solid waste (e.g., manure, compost) has more influence on mineralization than that of liquid waste [29,68], which justifies the choice of a model that integrates these driving factors into the I_Sys indicator [40].

4.3. Correlations between Vn, Soil Properties, EON and I_Sys

The study confirms the significant correlation between Vn_{mean} and SON, which has long been identified as an important variable that influences mineralization [9,12,28,69–71].

However, SON explained much less variance in Vn_{mean} in our dataset ($R^2 = 0.10$) than in the study of Clivot et al. [12] ($R^2 = 0.26$) or the meta-analysis of Ros et al. [31] ($R^2 = 0.40$). This difference may be explained by the non-linear relation between SON and Vn_{mean}, in which mineralization plateaus at high SON (Figure 5a), due to a larger proportion of stable organic matter in soils in the network. In fact, Vn_{mean} was much more strongly correlated with SMB than with SON, thus confirming that SMB is a valuable indicator of soil N availability [23,72] and is more important for nutrient turnover and availability to plants than SON [73].

Vn_{mean} was also more strongly correlated with POM-N than with SON, thus confirming the utility of measuring POM-N to estimate mineralization, since POM-N can be considered an unprotected potential source of N. Some studies identify POM-N as a strong predictor of mineralization [14] or one of its main drivers [21], while other studies indicate only that the N available from it depends greatly on its chemical composition and C:N ratio [17].

Among important soil properties, soil texture has often been observed to influence the availability of substrate for mineralization, since high clay content decreases SON decomposition by better protecting SON chemically and physically [9,10,74]. We observed a weakly significant negative correlation between Vn and clay content; the correlation itself was also weak, perhaps due to a relatively small range of clay contents in the soils studied. In contrast, we observed a stronger positive correlation between Vn_{mean} and the coarse sand content, since the soils with higher SON also had higher coarse sand contents.

Finally, our results confirm the utility of the EON method we used, since its correlation with Vn_{mean} was much stronger than that between SON and Vn_{mean}. However, the variance explained by EON ($R^2 = 0.23$) was much lower than that observed by Gianello and Bremner [48] ($R^2 = 0.81$), Schomberg et al. [75] ($R^2 = 0.58$), and Ros et al. [31] ($R^2 = 0.50$). This can be explained in part by the fact that these strong correlations were observed from laboratory measurements of mineralization, while mineralization and these indicators usually have weaker correlations when mineralization is measured under field conditions [31].

4.4. Modeling the N Mineralization Rate

Evaluation of six process-based models of European cropping systems led Yin et al. [76] to conclude that all of the models had difficulties predicting both the mean of and variance in soil N mineralization. It is clear that predicting Mn remains a challenge, due to the complexity of N cycling, which demonstrates why even the best models can explain only 65–80% of the variance [12,77]. Model 2, calibrated with our dataset, lies in this range. We also observed that a model parameterized only with soil properties and the I_Sys indicator was less accurate than the "soil" model of Clivot et al. [12] ($R^2 = 0.61$). It confirms, in the present study, the utility of using mineralization indicators as input variables of models, obtained either by measuring EON [69] or by incubation measurements [5,77].

SMB is rarely used to predict mineralization because it is relatively labor intensive, and few laboratories measure it routinely. Clivot et al. [12] concluded that SMB improved prediction of mineralization little, if SON had already been included as an input variable; this can be explained by the weaker correlation between the mineralization rate and SMB in their study than that with SON ($r = 0.54$ vs. 0.62, respectively), even though both were highly significant. In contrast, Vn_{mean} and SMB were more strongly correlated in our dataset than Vn_{mean} and SON ($r = 0.48$ vs. 0.31, respectively), thus replacing SON with SMB increased the accuracy of the GAMs.

The I_Sys indicator significantly increased the accuracy of mineralization predictions and represents an original contribution of this study. The selection of I_Sys in the final GAM confirms the influence, in the medium term, of crop rotations and the application of organic waste on mineralization, while also providing information that complements that provided by SMB and EON. The lack of correlation between I_Sys and these two variables leads to the hypothesis that EON is relatively insensitive to the cropping history over the medium term but is strongly influenced by the geochemical background and the cropping history over the long term, as shown by its strong correlation with SON (r = 0.64).

5. Conclusions

The experimental design of this study, based on repeatedly measuring Mn for three consecutive years after two years of unfertilized maize, ultimately placed it in the best possible conditions for estimating mineralization of SON. The measurements confirmed the high variability in mineralization, which lay near the top of the range reported in the literature. It can be explained by the combined effects of soil types, with variable but generally high SON content; cropping systems representative of livestock-production regions, with regular application of organic waste and the frequent presence of grassland in rotations; and the semi-oceanic climate, which favors mineralization.

The drivers of mineralization were those identified in the literature: SON, POM-N, SMB, EON, and soil texture. Original results of this study include (i) experimental evidence of the influence of cropping history in the medium term on mineralization, and (ii) the fact that the I_Sys indicator provides information complementary to EON. This result strengthens the hypothesis of Ros et al. [31] that mineralizable N cannot be predicted from a single soil test alone, but instead requires a combination of components including EON and site-specific information, such as land-use and soil properties.

The modeling approach identified the most influential measured variables and showed that the proportion of variance explained by a model based only on basic soil properties and the I_Sys indicator (R^2 = 0.47) was not sufficient to consider the model operational. The accuracy of the model increased greatly when SMB and EON were included, but the proportion of the variance explained by model 2 remained relatively moderate, despite the supplemental information provided by the model's covariates: basic soil properties, chemical and biological indicators, and a land-use indicator.

Author Contributions: Conceptualization, T.M. and Y.L.; methodology, T.M., Y.L., L.B., B.M. and N.B.; software, L.B.; validation, T.M., Y.L. and L.B.; formal analysis, T.M., P.G., B.L. and L.B.; investigation, T.M. and L.B.; resources, T.M. and L.B.; data curation, T.M., Y.L., L.B., B.M. and N.B.; writing—original draft preparation, T.M. and L.B.; writing—review and editing, T.M.; visualization, T.M. and L.B.; supervision, T.M.; project administration, T.M. and Y.L.; funding acquisition, T.M. and Y.L. All authors have read and agreed to the published version of the manuscript.

Funding: This study was financed by the Loire Bretagne Water Agency; the Regional Council of Brittany; the Departmental Councils of Côtes d'Armor, Finistère, and Morbihan; and the French government (DRAAF).

Institutional Review Board Statement: Not applicable.

Informed Consent Statement: Not applicable.

Data Availability Statement: The data analyzed in this article were published by *Data in Brief*: https://doi.org/10.1016/j.dib.2021.106795 (accessed on 1 April 2021). Raw data are deposited in the public repository Data INRAE. Data identification number: 10.15454/VYEYBK. Direct URL to data: https://doi.org/10.15454/VYEYBK (accessed on 19 January 2021).

Acknowledgments: This study was supported by partnerships with the SEMSE, AUREA, and Galys laboratories. The authors thank Michael Corson for proofreading the manuscript's English.

Conflicts of Interest: The authors declare no conflict of interest. The funders had no role in the design of the study; in the collection, analyses, or interpretation of data; in the writing of the manuscript, or in the decision to publish the results.

References

1. Makowski, D.; Wallach, D.; Meynard, J. Models of Yield, Grain Protein, and Residual Mineral Nitrogen Responses to Applied Nitrogen for Winter Wheat. *Agron. J.* **1999**, *91*, 377–385. [CrossRef]
2. Neeteson, J.J. Development of nitrogen fertilizer recommendations for arable crops in the Netherlands in relation to nitrate leaching. *Nutr. Cycl. Agroecosystems* **1990**, *26*, 291–298. [CrossRef]

3.   Meynard, J.; Justes, E.; Machet, J.; Recous, S. Nitrogen fertilizers for annuals. *Control. Nitrogen Conc. Agrosystems* **1997**, *83*, 183–199.
4.   Alvarez, R.; Steinbach, H.S.; Grigera, S.M.; Cartier, E.; Obregon, G.; Torri, S.; García, R. The Balance Sheet Method as a Conceptual Framework for Nitrogen Fertilization of Wheat in a Pampean Agroecosystem. *Agron. J.* **2004**, *96*, 1050–1057. [CrossRef]
5.   Romano, N.F.; Alvarez, R.; Bono, A.A.; Steinbach, H.S. Comparison of nitrogen fertilizer demand for wheat production between humid and semi-arid portions of the Argentinean Pampas using a mass balance method. *Arch. Agron. Soil Sci.* **2015**, *61*, 1409–1422. [CrossRef]
6.   Griffin, T.S.; Honeycutt, C.W.; Albrecht, S.L.; Sistani, K.R.; Torbert, H.A.; Wienhold, B.J.; Woodbury, B.L.; Hubbard, R.K.; Powei, J.M. Nationally coordinated evaluation of soil nitrogen mineralization rate using a standardized aerobic incubation protocol. *Commun. Soil Sci. Plant Anal.* **2008**, *39*, 257–268. [CrossRef]
7.   Stanford, G.; Smith, S.J. Nitrogen Mineralization Potentials of Soils. *Soil Sci. Soc. Am. J.* **1972**, *36*, 465–472. [CrossRef]
8.   Sharifi, M.; Zebarth, B.J.; Burton, D.L.; Grant, C.A.; Bittman, S.; Drury, C.F.; McConkey, B.G.; Ziadi, N. Response of Potentially Mineralizable Soil Nitrogen and Indices of Nitrogen Availability to Tillage System. *Soil Sci. Soc. Am. J.* **2008**, *72*, 1124–1131. [CrossRef]
9.   Delin, S.; Lindén, B. Relations Between Net Nitrogen Mineralization and Soil Characteristics Within an Arable Field. *Acta Agric. Scand. Sect. B Soil Plant Sci.* **2002**, *52*, 78–85. [CrossRef]
10.   Hassink, J. Effects of soil texture and grassland management on soil organic C and N and rates of C and N mineralization. *Soil Biol. Biochem.* **1994**, *26*, 1221–1231. [CrossRef]
11.   Parfitt, R.L.; Salt, G.J. Carbon and nitrogen mineralisation in sand, silt, and clay fractions of soils under maize and pasture. *Soil Res.* **2001**, *39*, 361. [CrossRef]
12.   Clivot, H.; Mary, B.; Valé, M.; Cohan, J.-P.; Champolivier, L.; Piraux, F.; Laurent, F.; Justes, E. Quantifying in situ and modeling net nitrogen mineralization from soil organic matter in arable cropping systems. *Soil Biol. Biochem.* **2017**, *111*, 44–59. [CrossRef]
13.   Kemmitt, S.; Wright, D.; Goulding, K.; Jones, D. pH regulation of carbon and nitrogen dynamics in two agricultural soils. *Soil Biol. Biochem.* **2006**, *38*, 898–911. [CrossRef]
14.   Luce, M.S.; Ziadi, N.; Zebarth, B.J.; Whalen, J.K.; Grant, C.A.; Gregorich, E.G.; Lafond, G.P.; Blackshaw, R.E.; Johnson, E.N.; O'Donovan, J.T.; et al. Particulate organic matter and soil mineral nitrogen concentrations are good predictors of the soil nitrogen supply to canola following legume and non-legume crops in western Canada. *Can. J. Soil Sci.* **2013**, *93*, 607–620. [CrossRef]
15.   Blanco-Canqui, H.; Francis, C.A.; Galusha, T.D. Does organic farming accumulate carbon in deeper soil profiles in the long term? *Geoderma* **2017**, *288*, 213–221. [CrossRef]
16.   Balesdent, J.; Besnard, E.; Arrouays, D.; Chenu, C. The dynamics of carbon in particle-size fractions of soil in a forest-cultivation sequence. *Plant Soil* **1998**, *201*, 49–57. [CrossRef]
17.   Whalen, J.K.; Bottomley, P.J.; Myrold, D.D. Carbon and nitrogen mineralization from light- and heavy-fraction additions to soil. *Soil Biol. Biochem.* **2000**, *32*, 1345–1352. [CrossRef]
18.   Gregorich, E.; Beare, M.; Mckim, U.; Skjemstad, J. Chemical and biological characteristics of physically uncomplexed organic matter. *Soil Sci. Soc. Am. J.* **2006**, *70*, 975–985. [CrossRef]
19.   Mirsky, S.B.; Lanyon, L.E.; Needelman, B.A. Evaluating Soil Management Using Particulate and Chemically Labile Soil Organic Matter Fractions. *Soil Sci. Soc. Am. J.* **2008**, *72*, 180–185. [CrossRef]
20.   Spargo, J.T.; Cavigelli, M.A.; Mirsky, S.B.; Maul, J.E.; Meisinger, J.J. Mineralizable soil nitrogen and labile soil organic matter in diverse long-term cropping systems. *Nutr. Cycl. Agroecosystems* **2011**, *90*, 253–266. [CrossRef]
21.   Bu, R.; Lu, J.; Ren, T.; Liu, B.; Li, X.; Cong, R. Particulate Organic Matter Affects Soil Nitrogen Mineralization under Two Crop Rotation Systems. *PLoS ONE* **2015**, *10*, e0143835. [CrossRef] [PubMed]
22.   Kader, M.A.; Sleutel, S.; Begum, S.A.; D'Haene, K.; Jegajeevagan, K.; De Neve, S. Soil organic matter fractionation as a tool for predicting nitrogen mineralization in silty arable soils. *Soil Use Manag.* **2010**, *26*, 494–507. [CrossRef]
23.   Franzluebbers, A.J.; Hons, F.M.; Zuberer, D.A. Long-Term Changes in Soil Carbon and Nitrogen Pools in Wheat Management Systems. *Soil Sci. Soc. Am. J.* **1994**, *58*, 1639. [CrossRef]
24.   Wienhold, B.J.; Halvorson, A.D. Nitrogen Mineralization Responses to Cropping, Tillage, and Nitrogen Rate in the Northern Great Plains. *Soil Sci. Soc. Am. J.* **1999**, *63*, 192–196. [CrossRef]
25.   Poudel, D.D.; Horwath, W.R.; Lanini, W.T.; Temple, S.R.; van Bruggen, A.H.C. Comparison of soil N availability and leaching potential, crop yields and weeds in organic, low-input and conventional farming systems in northern California. *Agric. Ecosyst. Environ.* **2002**, *90*, 125–137. [CrossRef]
26.   Sharifi, M.; Zebarth, B.J.; Burton, D.L.; Grant, C.A.; Porter, G.A. Organic Amendment History and Crop Rotation Effects on Soil Nitrogen Mineralization Potential and Soil Nitrogen Supply in a Potato Cropping System. *Agron. J.* **2008**, *100*, 1562–1572. [CrossRef]
27.   Constantin, J.; Beaudoin, N.; Laurent, F.; Cohan, J.-P.; Duyme, F.; Mary, B. Cumulative effects of catch crops on nitrogen uptake, leaching and net mineralization. *Plant Soil* **2011**, *341*, 137–154. [CrossRef]
28.   Mallory, E.B.; Griffin, T.S. Impacts of Soil Amendment History on Nitrogen Availability from Manure and Fertilizer. *Soil Sci. Soc. Am. J.* **2007**, *71*, 964–973. [CrossRef]
29.   Schroder, J.; Jansen, A.; Hilhorst, G. Long-term nitrogen supply from cattle slurry. *Soil Use Manag.* **2005**, *21*, 196–204. [CrossRef]
30.   Sharifi, M.; Zebarth, B.J.; Burton, D.L.; Grant, C.A.; Cooper, J.M. Evaluation of Some Indices of Potentially Mineralizable Nitrogen in Soil. *Soil Sci. Soc. Am. J.* **2007**, *71*, 1233–1239. [CrossRef]

31. Ros, G.H.; Temminghoff, E.J.M.; Hoffland, E. Nitrogen mineralization: A review and meta-analysis of the predictive value of soil tests. *Eur. J. Soil Sci.* **2011**, *62*, 162–173. [CrossRef]

32. McDonald, N.T.; Watson, C.J.; Lalor, S.T.J.; Laughlin, R.J.; Wall, D.P. Evaluation of Soil Tests for Predicting Nitrogen Mineralization in Temperate Grassland Soils. *Soil Sci. Soc. Am. J.* **2014**, *78*, 1051–1064. [CrossRef]

33. Scheu, S. Linkages between tree diversity, soil fauna and ecosystem processes. In *Forest Diversity and Function*; Springer: Berlin/Heidelberg, Germany, 2005; Volume 176, pp. 211–233.

34. Mary, B.; Beaudoin, N.; Justes, E.; Machet, J.M. Calculation of nitrogen mineralization and leaching in fallow soil using a simple dynamic model: N mineralization and leaching in fallow soil. *Eur. J. Soil Sci.* **1999**, *50*, 549–566. [CrossRef]

35. Oorts, K.; Laurent, F.; Mary, B.; Thiebeau, P.; Labreuche, J.; Nicolardot, B. Experimental and simulated soil mineral N dynamics for long-term tillage systems in northern France. *Soil Tillage Res.* **2007**, *94*, 441–456. [CrossRef]

36. Engels, T.; Kuhlmann, H. Effect of the rate of N fertilizer on apparent net mineralization of N during and after cultivation of cereal and sugar beet crops. *Z. Pflanzenernähr. Bodenkd.* **1993**, *156*, 149–154. [CrossRef]

37. Alvarez, R.; Steinbach, H.S. A review of the effects of tillage systems on some soil physical properties, water content, nitrate availability and crops yield in the Argentine Pampas. *Soil Tillage Res.* **2009**, *104*, 1–15. [CrossRef]

38. Rodrigo, A.; Recous, S.; Neel, C.; Mary, B. Modelling temperature and moisture effects on C-N transformations in soils: Comparison of nine models. *Ecol. Model.* **1997**, *102*, 325–339. [CrossRef]

39. Rémy, J.; Hébert, J. Le devenir des engrais azotés dans le sol. *Acad. Agric. Fr.* **1977**, *63*, 700–714.

40. Morvan, T.; Lambert, Y.; Germain, P.; Beff, L. A dataset from a 3-year network of field measurements of soil organic nitrogen mineralization under a mild oceanic temperate climate. *Data Brief* **2021**, *35*, 106795. [CrossRef]

41. Brisson, N.; Mary, B.; Ripoche, D.; Jeuffroy, M.; Ruget, F.; Nicoullaud, B.; Gate, P.; Devienne-Barret, F.; Antonioletti, R.; Durr, C.; et al. STICS: A generic model for the simulation of crops and their water and nitrogen balances. I. Theory and parameterization applied to wheat and corn. *Agronomie* **1998**, *18*, 311–346. [CrossRef]

42. Krom, M.D. Spectrophotometric determination of ammonia: A study of a modified Berthelot reaction using salicylate and dichloroisocyanurate. *Analyst* **1980**, *105*, 305–316. [CrossRef]

43. Sims, J.R.; Jackson, G.D. Rapid analysis of soil nitrate with chromotropic acid. *Soil Sci. Soc. Am. J.* **1970**, *35*, 603–606. [CrossRef]

44. AFNOR. Norme NF X 31-130, Méthodes chimiques, Détermination de la capacité d'échange cationique (CEC) et des cations extractibles. In *Normes Nationales et Documents Normatifs Nationaux*; Qualité des Sols: Paris, France, 1999.

45. AFNOR. Norme NF X 10390, Détermination du pH. In *Normes Nationales et Documents Normatifs Nationaux*; Qualité des Sols: Paris, France, 2005.

46. AFNOR. Norme NF X 31-107, Détermination de la distribution granulométrique des particules du sol, Méthode à la pipette. In *Normes Nationales et Documents Normatifs Nationaux*; Qualité du Sol: Paris, France, 2003.

47. Vance, E.D.; Brookes, P.C.; Jenkinson, D.S. An extraction method for measuring soil microbial biomass C. *Soil Biol. Biochem.* **1987**, *19*, 703–707. [CrossRef]

48. Gianello, C.; Bremner, J.M. A rapid steam distillation method of assessing potentially available organic nitrogen in soil. *Commun. Soil Sci. Plant Anal.* **1988**, *19*, 1551–1568. [CrossRef]

49. Hirte, J.; Leifeld, J.; Abiven, S.; Mayer, J. Maize and wheat root biomass, vertical distribution, and size class as affected by fertilization intensity in two long-term field trials. *Field Crops Res.* **2018**, *216*, 197–208. [CrossRef]

50. R Core Team. R: A Language and Environment for Statistical Computing. 2021. Available online: www.R-Project.org (accessed on 18 May 2021).

51. Louis, B.P.; Maron, P.-A.; Menasseri-Aubry, S.; Sarr, A.; Leveque, J.; Mathieu, O.; Jolivet, C.; Leterme, P.; Viaud, V. Microbial Diversity Indexes Can Explain Soil Carbon Dynamics as a Function of Carbon Source. *PLoS ONE* **2016**, *11*, e0161251. [CrossRef] [PubMed]

52. Wood, S.N. Fast stable restricted maximum likelihood and marginal likelihood estimation of semiparametric generalized linear models. *J. R. Stat. Soc. Ser. B Stat. Methodol.* **2011**, *73*, 3–36. [CrossRef]

53. Wallach, D. Working with Dynamic Crop Models. In *Evaluation, Analysis, Parameterization and Applications*; Elsevier: Amsterdam, The Netherlands, 2006.

54. Bellon-Maurel, V.; Fernandez-Ahumada, E.; Palagos, B.; Roger, J.-M.; McBratney, A. Critical review of chemometric indicators commonly used for assessing the quality of the prediction of soil attributes by NIR spectroscopy. *TrAC Trends Anal. Chem.* **2010**, *29*, 1073–1081. [CrossRef]

55. Lemaire, G.; Meynard, J. Part 1: On the Critical N Concentration in Agricultural Crops. Use of the Nitrogen Nutrition Index for the Analysis of Agronomical Data. In *Diagnosis of the Nitrogen Status in Crops*; Springer: Berlin, Germany, 1997; pp. 45–55.

56. Rozas, H.R.S.; Echeverría, H.E.; Barbieri, P.A. Nitrogen Balance as Affected by Application Time and Nitrogen Fertilizer Rate in Irrigated No-Tillage Maize. *Agron. J.* **2004**, *96*, 1622–1631. [CrossRef]

57. Bono, A.; Alvarez, R. Nitrogen mineralization in a coarse soil of the semi-arid Pampas of Argentina. *Arch. Agron. Soil Sci.* **2013**, *59*, 259–272. [CrossRef]

58. Lado-Monserrat, L.; Lull, C.; Bautista, I.; Lidón, A.; Herrera, R. Soil moisture increment as a controlling variable of the "Birch effect". Interactions with the pre-wetting soil moisture and litter addition. *Plant Soil* **2014**, *379*, 21–34. [CrossRef]

59. Franzluebbers, A.J. Microbial activity in response to water-filled pore space of variably eroded southern Piedmont soils. *Appl. Soil Ecol.* **1999**, *11*, 91–101. [CrossRef]

60. Delphin, J.-E. Estimation of nitrogen mineralization in the field from an incubation test and from soil analysis. *Agronomie* **2000**, *20*, 349–361. [CrossRef]
61. Carpenter-Boggs, L.; Pikul, J.L.; Vigil, M.F.; Riedell, W.E. Soil Nitrogen Mineralization Influenced by Crop Rotation and Nitrogen Fertilization. *Soil Sci. Soc. Am. J.* **2000**, *64*, 2038–2045. [CrossRef]
62. Culman, S.W.; Snapp, S.S.; Green, J.M.; Gentry, L.E. Short- and Long-Term Labile Soil Carbon and Nitrogen Dynamics Reflect Management and Predict Corn Agronomic Performance. *Agron. J.* **2013**, *105*, 493–502. [CrossRef]
63. Osterholz, W.R.; Liebman, M.; Castellano, M.J. Can soil nitrogen dynamics explain the yield benefit of crop diversification? *Field Crops Res.* **2018**, *219*, 33–42. [CrossRef]
64. Zhang, X.; Fang, Q.; Zhang, T.; Ma, W.; Velthof, G.L.; Hou, Y.; Oenema, O.; Zhang, F. Benefits and trade-offs of replacing synthetic fertilizers by animal manures in crop production in China: A meta-analysis. *Glob. Change Biol.* **2020**, *26*, 888–900. [CrossRef]
65. Liu, S.; Wang, M.; Yin, M.; Chu, G.; Xu, C.; Zhang, X.; Abliz, B.; Tang, C.; Wang, D.; Chen, S. Fifteen years of crop rotation combined with straw management alters the nitrogen supply capacity of upland-paddy soil. *Soil Tillage Res.* **2022**, *215*, 105219. [CrossRef]
66. Verloop, J.; Hilhorst, G.J.; Oenema, J.; Van Keulen, H.; Sebek, L.B.J.; Van Ittersum, M.K. Soil N mineralization in a dairy production system with grass and forage crops. *Nutr. Cycl. Agroecosystems* **2014**, *98*, 267–280. [CrossRef]
67. Crystal-Ornelas, R.; Thapa, R.; Tully, K.L. Soil organic carbon is affected by organic amendments, conservation tillage, and cover cropping in organic farming systems: A meta-analysis. *Agric. Ecosyst. Environ.* **2021**, *312*, 107356. [CrossRef]
68. Schroder, J.J.; Uenk, D.; Hilhorst, G.J. Long-term nitrogen fertilizer replacement value of cattle manures applied to cut grassland. *Plant Soil* **2007**, *299*, 83–99. [CrossRef]
69. Dessureault-Rompré, J.; Zebarth, B.J.; Burton, D.L.; Georgallas, A. Predicting soil nitrogen supply from soil properties. *Can. J. Soil Sci.* **2015**, *95*, 63–75. [CrossRef]
70. Griffin, G.F.; Laine, A.F. Nitrogen Mineralization in Soils Previously Amended with Organic Wastes 1. *Agron. J.* **1983**, *75*, 124–129. [CrossRef]
71. Cabrera, M.L.; Kissel, D.E. Potentially Mineralizable Nitrogen in Disturbed and Undisturbed Soil Samples. *Soil Sci. Soc. Am. J.* **1988**, *52*, 1010–1015. [CrossRef]
72. Hassink, J. Effect of soil texture on the size of the microbial biomass and on the amount of c and n mineralized per unit of microbial biomass in dutch grassland soils. *Soil Biol. Biochem.* **1994**, *26*, 1573–1581. [CrossRef]
73. Moharana, P.C.; Sharma, B.M.; Biswas, D.R.; Dwivedi, B.S.; Singh, R.V. Long-term effect of nutrient management on soil fertility and soil organic carbon pools under a 6-year-old pearl millet–wheat cropping system in an Inceptisol of subtropical India. *Field Crops Res.* **2012**, *136*, 32–41. [CrossRef]
74. Bechtold, J.S.; Naiman, R.J. Soil texture and nitrogen mineralization potential across a riparian toposequence in a semi-arid savanna. *Soil Biol. Biochem.* **2006**, *38*, 1325–1333. [CrossRef]
75. Schomberg, H.H.; Wietholter, S.; Griffin, T.S.; Reeves, D.W.; Cabrera, M.L.; Fisher, D.S.; Endale, D.M.; Novak, J.M.; Balkcom, K.S.; Raper, R.L.; et al. Assessing Indices for Predicting Potential Nitrogen Mineralization in Soils under Different Management Systems. *Soil Sci. Soc. Am. J.* **2009**, *73*, 1575–1586. [CrossRef]
76. Yin, X.; Kersebaum, K.-C.; Beaudoin, N.; Constantin, J.; Chen, F.; Louarn, G.; Manevski, K.; Hoffmann, M.; Kollas, C.; Armas-Herrera, C.M.; et al. Uncertainties in simulating N uptake, net N mineralization, soil mineral N and N leaching in European crop rotations using process-based models. *Field Crops Res.* **2020**, *255*, 107863. [CrossRef]
77. Alvarez, R.; Steinbach, H.S. Modeling Apparent Nitrogen Mineralization under Field Conditions Using Regressions and Artificial Neural Networks. *Agron. J.* **2011**, *103*, 1159–1168. [CrossRef]

Nitrogen

Article

Autumn Tillage Reduces the Effect of Plant Cover on Topsoil Nitrogen Leaching

Jorge F. Miranda-Vélez [1,*] **and Iris Vogeler** [1,2]

[1] Department of Agroecology, Aarhus University, 8830 Tjele, Denmark; iris.vogeler@agro.au.dk
[2] Grass and Forage Science/Organic Agriculture, Christian-Albrecht University of Kiel, 24118 Kiel, Germany
* Correspondence: jorge_mv@agro.au.dk

Abstract: Keeping cover crops to reduce nitrogen leaching often conflicts with timing tillage operations before the soil becomes un-trafficable during winter, while leaving cover crops in the field until spring raises concerns over pre-emptive competition with the following crop. Therefore, farmers may resort to tilling their fields in autumn after letting cover crops remain in the fields for only a short period of time. We explore the effects of this practice in a laboratory lysimeter setting by analyzing the leaching of nitrate from intact topsoil cores. Cores were extracted from no-till (NT) plots and plots tilled in autumn (AuT), in areas kept bare (B) and with volunteer winter rye plant cover (V) after harvest. Nitrate breakthrough curves show that V significantly reduced N leaching by 61% relative to B in NT, but did not have a significant effect in AuT. Dissection of leached cores and undisturbed reference cores indicated a significant removal of mineral N from the soil during the lysimeter experiment for all treatments except V in NT. This indicates that volunteer cover removed a crucial amount of leachable N and suggests that tillage counteracted the effect of V in AuT, likely due to a combination of reduced uptake and re-mineralization of N in cover crop residue.

Keywords: nitrogen; nitrogen leaching; autumn tillage; no-till; lysimeter

Citation: Miranda-Vélez, J.F.; Vogeler, I. Autumn Tillage Reduces the Effect of Plant Cover on Topsoil Nitrogen Leaching. *Nitrogen* **2022**, *3*, 186–196. https://doi.org/10.3390/nitrogen3020014

Academic Editor: Jacynthe Dessureault-Rompré

Received: 14 March 2022
Accepted: 11 April 2022
Published: 13 April 2022

Publisher's Note: MDPI stays neutral with regard to jurisdictional claims in published maps and institutional affiliations.

1. Introduction

Keeping plant cover between harvest and the sowing of the subsequent cash crop, i.e., cover crops, has many advantages in agriculture. Cover crops can be used as a natural control against weeds [1] and to reduce nitrogen leaching after harvest [2,3]. Additionally, cover crops are known to protect agricultural soil from erosion, improve soil structure and fertility, and stimulate carbon storage in the soil [4–6].

Concerns over pre-emptive competition over nutrients [7] and water [8], as well as poor field trafficability in early spring and autumn spreading of animal manures [9], often motivate farmers to limit the residence time of winter cover crops. Thus, cover crops successfully established after harvest may be terminated in the same autumn, depending on the farmer's judgement. In the Nordic countries, for instance, the use of post-harvest cover crops is either compulsory with area requirements depending on management (Denmark) or is encouraged and subsidized (Sweden, Norway and Finland). However, farmers are typically free to till their fields and terminate the cover crops after 20 October in Denmark and Sweden and after 1 October in Finland. Only in Norway is the associated subsidy contingent on spring termination of the cover crops [10].

Cover crops in temperate climates extend roots and take up soil N during a relatively short growth period before being killed by frost or terminated by the farmer. During this period, however, post-harvest N uptake by cover crops in temperate climates frequently represents over 20% of the applied fertilizer N, resulting in a 35% average reduction in soil mineral N in autumn and reducing nitrate leaching by 40 to 70% [11,12]. Upon cover crop termination, litter and root breakdown in the soil results in a gradual re-release of a significant fraction of the taken-up N. The timing and extent of this process is affected by a number of factors—e.g., type of cover crop, degree of incorporation, air and soil temperature, precipitation, and termination time—and is key to the overall effect of the cover crops [13]. If a significant amount of soil N is taken up by the cover crops and re-mineralized during the following growth season, it will be available for uptake by the cash crop and less fertilizer need to be applied to the soil. Indeed, Danish regulations require a fertilizer reduction of up to 25 kg N ha^{-1} following cover crops in order to utilize the N "carried over" by the cover crops [10]. However, if the time for soil N uptake by cover crops is short and/or re-mineralization of cover crop N takes place during a period of fallow with high precipitation, it will be at high risk of leaching [14,15]. These considerations, unfortunately, often conflict with farmer's habits and their concerns involving weed control (including the cover crop) and reduced biomass and grain yields on the following cash crop [10,16,17].

This study explores the effect of autumn tillage on the capacity of cover crops to reduce nitrogen leaching. We compared the effect of autumn inversion tillage (AuT) against no-till (NT) on the reduction of nitrogen leaching by volunteer winter rye as cover crops in a laboratory lysimeter setting. We expect that the effectiveness of the volunteer cover crops will be diminished by autumn tillage for two reasons: (1) shortening of the time available for the volunteers to take up N from the soil and (2) re-mineralization of nitrogen in the incorporated volunteer residue. Thus, we hypothesize that soil mobile N contents will be high in AuT regardless of cover crop treatment, and that leached N recovery from AuT cores with volunteer cover crops will be greater than in corresponding NT cores.

2. Materials and Methods

2.1. Field Operations and Sampling

Sampling took place in the CENTS long-term rotation and tillage experiment at the department of Agroecology, Aarhus University Flakkebjerg (55°19′ N, 11°23′ E) [18]. The soil is a sandy loam with 14.7% clay, 13.7% silt and 69.6% sand contents, average bulk density of 1.53 Mg m^{-3} and an average organic matter content of 2% [19]. Soil pH (water suspension) at the time of sampling ranged between 7.4 and 8.3, with a mean value of 7.8. The long-term experiment consists of a randomized split-plot design with four crop rotations (R1–R4) as main plots and four levels of tillage (direct sowing, shallow harrowing, deep harrowing and inversion tillage) as sub-plots, arranged in four replicate blocks [20]. For this study, sampling was restricted to the R2 rotation, which consists of 3 years of winter barley (*Hordeum vulgare*) followed by 1 year winter rape (*Brassica napus*) and 2 years winter wheat (*Triticum aestivum*), all with straw retention after harvest [18]. The last crop before sampling was substituted with winter rye (*Secale cereale*) followed by fodder radish (*Raphanus sativus*) as a winter cover crop. Sampling was also restricted to direct sowing (NT) and inversion tillage (AuT). Finally, only three of the four experimental blocks in the long-term experiment were used for sampling and are considered here as replicate samples.

The NT treatment consisted of sowing by direct drilling using a single-disk drill (2002 to spring 2006) and later a single chisel coulter drill (autumn 2006–present), and straw retention at harvest. The AuT treatment consisted of seeding with a drag coulter seed drill, straw retention at harvest and inversion ploughing followed by rolling before sowing [20]. On the year of sampling, inversion tillage in the AuT sub-plots was carried out in late October.

In late August, glyphosate (1 L ha^{-1} Roundup Bio) was applied to the NT and AuT sub-plots in the R2 rotation, killing the established mixture of cover crops, volunteers and weeds. This allowed the center portion of each sub-plot to remain bare (B), while the edges of the sub-plots were re-established exclusively with winter rye volunteers (V). Clippings carried out in the V portions of AuT showed that a growth of approximately 600 kg ha^{-1} DW in aboveground winter rye biomass at the time of tillage in October.

In early December, two intact topsoil cores (Ø = 20 cm, h = 20 cm) were extracted side by side from the center (B) and edge (V) of each AuT and NT subplot. Core extraction was carried out by slowly pressing a steel cylinder into the soil with a hydraulic press, and subsequently digging around the buried cylinder with spades to manually retrieve the cores. This resulted in two separate sets of twelve samples covering four treatment combinations (AuT/NT × B/V) per block. One set of cores was designated as reference for field conditions (Reference), while the second set of cores was designated as experimental set to undergo simulated precipitation in the laboratory (lab-rain). All soil cores were transported back from the field and put in storage at 2–4 °C on the same day as extracted, awaiting sample preparation.

Additionally, composite topsoil samples (0–20 cm depth) were taken from the core sampling areas of all tillage and cover crop treatment combinations. Representative sub-samples of loose soil were then used for total soil C and total N content analysis in a Vario Max Cube organic elemental analyzer (Elementar, Germany).

2.2. Sample Handling and Preparation

The soil water content of all cores, both Reference and lab-rain, was equalized before the leaching experiment. Each intact core was moved into a 5 °C temperature-controlled room and placed on a ceramic plate inside a ~40 L plastic tub and slowly saturated with a simulated soil solution ($CaCl_2$ 0.05 M) (Figure 1A). After 7 days of saturation, each core was slowly drained to a pressure of −10 hPa using the same ceramic plates and a controlled vacuum system. Upon reaching equilibrium with the vacuum system, lab-rain cores were transferred to a benchtop lysimeter for a leaching experiment. The Reference cores, in turn, were removed from the ceramic plates upon reaching equilibrium at −10 hPa, and kept lidded at 5 °C for the duration of the leaching experiment.

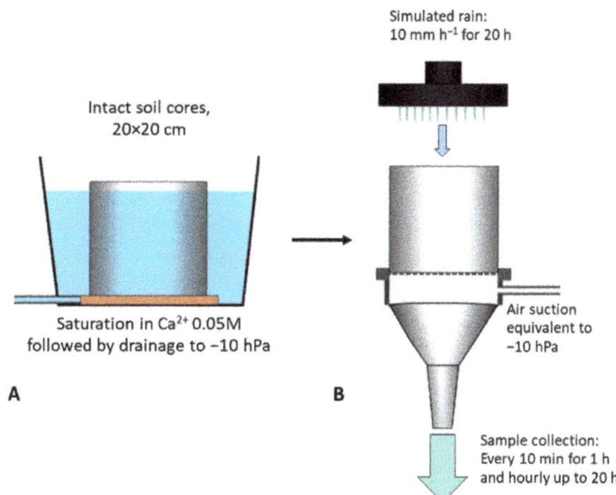

Figure 1. Sample preparation (**A**) and leaching experiment (**B**) setup. Preparation consisted of saturating intact soil cores in a simulated soil solution followed by drainage and equilibration at −10 hPa. In the leaching experiment, the soil cores were irrigated with simulated rain at a rate of 10 mm h⁻¹ for 20 h, while leachate was collected.

2.3. Leaching Experiment

The leaching experiment consisted of 200 mm of simulated precipitation, administered in a single course of 20 h at a constant rate of 10 mm h⁻¹ on individual benchtop lysimeters (Figure 1B). The lysimeters were fitted with rotating rain heads utilizing blunt needles as dripping nozzles. In order to prevent soil dispersion during the leaching experiment, simulated rainwater consisted of a weak salt solution (1.76 mg L⁻¹ $CaCl_2·2H_2O$ + 3.05 mg L⁻¹ $MgCl_2·6H_2O$ + 7.07 mg L⁻¹ NaCl), similar to that used in other leaching studies (e.g., [21,22]). The total simulated precipitation, 200 mm, corresponds approximately to the total precipitation recorded by the meteorological station at AU Flakkebjerg (211 mm) during the months of October, November and December of 2019. For comparison, the total normal autumn precipitation (September–November, 1961–1990) in Denmark is 228 mm [23]. The leachate from each core was collected in pre-weighed plastic bottles using automatic rotating carrousels at 10-min intervals during the first hour and at 1-h intervals until termination of the experiment.

The nitrate in the collected leachate was quantified by ion chromatography using a Metrosep cation resin suppressor (Metrohm, Switzerland) and an A Supp 5 anion exchange column followed by an electric conductivity detector, with a carbonate buffer as eluent. Leachate samples were filtered using cartridge polyether sulfone (PES) filters (pore size 0.22 μm) before injection.

Following the leaching experiment, both the Reference and lab-rain soil cores were dissected in four layers, corresponding to depths of 0–5 cm, 5–10 cm, 10–15 cm and 15–20 cm, relative to the metal core casings. Each layer in each core was weighed immediately after dissection and placed in an airtight plastic bag, then stored at 2 °C awaiting sub-sampling.

Two representative soil sub-samples (~50 g) were taken from each layer in each core. The first set dried at 105 °C for 24 h for water content determination at the time of subsampling. The second set of subsamples was weighed and placed in vials for NO_3^- and NH_4^+ (mineral N) extraction. Mineral N extraction consisted of suspending the soil samples in 0.2 L of KCl 1M solution and mixing for 30 min in a rotary shaker at 20 rpm, then filtering the supernatant using ashless paper filters. The collected extract was then frozen at -20 °C awaiting analysis. Nitrogen as nitrate and as ammonium in the extracts were quantified colorimetrically in a Seal Analytical AA500 auto-analyzer as described by Best [24] and Crooke and Simpson [25], respectively.

2.4. Calculations and Statistical Analysis

All data handling, visualization and analysis was carried out using R version 4.1.0 "Camp Pontanezen", released in May 2021 [26].

Total C, total N and pH measurements were analyzed using linear mixed models with tillage and cover crop treatments as main effects and field block as random effect (packages lme4 and lmerTest [27,28]).

Nitrate leaching (mg) was calculated as the product of the collected volume of leachate at each sampling time and the leachate NO_3^- concentration as determined by ion chromatography. The total leached N as NO_3^- (N_{NO3}, mg N) was calculated as the sum of the leached NO_3^- over the course of the leaching experiment, multiplied by the ratio of the atomic mass of N to the molar mass of the NO_3^- ion. The mass of mineral N (Nmin, mg N) extracted from soil subsamples was calculated as the sum of N as NO_3^- and NH_4^+. Total Nmin was then calculated for each dissection layer and full cores using the wet mass at dissection and the determined water content.

Differences in N_{NO3} leaching were analyzed using linear mixed models, with total leached N_{NO3} as response variable, tillage and plant cover as main effects and field block as random effect. Model estimates and 95% confidence intervals for leached N_{NO3} were obtained from model marginal means (package emmeans [29]) and significant differences were evaluated by multiple pairwise comparison among all tillage and plant cover treatment combinations (package multcomp [30]).

The effects of tillage and plant cover on the differences between the total Nmin content of the Reference and lab-rain cores was evaluated using linear mixed models with experimental group (i.e., Reference or lab-rain), tillage and plant cover as main effects and field block as random effect. Here, all interactions including the three-way interaction were explicitly preserved in the model in order to evaluate the influence of tillage and cover crop treatments on the difference between Reference and lab-rain cores. Model estimates and 95% confidence intervals were obtained from model marginal means and significant differences between Reference and lab-rain were evaluated by multiple pairwise comparisons grouped by tillage and plant cover treatment combinations.

3. Results and Discussion

3.1. Breakthrough Curves and N_{NO3} Leaching

The NO_3^- breakthrough (Figure 2(top)) and cumulative N_{NO3} leaching (Figure 2(bottom)) curves show different effects of cover crops on N leaching from the topsoil, depending on tillage. In the NT treatment, there is a clear difference due to plant cover in both leachate NO_3^- concentrations and total N_{NO3} leached over the course of the experiment. Specifically, NO_3^- concentrations were lower and there was less N_{NO3} leached under volunteer plant cover, compared to bare soil. In AuT, the difference between B and V was much more subtle for both NO_3^- concentrations and accumulated N_{NO3} leaching throughout the experiment. Additionally, NT had an overall effect on the elution of nitrate from the intact soil cores, independent of plant cover, where NO_3^- peaked earlier and more sharply in NT compared to AuT. This, in turn, indicates increased preferential flow due to better-developed macropore flow pathways [31] in NT.

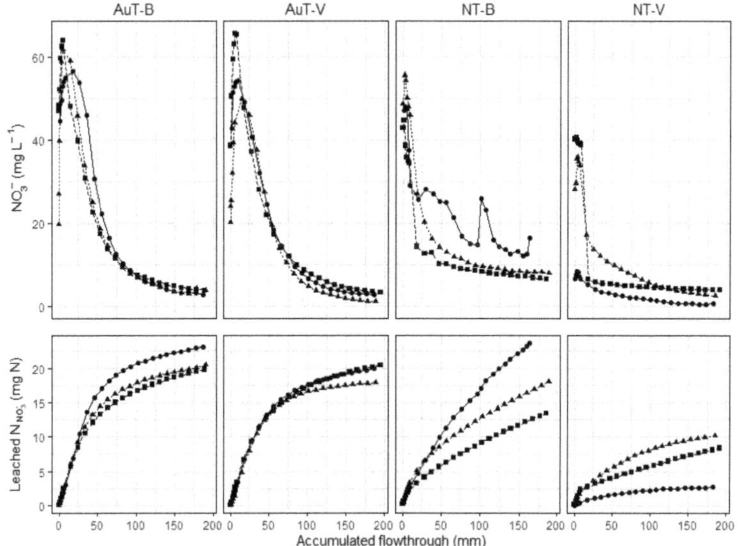

Figure 2. Nitrate concentration breakthrough (**top**) and accumulated leaching of NO_3^- N (**bottom**) curves from intact soil cores subjected to ~200 mm of simulated rain in a laboratory lysimiter. The cores belonged to two tillage treatments, no-till (NT) and autumn inversion tillage (AuT), as well as two cover crop treatments, bare fallow (B) and winter rye volunteers (V). Symbols indicate different field experiment blocks.

Statistical analysis of total N_{NO3} leaching showed significant effects by both plant cover (F = 10.98, df = 8, *p* = 0.010) and tillage (F = 14.72, df = 8, *p* = 0.005), as well as a significant interaction between plant cover and tillage (F = 5.88, df = 8, *p* = 0.041). Pairwise comparisons (Table 1) show that N_{NO3} leaching under simulated precipitation was significantly lower in V compared to B in NT. The reduction corresponds to 11.3 mg N, i.e., 61% of the N_{NO3} leached in the bare soil treatment of NT. In contrast, no significant differences were found between cover crop treatments in AuT, with very similar total N_{NO3} leaching amounts of approximately 20 mg N per core. These results indicate, firstly, that the volunteer treatment in NT significantly and considerably reduced N losses during the leaching experiment. Secondly, this effect was completely lost in the volunteer treatment of autumn-tilled samples. Importantly, there was also no significant difference between N_{NO3} leaching between the bare fallow treatment in AuT and the bare fallow treatment in NT, indicating that tillage treatment alone did not reduce total N leaching from the intact cores.

Finally, all treatments have the same crop history, having been consistently kept under the same crop rotation since the establishment of the long-term trial in 2002. It is therefore highly unlikely that the reduction was caused by factors other than the establishment of volunteers and their interaction with tillage in the year of sampling.

Table 1. Nitrogen as NO_3^- (N_{NO3}) leached from intact topsoil cores subjected to 200 mm of simulated rain. Mean values and 95% confidence intervals (CI) were obtained from linear mixed effects models with tillage and plant cover as main effects and experimental block as random effect.

Tillage	Plant Cover	Mean Total Recovered N_{NO3} (mg N)	95% CI (mg N)	Group [1]
NT	B	18.4	12.09–24.62	a
NT	V	7.1	0.83–13.36	b
AuT	B	21.1	14.86–27.39	a
AuT	V	19.4	13.12–25.64	a

[1] Multiple pairwise comparisons significance threshold, $p = 0.05$.

As mentioned earlier, cover crop residence time is known to increase the overall effect of cover crops in reducing N leaching, with significantly lower N leaching with spring incorporation relative to autumn incorporation [10]. However, our results suggest that a difference in autumn cover crop residence of as little as one month can significantly hinder the efficacy of cover crops in reducing N leaching. Indeed, Sieling [17] highlight that N-uptake by cover crops is intrinsically linked to dry matter accumulation and therefore to intercepted photosynthetically active radiation. This means that cover crop residence is of particular importance for N uptake in the autumn when days are still relatively long (and warm). According to Vos and Van Der Putten [32], cover crops in autumn very broadly accumulate dry matter at a rate of 1.12 g per MJ intercepted global radiation. This figure allows us to make a broad calculation of the effect of autumn tillage on cover crops. Based on 65.5 MJ m^{-2} total global radiation measured in November at Flakkebjerg station and an aboveground N content of 3.76% measured in plant clippings, the maximum potential N uptake by cover crops in AuT was 87 mg N per core lower than in NT. As discussed before, there are many other factors at play in determining the effect of cover crops in reducing N leaching after harvest, among them the fate of taken-up N once the cover crop is terminated. However, in terms of N uptake alone, prolonged field residence in early and mid-autumn is likely only second in importance to successful establishment of cover crops.

3.2. Soil Nmin Content

The mass of Nmin at the four dissected depths (Figure 3) was expectedly different in Reference and lab-rain cores, with clear depletion of soil mineral N after 200 mm of simulated rain in all treatments. Likewise, total mineral N contents in Reference and lab-rain cores show significant N losses during the leaching experiment in all tillage and plant cover treatments, except for NT-V (Table 2). These losses, approximately 22 mg N per core on average, closely resemble the total N_{NO3} amounts recovered in the leachate for B in NT and both B and V in AuT.

Pairwise comparisons show that Nmin contents did not differ significantly between bare fallow and volunteer cores for either Reference or lab-rain cores in AuT. In contrast, pairwise comparison of NT Reference cores revealed a significantly higher total Nmin amount in B compared to V (13.6 mg N, SE = 4.96, $p = 0.017$) and, given that the volunteer cover crops were not incorporated in NT, this difference is attributable exclusively to a reduced cover crop N uptake. Interestingly, the amount of Nmin lost from NT-V cores was considerably smaller than in the AuT-B and AuT-V treatments. This supports our hypothesis that, through a combination of reduced uptake and re-mineralization, autumn tillage can hinder the intended function of cover crops.

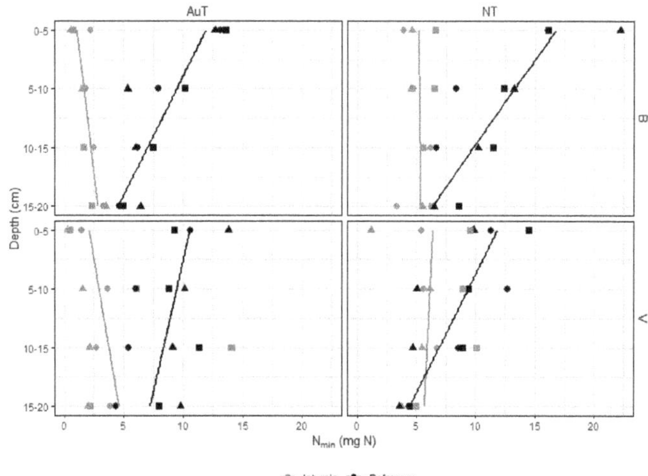

Figure 3. Mineral nitrogen (Nmin) amounts determined in four 5 cm layers in each intact core. Lab-rain cores underwent 200 mm of simulated rain in a laboratory lysimeter before dissection, while Reference cores were dissected without simulated rain. AuT and NT represent, respectively, inversion tillage carried out in late October and no-till. B and V represent bare fallow and winter rye volunteers as cover crops, respectively. The solid lines indicate linear interpolations of Nmin changes with depth.

Table 2. Total mineral nitrogen (Nmin) in intact cores dissected after 200 mm of simulated rain in laboratory lysimeters (lab-rain) and without simulated rain (Reference). AuT represents cover crop termination by inversion tillage approximately one month before sampling. NT represents no-till and continued cover crop residence until sampling. B and V represent bare fallow and winter rye volunteers as cover crops, respectively. Model estimates and 95% confidence intervals (CI) for Nmin were obtained from linear mixed effects models with experimental set tillage and plant cover as main effects and experimental block as random effect.

Experimental Set	Tillage	Plant Cover	Total Nmin (mg N)	95% CI (mg N)	Group [1]
Reference			46.14	35.15–56.1	a
Lab-rain	NT	B	21.15	11.17–31.1	b
Difference			24.99		
Reference			32.55	22.57–42.5	a
Lab-rain	NT	V	24.16	14.17–34.1	a
Difference			8.39		
Reference			33.02	23.04–43.0	a
Lab-rain	AuT	B	7.98	−2.01–18.0	b
Difference			25.05		
Reference			35.75	25.76–45.7	a
Lab-rain	AuT	V	13.57	3.59–23.6	b
Difference			22.17		

[1] Grouping by pairwise comparisons restricted to Reference and lab-rain treatments of the same tillage and plant cover treatments. Significance threshold, $p = 0.05$.

Unexpectedly, lab-rain NT cores retained on average approximately 5 mg N per dissected layer after the leaching experiment. In contrast, lab-rain AuT cores retained an average N mass per layer of 2 mg N in B and 3.4 mg N in V (Figure 3). The difference between AuT and NT lab-rain cores is likely due to a bypass effect, where soil solutes dispersed in the matrix are partially protected from leaching in soils experiencing heavy precipitation. This effect has been found to be more prevalent in no-till and reduced till soils, as resident solutes in the matrix are increasingly bypassed by macropore water flow [33,34]. Importantly, this effect appears to be independent from cover crops treatment in our results, with no significant difference between B and V lab-rain total N content, suggesting that the soil N which is most strongly bypassed by macropore flow at high precipitation rates, is also poorly available for uptake by cover crops. Further research into the plant availability of matrix-associated Nmin in NT systems is necessary, particularly in the context of preferential exploration of pre-existing macropores by plant roots in more compact soils [35].

Organic elemental analysis of topsoil samples showed no significant main effects of cover crop or tillage on either total C or total N contents (mean values 15.13 mg C g^{-1} and 1.37 mg N g^{-1}), in spite of near-significant trends of greater total C (3.6 mg C g^{-1}, SE = 1.58, $p = 0.052$) and total N (0.29 mg N g^{-1}, SE = 0.149, $p = 0.088$) in NT compared to AuT. The trend of greater total C content in NT resembles results by Gómez-Muñoz et al. [19], who found significantly higher soil total C contents in the upper 25 cm of NT soil in the same field experiment. However, lack of significant differences in our total C and N measurements and the fact that leaching was high in NT-B as well as AuT-B and AuT-V suggest that the results from the leaching experiment are not related to any underlying differences in total C and N soil contents. In Denmark, inversion tillage is typically carried out with a plough depth of 20–30 cm. It is therefore possible some volunteer plant material was buried below the sampling depth of the soil cores, in which case a portion of the corresponding re-mineralization of cover crop N would not have been captured by the leaching experiment. However, examination of the mineral N contained in the dissection layers at different depths (Figure 3) shows that Nmin is more evenly distributed across depths in AuT-V compared to AuT-B, suggesting that cover crop residues were mixed throughout the plough layer and the taken-up N had begun to re-mineralize in the one month span between tillage and sampling. Indeed, incubation studies have previously found that between 20% and 60% of the total N content in cover crop residues can mineralize within 1 month of incorporation at temperatures similar to those found in the field in autumn and winter [36,37].

3.3. Considerations on N Losses at Field Scale

Nutrient losses from the topsoil are of great relevance both for soil fertility and diffuse nutrient emissions from agriculture. Firstly, the root mass of many cash crops is strongly concentrated in the upper layers of the soil [38,39], and spring crops tend to proliferate new roots faster in the upper soil layers at early stages of development [40]. Secondly, the roots of many common cover crops, including winter rye, primarily explore the topsoil in the autumn [41], resulting often in greatest root growth in the upper 20 cm of the soil [42]. Thus, although autumn N leaching in the topsoil does not constitute a removal of said N from the soil column, it does reduce N availability for the following cash crop and increases the risk of diffuse N emissions into the environment.

Our breakthrough curve results indicate topsoil leaching losses of 2.2 kg N ha^{-1} in NT-V and approximately 6.2 kg N ha^{-1} in all other treatments, while soil core dissections show leaching losses of approximately 2.7 kg N ha^{-1} and between 7 and 8 kg N ha^{-1}, respectively. These losses are much smaller than those commonly reported from field and outdoor lysimeter trials, which tend to average approximately 50 kg N ha^{-1} and can be as high as 98 kg N ha^{-1} (e.g., [11,12]). This is due primarily to the use of core samples rather than full soil columns, which limits the observations to the upper 20 cm of the soil. Additionally, it is likely some N was leached from the topsoil before sampling. The weather station at Flakkebjerg registered a total precipitation of 266 mm during the uncommonly wet 2019 autumn in Denmark [43], which would have carried significant amounts of NO$_3^-$ into the subsoil, below the sampling depth of our soil cores. Finally, it is also likely that some NO$_3^-$ was removed during core saturation and subsequent draining in the laboratory, although the total removal of N by drained water would have been limited as the volumetric water content of the cores was high after sample preparation (at a pressure head of -10 hPa). In either case, given that all cores underwent the exact same sample preparation procedure regardless of treatment, the differences between treatments (or lack thereof) remain informative in spite of the reduced total leaching amounts.

We acknowledge the important distinction between N leached from the intact cores in this study, and N leaching in the field. Extrapolating mesocosm and laboratory results to the field scale is not entirely straightforward, given the natural variability of the soil and the limited representation of this variability that a small sample can provide. However, Valkama et al. [12] found no significant differences in results from field and lysimeter analyses in a meta-analysis of N leaching losses that included 13 field experiments and 6 lysimeter experiments. Furthermore, the representative quality of other measurements, e.g., suction cup or tile drain measurements, has also been questioned as soil N content and actual water drainage in the field remain difficult to determine [44], forcing studies to extrapolate or model some part of their results. Thus, exploiting the increased opportunities for controlled drainage and soil analysis afforded by laboratory-scale analyses remains valuable in the study of N leaching in agriculture.

Author Contributions: Conceptualization, J.F.M.-V. and I.V.; methodology, J.F.M.-V. and I.V.; validation, I.V.; formal analysis, J.F.M.-V.; investigation, J.F.M.-V. and I.V.; data curation, J.F.M.-V.; writing—original draft preparation, J.F.M.-V.; writing—review and editing, I.V.; visualization, J.F.M.-V.; supervision, I.V.; project administration, I.V.; funding acquisition, I.V. All authors have read and agreed to the published version of the manuscript.

Funding: This research was funded by the Aarhus University Research Foundation (AUFF) Starting Grant.

Institutional Review Board Statement: Not applicable.

Informed Consent Statement: Not applicable.

Data Availability Statement: The data analyzed and reported in this study is openly available in FigShare at doi:10.6084/m9.figshare.19354721.

Acknowledgments: We would like to thank Lars Juhl Munkholm and Elly Møller Hansen for their help in the execution of this project, and for granting access to the long-term reduced tillage experiment at Aarhus University Flakkebjerg. Additionally, we would like to thank Michael Koppelgaard, Stig Rassmussen, Karen B. Heinager and Eugene Driessen for their technical assistance in the field and laboratory.

Conflicts of Interest: The authors declare no conflict of interest.

References

1. Creamer, N.G.; Bennett, M.A.; Stinner, B.R.; Cardina, J.; Regnier, E.E. Mechanisms of weed suppression in cover crop-based production systems. *HortScience* **1996**, *31*, 410–413. [CrossRef]
2. Askegaard, M.; Olesen, J.E.; Rasmussen, I.A.; Kristensen, K. Nitrate leaching from organic arable crop rotations is mostly determined by autumn field management. *Agric. Ecosyst. Environ.* **2011**, *142*, 149–160. [CrossRef]
3. Macdonald, A.J.; Poulton, P.R.; Howe, M.T.; Goulding, K.W.T.; Powlson, D.S. The use of cover crops in cereal-based cropping systems to control nitrate leaching in SE England. *Plant Soil* **2005**, *273*, 355–373. [CrossRef]
4. Dabney, S.M.; Delgado, J.A.; Reeves, D.W. Using winter cover crops to improve soil and water quality. *Commun. Soil Sci. Plant Anal.* **2001**, *32*, 1221–1250. [CrossRef]
5. Bronick, C.J.; Lal, R. Soil structure and management: A review. *Geoderma* **2005**, *124*, 3–22. [CrossRef]
6. Thorup-Kristensen, K. The effect of nitrogen catch crop species on the nitrogen nutrition of succeeding crops. *Fertil. Res.* **1994**, *37*, 227–234. [CrossRef]
7. Thorup-Kristensen, K. The Effect of Nitrogen Catch Crops On the Nitrogen Nutrition of A Succeeding Crop: I. Effects Through Mineralization And Pre-Emptive Competition. *Acta Agric. Scand. Sect. B Soil Plant Sci.* **1993**, *43*, 74–81. [CrossRef]
8. Stipešević, B.; Kladivko, E.J. Effects of winter wheat cover crop desiccation times on soil moisture, temperature and early maize growth. *Plant Soil Environ.* **2005**, *51*, 255–261. [CrossRef]
9. Chambers, B.J.; Smith, K.A.; Pain, B.F. Strategies to encourage better use of nitrogen in animal manures. *Soil Use Manag.* **2000**, *16*, 157–166. [CrossRef]

10. Aronsson, H.; Hansen, E.M.; Thomsen, I.K.; Liu, J.; Øgaard, A.F.; Känkänen, H.; Ulén, B. The ability of cover crops to reduce nitrogen and phosphorus losses from arable land in southern Scandinavia and Finland. *J. Soil Water Conserv.* **2016**, *71*, 41–55. [CrossRef]

11. Tonitto, C.; David, M.B.; Drinkwater, L.E. Replacing bare fallows with cover crops in fertilizer-intensive cropping systems: A meta-analysis of crop yield and N dynamics. *Agric. Ecosyst. Environ.* **2006**, *112*, 58–72. [CrossRef]

12. Valkama, E.; Lemola, R.; Känkänen, H.; Turtola, E. Meta-analysis of the effects of undersown catch crops on nitrogen leaching loss and grain yields in the Nordic countries. *Agric. Ecosyst. Environ.* **2015**, *203*, 93–101. [CrossRef]

13. Thorup-Kristensen, K.; Nielsen, N.E. Modelling and measuring the effect of nitrogen catch crops on the nitrogen supply for succeeding crops. *Plant Soil* **1998**, *203*, 79–89. [CrossRef]

14. Thorup-Kristensen, K.; Dresbøll, D.B. Incorporation time of nitrogen catch crops influences the N effect for the succeeding crop. *Soil Use Manag.* **2010**, *26*, 27–35. [CrossRef]

15. Känkänen, H.; Kangas, A.; Mela, T. Timing incorporation of different green manure crops to minimize the risk of nitrogen leaching. *Agric. Food Sci.* **1998**, *7*, 553–567. [CrossRef]

16. Böldt, M.; Taube, F.; Vogeler, I.; Reinsch, T.; Kluß, C.; Loges, R. Evaluating Different Catch Crop Strategies for Closing the Nitrogen Cycle in Cropping Systems—Field Experiments and Modelling. *Sustainability* **2021**, *13*, 394. [CrossRef]

17. Sieling, K. Improved N transfer by growing catch crops—A challenge. *J. Für Kult.* **2019**, *71*, 145–160. [CrossRef]

18. Hansen, E.M.; Munkholm, L.J.; Olesen, J.E.; Melander, B. Nitrate Leaching, Yields and Carbon Sequestration after Noninversion Tillage, Catch Crops, and Straw Retention. *J. Environ. Qual.* **2015**, *44*, 868–881. [CrossRef]

19. Gómez-Muñoz, B.; Jensen, L.S.; Munkholm, L.; Olesen, J.E.; Hansen, E.M.; Bruun, S. Long-term effect of tillage and straw retention in conservation agriculture systems on soil carbon storage. *Soil Sci. Soc. Am. J.* **2021**, *85*, 1465–1478. [CrossRef]

20. Hansen, E.M.; Munkholm, L.J.; Melander, B.; Olesen, J.E. Can non-inversion tillage and straw retainment reduce N leaching in cereal-based crop rotations? *Soil Tillage Res.* **2010**, *109*, 1–8. [CrossRef]

21. Katuwal, S.; Norgaard, T.; Moldrup, P.; Lamandé, M.; Wildenschild, D.; de Jonge, L.W. Linking air and water transport in intact soils to macropore characteristics inferred from X-ray computed tomography. *Geoderma* **2015**, *237–238*, 9–20. [CrossRef]

22. Paradelo, M.; Katuwal, S.; Moldrup, P.; Norgaard, T.; Herath, L.; de Jonge, L.W. X-ray CT-Derived Soil Characteristics Explain Varying Air, Water, and Solute Transport Properties across a Loamy Field. *Vadose Zone J.* **2016**, *15*, 1–13. [CrossRef]

23. Rubek, F. Sammendrag af Vinter 2020–2021. 2021. Available online: https://www.dmi.dk/fileadmin/user_upload/Afrapportering/Seasonsammendrag/Sammendrag_af_vinter_2020-2021.pdf (accessed on 3 March 2022).

24. Best, E.K. An Authomated Method for Determining Nitrate-Nitrogen in Soil Extracts. *Qld. J. Agric. Anim. Sci.* **1976**, *33*, 161–166.

25. Crooke, W.M.; Simpson, W.E. Determination of ammonium in Kjeldahl digests of crops by an automated procedure. *J. Sci. Food Agric.* **1971**, *22*, 9–10. [CrossRef]

26. R Core Team. *R: A Language and Environment for Statistical Computing (Version 4.1.0)*; R Foundation for Statistical Computing: Vienna, Austria, 2021. Available online: https://www.r-project.org/ (accessed on 3 March 2022).

27. Bates, D.; Mächler, M.; Bolker, B.; Walker, S. Fitting linear mixed-effects models using lme4. *J. Stat. Softw.* **2015**, *67*, 1–48. [CrossRef]

28. Kuznetsova, A.; Brockhoff, P.B.; Christensen, R.H.B. lmerTest Package: Tests in Linear Mixed Effects Models. *J. Stat. Softw.* **2017**, *82*, 1–26. [CrossRef]

29. Lenth, R. Emmeans: Estimated Marginal Means, AKA Least-Squares Means (R Package Version 1.6.2-1). 2021. Available online: https://cran.r-project.org/package=emmeans (accessed on 3 March 2022).

30. Hothorn, T.; Bretz, F.; Westfall, P. Simultaneous inference in general parametric models. *Biom. J.* **2008**, *50*, 346–363. [CrossRef]

31. Soto-Gómez, D.; Pérez-Rodríguez, P.; Vázquez-Juiz, L.; López-Periago, J.E.; Paradelo, M. Linking pore network characteristics extracted from CT images to the transport of solute and colloid tracers in soils under different tillage managements. *Soil Tillage Res.* **2018**, *177*, 145–154. [CrossRef]

32. Vos, J.; van der Putten, P.E.L. Field observations on nitrogen catch crops. I. Potential and actual growth and nitrogen accumulation in relation to sowing date and crop species. *Plant Soil* **1997**, *195*, 299–309. [CrossRef]

33. Gerke, H.H. Preferential flow descriptions for structured soils. *J. Plant Nutr. Soil Sci.* **2006**, *169*, 382–400. [CrossRef]

34. Shipitalo, M.J.; Dick, W.A.; Edwards, W.M. Conservation tillage and macropore factors that affect water movement and the fate of chemicals. *Soil Tillage Res.* **2000**, *53*, 167–183. [CrossRef]

35. Atkinson, J.A.; Hawkesford, M.J.; Whalley, W.R.; Zhou, H.; Mooney, S.J. Soil strength influences wheat root interactions with soil macropores. *Plant Cell Environ.* **2020**, *43*, 235–245. [CrossRef] [PubMed]

36. Miranda-Vélez, J.F.; Vogeler, I. Exploring Temperature-Related Effects in Catch Crop Net N Mineralization Outside of First-Order Kinetics. *Nitrogen* **2021**, *2*, 110–127. [CrossRef]

37. Thomsen, I.K.; Elsgaard, L.; Olesen, J.E.; Christensen, B.T. Nitrogen release from differently aged *Raphanus sativus* L. nitrate catch crops during mineralization at autumn temperatures. *Soil Use Manag.* **2016**, *32*, 183–191. [CrossRef]

38. Gan, Y.T.; Liang, B.C.; Liu, L.P.; Wang, X.Y.; McDonald, C.L. C:N ratios and carbon distribution profile across rooting zones in oilseed and pulse crops. *Crop Pasture Sci.* **2011**, *62*, 496–503. [CrossRef]

39. Munkholm, L.J.; Hansen, E.M.; Olesen, J.E. The effect of tillage intensity on soil structure and winter wheat root/shoot growth. *Soil Use Manag.* **2008**, *24*, 392–400. [CrossRef]

40. Smit, A.L.; Groenwold, J. Root characteristics of selected field crops: Data from the Wageningen Rhizolab (1990–2002). *Plant Soil* **2005**, *272*, 365–384. [CrossRef]

41. Kemper, R.; Bublitz, T.A.; Müller, P.; Kautz, T.; Döring, T.F.; Athmann, M. Vertical Root Distribution of Different Cover Crops Determined with the Profile Wall Method. *Agriculture* **2020**, *10*, 503. [CrossRef]
42. Guo, B.-B.; Liu, B.-C.; He, L.; Wang, Y.-Y.; Feng, W.; Zhu, Y.-J.; Jiao, N.-Y.; Wang, C.-Y.; Guo, T.-C. Root and nitrate-N distribution and optimization of N input in winter wheat. *Sci. Rep.* **2019**, *9*, 18018. [CrossRef]
43. Rubek, F. Efterår 2019. Available online: https://www.dmi.dk/fileadmin/user_upload/Afrapportering/Seasonsammendrag/Sammendrag_2019_efteraar.pdf (accessed on 3 March 2022).
44. Ramos, C.; Kücke, M. A Review of Methods for Nitrate Leaching Measurement. *Acta Hortic.* **2001**, *563*, 259–266. [CrossRef]

Nitrogen

Article

Winter Malting Barley Growth, Yield, and Quality following Leguminous Cover Crops in the Northeast United States

Arthur Siller [1,*], Heather Darby [2], Alexandra Smychkovich [1] and Masoud Hashemi [1,*]

[1]   Stockbridge School of Agriculture, University of Massachusetts, Amherst, MA 01003, USA;
     asmychko@umass.edu
[2]   Department of Plant and Soil Science, University of Vermont, St. Albans, VT 05478, USA;
     Heather.Darby@uvm.edu
*   Correspondence: asiller@umass.edu (A.S.); masoud@umass.edu (M.H.)

Abstract: There is growing interest in malting barley (*Hordeum vulgare* L.) production in the Northeastern United States. This crop must meet high quality standards for malting but can command a high price if these quality thresholds are met. A two-year field experiment was conducted from 2015 to 2017 to evaluate the impact of two leguminous cover crops, sunn hemp (*Crotalaria juncea* L.) and crimson clover (*Trifolium incarnatum* L.), on subsequent winter malting barley production. Four cover crop treatments—sunn hemp (SH), crimson clover (CC), sunn hemp and crimson clover mixture (SH + CC), and no cover crop (NC)—were grown before planting barley at three seeding rates (300, 350, and 400 seeds m^{-2}). SH and SH + CC produced significantly more biomass and residual nitrogen than the CC and NC treatments. Higher barley seeding rates led to higher seedling density and winter survival. However, the subsequent spring and summer barley growth metrics, yield, and malting quality were not different in any of the treatments. There is much left to investigate in determining the best malting barley production practices in the Northeastern United States, but these results show that winter malting barley can be successfully integrated into crop rotations with leguminous plants without negative impacts on barley growth, yield, and grain quality.

Keywords: winter malting barley; malting quality indices; summer cover crops; sunn hemp; crimson clover; seeding rate; nitrogen management

Citation: Siller, A.; Darby, H.; Smychkovich, A.; Hashemi, M. Winter Malting Barley Growth, Yield, and Quality following Leguminous Cover Crops in the Northeast United States. *Nitrogen* **2021**, *2*, 415–427. https://doi.org/10.3390/nitrogen2040028

Academic Editor: Jacynthe Dessureault-Rompré

Received: 1 September 2021
Accepted: 5 October 2021
Published: 8 October 2021

Publisher's Note: MDPI stays neutral with regard to jurisdictional claims in published maps and institutional affiliations.

1. Introduction

Winter malting barley (*Hordeum vulgare* L.) is an emerging crop in the Northeastern United States [1,2]. Although it has the potential to be a profitable crop in the region, achieving the high-quality grains needed for malting purposes is challenging because of the Northeast's humid environment and seasonal temperature extremes [3]. However, if the barley meets the malting quality standards, there is a price premium compared with feed grain [4] and the potential for additional local markets in the regional malting and brewing industry [5,6].

A successful malting barley crop must grow well, produce good yields of high-quality grain, and be harvested and stored correctly to maintain quality [7]. To be acceptable for malting, the barley grains should be large, low in protein, free of or very low in carcinogenic deoxynivalenol (DON) toxin [8], and sprout well during the malting process [9]. Farmers can successfully grow malting barley by combining three methods: (1) choosing a site-appropriate variety that will overwinter, resist locally common diseases, and remain upright after heading [10,11]; (2) correctly timing their harvest to avoid partial sprouting in the field and using forced air dryers if weather does not permit dry-down in the field [2,12]; and (3) using growing practices that have been shown to promote good malting quality [1,13].

Grain size is quantified by three metrics: test weight, percent plump, and percent thin. High test weight and high percent plump indicate that the kernels are large and relatively uniform, while high percent thin means that the kernels are small with little energy for the malting process [7]. For the grains to sprout well during malting, they should have greater than 95% germinative energy and they must have a falling number greater than 250 s, indicating that they have not pre-sprouted in the field or during storage. Deoxynivalenol (DON) content must be below 0.5 mg kg^{-1} while protein should be below 125 g kg^{-1} [10,12]. While regional weather variability may prevent farmers from achieving malt-quality harvests in every year [2], the growing body of research into winter malting barley production can minimize this production risk for growers.

Previous research has produced varietal recommendations for malting barley growing in the Northeast [10,11] as well as recommendations for planting dates and fertilization rates [13,14]. However, winter malting barley is still a relatively new and minor crop in the Northeast [15] and most scientific reports are related to the drier areas of western North America.

In addition to selecting appropriate varieties and awareness of agronomic recommendations, it is essential that farmers understand how malting barley interacts with other crops in their rotations [16] and how other growing practices could influence crop rotation decisions. Since nitrogen can affect many aspects of barley malting quality, nitrogen cycling is of particular interest when considering how to integrate malting barley into a larger crop rotation. High levels of nitrogen fertilization can increase malting barley yields and grain size [17,18] but can also lead to excessive protein content [13,17–20], lower nitrogen use efficiency [13,17,19,21], and lower falling number [13]. While leguminous crops can contribute substantial amounts of nitrogen to subsequent crops, it is unknown whether nitrogen from legumes would have the same effects on winter malting barley as soluble nitrogen sources.

Since barley has the potential for producing a large number of tillers in the spring, the final yield does not respond linearly to increased planting density [22,23]. However, higher seeding rates may counteract the effect of excess nitrogen since higher seeding rates can reduce protein concentration and grain size [20,24]. The impact of crop rotation patterns may also be more noticeable in the fall when the plants are small [23] and different seeding rates may be differentially productive following a nitrogen-producing legume than a summer fallow.

Sunn hemp (*Crotalaria juncea* L.) and crimson clover (*Trifolium incarnatum* L.) are grown in the Northeast as summer forages or cover crops and can fit well into short growing periods before winter barley planting in the fall [25,26]. Farther north in eastern North America, Darby et al. [23] reported that barley yield was lower following sunn hemp but its malting quality was not affected, while crimson clover did not impact barley growth relative to summer fallow. In western North America, winter malting barley has also performed well following peas (*Pisum sativum* L.) and canola (*Brassica napus* L.) [16,17] but these crops are not commonly grown in the Northeast, and local rotation recommendations are needed. Whether grown as forages or cover crops, sunn hemp and crimson clover can have many impacts on agricultural productivity and ecosystem services. They can protect the soil from erosion [27], contribute organic matter to the soil [26], reduce insect pest damage [28,29], provide income if harvested as a forage [26,27], and add plant-available nitrogen to the soil [27,30].

In many rotations, the nitrogen contribution from sunn hemp or crimson clover would be beneficial to the following crop, but this may not be the case for winter malting barley if nitrogen from the preceding crop leads to excessive grain protein or otherwise reduces malting quality. Alternatively, if nitrogen from leguminous cover crops does not negatively affect malting barley, this would suggest that farmers can plant winter malting barley with minimal concern about excess nitrogen contributions from preceding crops.

The current experiment examines how integrating legumes into winter malting barley cultivation can affect grain yield and quality, and whether the barley seeding rate changes these effects. This knowledge will complement previous varietal assessment and agronomic management recommendations to help farmers improve their profitability and integrate winter malting barley into their overall farm systems.

2. Materials and Methods

Experimental Site: A two-year field experiment was performed at the University of Massachusetts Agricultural Experiment Station Farm in South Deerfield, MA (42° N, 73° W) on fine Hadley loam soil. In both years, the experimental crops were grown after summer corn silage (*Zea Mays* L.) and winter fallow. The top 15 cm of soil was analyzed before cover cropping each year and the fields were amended with lime, sulfur, and potassium as recommended by the University of Massachusetts Soil and Plant Nutrient Testing Laboratory (Amherst, MA) for barley production. The experiment site was prepared using disk tillage immediately before the first planting date. Extreme weather events did not appear to influence the experiment in either year (Table 1).

Table 1. Weather Data for the experimental site in 2015 and 2016.

Year	Month	Avg Temp (°C)	Departure from Avg. *	Max Temp (°C)	Departure from Avg.	Min Temp (°C)	Departure from Avg.	Total Precipitation (cm)	Departure from Avg.
2015	July	21.1	−1.3	32.7	4.0	11.3	−4.8	8.4	−2.1
	August	21.1	−0.3	32.5	4.6	11.3	−3.7	6.4	−3.9
	September	18.3	0.9	33.0	8.8	4.9	−5.7	16.3	4.9
	October	9.2	−1.4	23.3	6.3	−7.4	−11.8	5.6	−7.4
	November	6.2	1.7	23.1	12.7	−8.9	−7.5	5.1	−3
	December	4.0	4.8	16.4	12.2	−5.5	0.4	11.9	1.3
2016	January	−2.7	1.6	11.0	9.7	−15.5	−5.5	3.8	−3.7
	February	−1.9	1.3	14.9	12.0	−26.1	−16.8	10.4	2.4
	March	4.7	3.2	25.5	18.0	−8	−3.5	8.4	0.1
	April	7.4	−0.8	26.2	11.2	−11	−12.5	5.3	−4.5
	May	14.2	−0.2	32.6	11.6	−1.7	−9.5	6.6	−2.7
	June	19.1	−0.2	30.9	5.5	5.3	−7.7	3.6	−8.7
	July	22.3	−0.1	34.4	5.7	9.9	−6.1	4.3	−6.2
	August	22.2	0.8	33.6	5.7	8.9	−6.1	4.6	−5.6
	September	17.7	0.3	30.6	6.4	2.7	−8	9.3	−2
	October	10.3	−0.4	23.8	6.9	−4	−8.4	5.4	−7.6
	November	4.2	−0.3	19.0	8.6	−6.5	−5.1	8.2	0.1
	December	−1.6	−0.7	11.4	7.2	−19.2	−13.3	7.7	−2.9
2017	January	−1.3	3.0	13.6	12.2	−18.7	−8.7	7.0	−0.5
	February	−0.3	2.8	20.8	17.8	−18.6	−9.2	3.8	−4.2
	March	−0.8	−2.4	15.4	7.9	−14.4	−10	4.0	−4.3
	April	10.2	2.0	29.0	14.0	−3.7	−5.2	11.1	1.3
	May	13.0	−1.4	33.3	12.3	0.2	−7.6	8.2	−1.1
	June	18.8	−0.4	34.4	9.0	4.7	−8.3	11.8	−0.5
	July	20.6	−1.8	32.3	3.6	10.7	−5.4	5.7	−4.8

* Average weather data from Amherst, MA—eight miles from South Deerfield. Averages are based on the years 2001–2020.

Experimental Layout: Four replications of each treatment were planted in a randomized complete block design, with cover crop species and barley seeding rate as fixed main effects. The 12 treatments consisted of balanced combinations of three barley seeding rates (300, 350, and 400 seeds m^{-2}) and four summer cover crop species, including sunn hemp (SH), crimson clover (CC), sunn hemp and crimson clover (SH + CC), and summer fallow with no cover crop (NC).

Field Management and Assessments: Summer cover crops were planted on 19 July 2015 and 11 August 2016. The cover crops were planted at the following rates. SH: 33.6 kg ha^{-1} sunn hemp, CC: 20.2 kg ha^{-1} crimson clover, SH + CC: 16.8 kg ha^{-1} sunn hemp and 16.8 kg ha^{-1} crimson clover. Cover crop aboveground biomass was sampled from two 0.5 m^2 sections on 8 September 2015 and 15 September 2016. Cover crop nitrogen content was calculated from crude protein using near-infrared spectroscopy (NIR) (Inframatic 8600, Perten Instruments). Cover crop biomass analysis included weeds growing with cover crops. Cover crops were flail mowed and terminated using a rototiller on 15 September 2015 and 16 September 2016.

Wintmalt, a 2-row malting barley, was planted on 25 September 2015 and 30 September 2016. Wintmalt is the common winter barley grown in New England. Cover crops and barley seeds were planted two cm deep, using a custom-made plot-size cone seed drill with 17.8 cm between rows.

Barley stands were counted on 16 October 2015 and October 2016. However, stand count data from 2016 was lost and the reported barley stand counts are based solely on data from 2015. Fall soil nitrate was measured immediately following the first hard frost on 20 October 2015 and 17 November 2016. Winter survival was not assessed in the spring of 2016 and was measured on 28 April 2017 using a 0–10 scale with 10 as complete survival and 0 as totally winter killed. Soil samples were collected to assess spring soil nitrate using five 6-inch-deep cores per plot, air dried, and soil nitrate content was determined using a LaChat QuickChem 8500 Series 2 Flow Injection Analysis System [31]. An amount of 28 kg ha^{-1} nitrogen was applied as calcium ammonium nitrate on 15 April in both 2016 and 2017.

Foliar disease was estimated on 10 July 2017 as a percentage of leaf surface area infected using the disease guides in the American Phytopathological Society's "A Manual of Assessment Keys for Plant Disease" [32]. Due to rapid drought-induced foliar desiccation, foliar diseases were not measured in 2016. Heading date was declared when half of the tillers had emerged heads and is reported as Julian date. Plant height was measured on 24 June 2016 and 10 July 2017 while lodging was assessed on 12 July 2016 and 10 July 2017. Lodging was visually evaluated on a 0–10 scale with 0 as no lodging and 10 as completely lodged.

Harvest and Laboratory Analyses: Barley was harvested on 19 July 2016 and 17 July 2017 using an ALMACO SPC20 plot combine. A subsample of the grain was dried in a forced air oven at 38 °C to preserve kernel integrity. Germinative energy, test weight, and 1000-kernel weight were determined using ASBC methods Barley–3A, Barley–2B, and Barley–2 [9]. 2017 grain samples could not be analyzed immediately following grain harvest and the samples had to be stored in a walk-in cooler for several years before analysis. As a result, germinative energy was considerably lower for these samples and, although their inclusion would not change the statistical results, they are not included in our results and analysis. Malting quality was assessed at the E.E. Cummings Crop Testing Laboratory at the University of Vermont (Burlington, VT). Crude protein content as a proportion of dry matter was measured with a Perten Inframatic 8600 Flour Analyzer, and falling number was assessed using the AACC Method 56–81B [33] on a Perten FN 1500 Falling Number Machine. DON content was evaluated in the subsamples using the NEOGEN Corp. Veratox DIN 2/3 Quantitative Test with a limit of detection of 0.1 mg kg^{-1}.

Statistical Analysis: Data were analyzed using the permlmer and lmer functions in the predictmeans [34] and lme4 [35] packages of R statistical software [36]. Permutation tests were used to assess the impact of the main fixed effects of *cover crop type* and *barley seeding rate* as well as their interaction at a significance level of $p \leq 0.05$. This non-parametric method was used to account for non-normal distribution of residuals and heterogeneous variance in many response variables. Bonferroni adjusted *t*-tests were used to make pairwise comparisons between all *cover crop type* treatments and orthogonal polynomial regression was used to assess the continuous effect of *barley seeding rate*. Because variance between groups was homogeneous, pooled standard deviations were used to calculate pairwise *t*-tests for *fall soil nitrate* and *barley grain falling number* while non-pooled standard deviations were used in the analysis of *cover crop biomass* and *cover crop nitrogen content* to account for heterogeneous variance. The random variables of *year* and *block* were combined into one random variable and data from both years were combined and analyzed collectively as eight replications.

3. Results

3.1. Cover Crop Biomass and Nitrogen Content

Although nitrogen concentrations in aerial parts of the cover crops were not significantly different, due to higher biomass production sunn hemp (SH) and sunn hemp-crimson clover mixture (SH + CC) produced significantly more biomass than either monocrop crimson clover (CC) or weedy fallow (NC) (Table 2, Figure 1). Therefore, their aboveground biomass residues added more nitrogen to the soil when incorporated before barley planting (Table 2, Figure 2). SH produced the highest amount of aboveground dry matter (2.90 t ha^{-1}), thus contributing the highest nitrogen (78.8 kg ha^{-1}). Both local weeds and CC treatments produced considerably lower dry matter (0.87 and 0.59 t ha^{-1}, respectively); therefore, their nitrogen contributions to the cropping system were minimal compared with SH and cover crop mixture (Figure 3). Interestingly, the local weeds in NC plots contained significantly more nitrogen (37.3 g kg^{-1}) than in monoculture and mix legumes (Figure 2). The results indicate that the local weeds were more aggressive and efficient in taking up nitrogen from the soil. The effective nutrient uptake by weeds from the soil likely demonstrates their capacity to adapt to changes in edaphoclimatic conditions during the growing season [37,38]. To our surprise, CC performed poorly in terms of atmospheric nitrogen fixation and accumulation in plants' aerial tissues. The failure of CC to fix nitrogen could be due to the poor rhizobium activity and thus requires further investigation.

Table 2. Mean cover crop dry matter biomass and nitrogen content following different cover crops in South Deerfield MA in 2014–2015 and 2015–2016.

Cover Crop Type	Cover Crop Dry Matter Yield (t ha^{-1})		Cover Crop Nitrogen Content (g kg^{-1})		Cover Crop Nitrogen Yield (kg ha^{-1})	
Sunn Hemp	2.90	a	27.1	b	78.7	a
Crimson Clover	0.59	b	29.4	b	17.5	c
Sunn Hemp and Crimson Clover	2.24	a	25.6	b	57.4	a
No Cover Crop	0.87	b	37.3	a	32.5	b
Overall Experiment Mean	1.82		27.3		49.6	
Effect Significance						
Cover Crop Type	***		*		***	

Note. *, $p \leq 0.05$; ***, $p \leq 0.001$; ns, non-significant according to non-parametric permutation tests. For significant effects, all pairwise comparisons were made using Bonferroni adjusted *t*-tests. Means followed by the same letter are not significantly different from each other ($p \leq 0.05$). Cover crop dry matter and nitrogen content includes weeds.

3.2. Soil Nitrate in the Fall and Spring

In the late fall, soil nitrate levels were nearly identical in CC (2.0 mg kg^{-1}) and NC (2.0 mg kg^{-1}) plots (Table 3). Soil nitrate was significantly higher in the SH plots (3.3 mg kg^{-1}), while SH + CC plots had intermediate soil nitrate (2.8 mg kg^{-1}) (Table 3). As expected, spring soil nitrate was substantially higher overall compared to fall soil nitrate and ranged from 5.9 to 7.2 mg kg^{-1}, depending on cover crop treatments. However, soil nitrate in spring was highly variable and the differences between treatments were not significant (Table 3).

3.3. Barley Seeding Rate Impact on Stand Establishment, Winter Survival, and Growth of Barley

Barley seedling populations increased with increased seeding rate (Table 4). There was a significant quadratic regression describing this relationship (Seedling Population = $- 0.0105x^2 + 8.1424x - 1211$), showing that there was a leveling off of barley population between 350 and 400 seeds m^{-1}. This suggests that higher seeding rates do not increase barley populations towards the higher end of this range (Figure 4). Higher seeding rates also significantly improved seedlings' winter survival (Winter Survival = 0.0313x + 86.25) but the difference was small and all seeding rate levels had better than 95 percent winter survival (Figure 5).

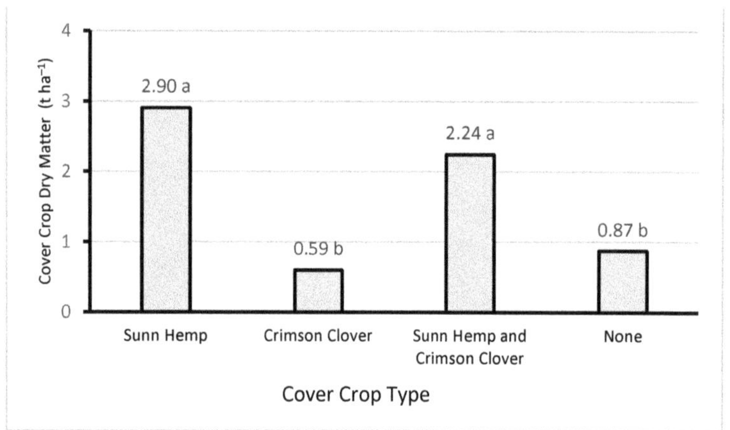

Figure 1. Mean aboveground cover crop biomass (t ha^{-1}) as a function of cover crop treatment. Means followed by the same letter are not significantly different from each other according to Bonferroni adjusted *t*-tests at $p \leq 0.05$.

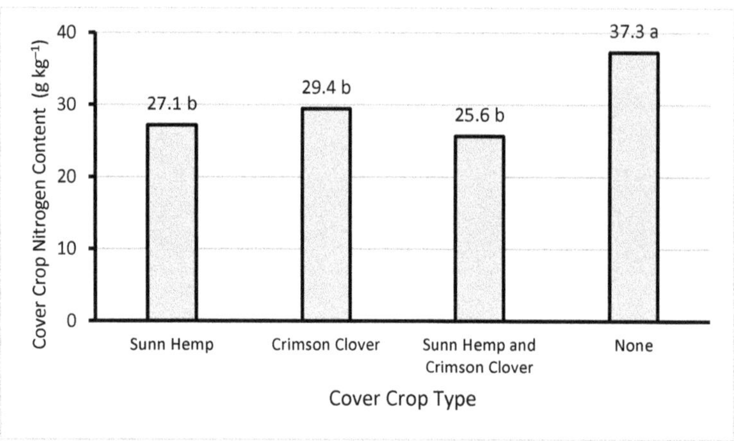

Figure 2. Cover crop biomass nitrogen content (g kg^{-1}) as a function of cover crop treatment. Means followed by the same letter are not significantly different from each other according to Bonferroni adjusted *t*-tests at $p \leq 0.05$.

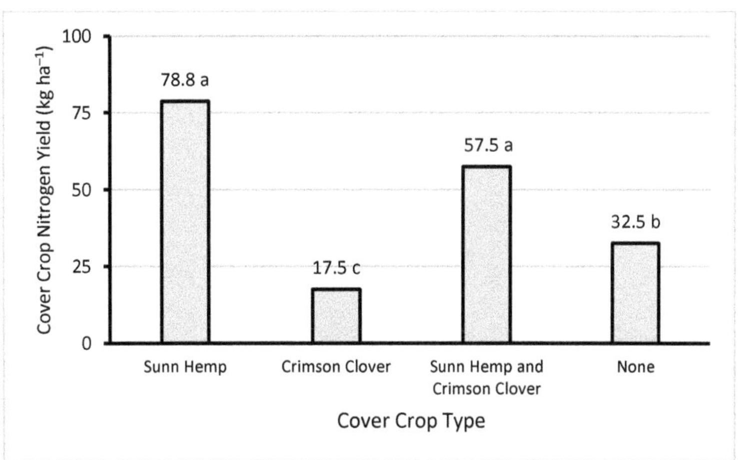

Figure 3. Mean aboveground cover crop nitrogen (kg ha^{-1}) as a function of cover crop treatment. Means followed by the same letter are not significantly different from each other according to Bonferroni adjusted *t*-tests at $p \leq 0.05$.

Table 3. Mean fall and spring soil nitrate following different cover crops in South Deerfield MA in 2014–2015 and 2015–2016.

Cover Crop Type	Fall Soil Nitrate (mg kg^{-1})		Spring Soil Nitrate (mg kg^{-1})
Sunn Hemp	3.3	a	7.2
Crimson Clover	2.0	b	6.7
Sunn Hemp and Crimson Clover	2.8	ab	6.3
No Cover Crop	2.1	b	5.9
Overall Experiment Mean			
	2.5		6.5
Effect Significance			
Cover Crop Type	**		ns

Note. **, $p \leq 0.01$; ns, non-significant according to non-parametric permutation tests. For significant effects, all pairwise comparisons were made using Bonferroni adjusted *t*-tests. Means followed by the same letter are not significantly different from each other ($p \leq 0.05$).

Table 4. Mean barley establishment, winter survival, and growth metrics at different barley seeding rates in South Deerfield MA in 2014–2015 and 2015–2016.

Barley Seeding Rate	Barley Population (Plants m^{-2})		Winter Survival (Percent)		Heading Date (Julian Day)	Height (cm)	Foliar Disease (Percent of Plants Affected)	Lodging
300 seeds m^{-2}	288.5	b	95.3	b	139.9	54.1	52.5	10.4
350 seeds m^{-2}	355.0	a	97.8	ab	139.0	56.6	47.5	11.8
400 seeds m^{-2}	369.1	a	98.4	a	139.6	54.9	47.5	10.5
Overall Experiment Mean								
	337.5		97.2		139.5	55.2	49.2	10.9
Effect Significance								
Barley Seeding Rate	***		**		ns	ns	ns	ns

Note. **, $p \leq 0.01$; ***, $p \leq 0.001$; ns, non-significant according to non-parametric permutation tests. For significant effects, all pairwise comparisons were made using Bonferroni adjusted *t*-tests. Means followed by the same letter are not significantly different from each other ($p \leq 0.05$). Barley seeding rate was also evaluated as a.continuous effect using orthogonal polynomial regression.

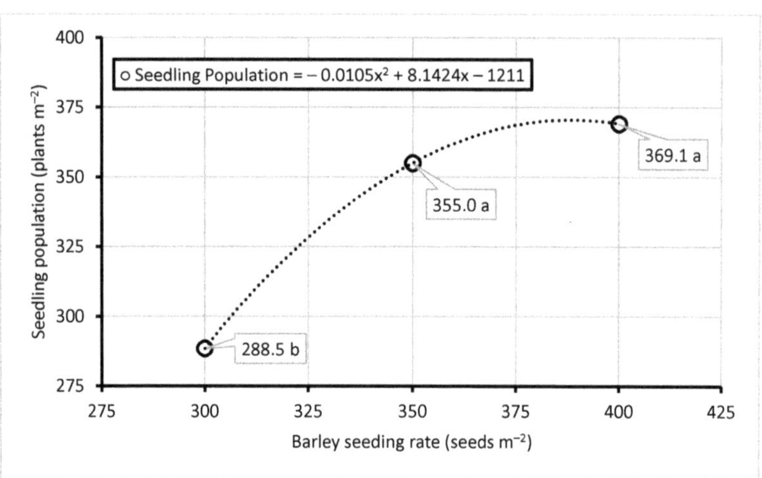

Figure 4. Mean seedling population (plants m^{-1}) as a function of barley seeding rate. Means followed by the same letter are not significantly different from each other according to Bonferroni adjusted *t*-tests at $p \leq 0.05$.

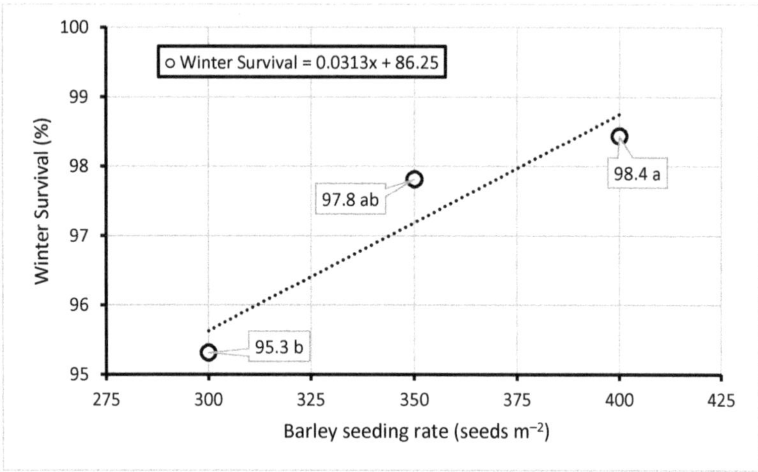

Figure 5. Mean winter survival (%) as a function of barley seeding rate. Means followed by the same letter are not significantly different from each other according to Bonferroni adjusted *t*-tests at $p \leq 0.05$.

Seeding rate did not have a significant impact on late vegetative and reproductive growth (Table 4). Seeding rate showed no influence on heading date or height of the barley plants, the severity of foliar disease, or the prevalence of crop lodging. All plots reached 50 percent heading within two days of each other and height averaged 55 cm across the experiment. Forty-nine percent of the leaves were affected by foliar disease and eleven percent of barley lodged before harvest across all treatments.

3.4. Influence of Cover Crop Species and Barley Seeding Rate on Barley Yield and Malting Quality Characteristics

There were no significant treatment effects on barley grain yield but there were two non-significant trends of note (Table 5). First, barley in SH and SH + CC treatments produced slightly more grain (3.74 and 3.67 t ha^{-1}, respectively) than either CC (3.39 t ha^{-1}) or NC (3.32 t ha^{-1}). Given that the cover crop treatment groups had very different organic matter and nitrogen contributions the previous fall, the small difference in yield could be related to the larger differences in soil condition at barley planting. Second, there was a non-significant yield improvement as seeding rate increased, possibly related to differences in stand quality due to differences in initial barley seedling population and winter survival. However, as noted above, the differences in yield were small and showed that all of the experimental treatments examined in this study can be similarly productive in malting barley cropping systems.

Table 5. Mean grain yield and quality metrics for winter malting barley following different cover crops and barley seeding rates in South Deerfield MA in 2014–2015 and 2015–2016.

Cover Crop Type	Barley Seeding Rate (seeds m^{-2})	Grain Yield (13.5% Moisture) (t ha^{-1})	Protein (0% Moisture) (g kg^{-1})	Test Weight (kg hl^{-1})	1000 Kernel Weight (g)	Germinative Energy (percent)	Falling Number (seconds)	DON (mg kg^{-1})
Sunn Hemp	300	3.53	118.4	58.68	46.40	80.38	233.9	0.20
	350	3.95	107.7	58.15	45.89	84.38	231.0	0.18
	400	3.75	110.0	58.06	45.55	79.75	234.4	0.10
Crimson Clover	300	3.15	114.8	58.66	47.00	85.00	237.3	0.17
	350	3.24	106.9	58.60	46.14	84.00	221.8	0.10
	400	3.77	114.3	57.90	46.70	80.00	227.0	0.20
Sunn Hemp and Crimson Clover	300	3.62	108.5	59.15	45.73	81.63	221.5	0.18
	350	3.85	111.0	58.28	47.13	89.13	212.1	0.15
	400	3.54	109.0	58.58	45.70	81.38	214.9	0.19
No Cover Crop	300	3.25	112.3	56.57	46.54	82.00	230.5	0.05
	350	3.23	111.1	58.85	46.43	79.38	211.5	0.15
	400	3.48	110.9	57.50	45.36	77.00	221.4	0.25
Cover Crop Type								
Sunn Hemp		3.74	112.0	58.30	45.95	81.50	233.1	0.16
Crimson Clover		3.39	111.9	58.39	46.61	83.00	228.3	0.15
Sunn Hemp and Crimson Clover		3.67	109.5	58.67	46.18	84.04	216.2	0.17
No Cover Crop		3.32	111.5	57.64	46.11	79.46	221.1	0.15
Barley Seeding Rate								
300 seeds m^{-2}		3.39	113.5	58.26	46.42	82.25	230.6	0.15
350 seeds m^{-2}		3.57	109.2	58.47	46.39	84.22	219.1	0.14
400 seeds m^{-2}		3.64	111.1	58.01	45.83	79.53	224.4	0.18
Overall Experiment Mean								
		3.53	111.2	58.25	46.21	82.00	224.6	0.16
Effect Significance								
Cover Crop Type		ns	ns	ns	ns	ns	*	ns
Barley Seeding Rate		ns	ns	ns	ns	ns	ns	ns
Cover Crop Type × Seeding Rate		ns	ns	ns	ns	ns	ns	ns

Note. *, $p \leq 0.05$; ns, non-significant according to non-parametric permutation tests. Although cover crop type had a significant effect on falling number, pairwise comparisons made using Bonferroni adjusted *t*-tests did not show significant differences between any cover crop types ($p \leq 0.05$). Cover crop type is evaluated as a discrete effect and barley seeding rate as a continuous effect.

As was the case with grain yield, there were no significant differences in malting barley grain quality among cover crops or barley seeding rates (Table 5). Across the experiment, mean test weight was 58.2 kg hl^{-1}, mean 1000 kernel weight was 46.2 g, mean germinative energy was 82 percent, mean protein content was 111.2 g kg^{-1}, mean falling number was 224 s, and mean DON content was 0.16 mg kg^{-1}. Overall, while the barley met some malting quality standards, it fell below others. Protein and DON content were in good ranges for brewing purposes (below 125 g kg^{-1} and 0.5 mg kg^{-1}, respectively). Test weight and falling number were a little lower than standard malting quality for these two indices (61.8 kg hl^{-1} and 250 s) while germinative energy was much lower than required for malting (95 percent). There is no specific standard for 1000 kernel weight. Relatively low germinative energy may have been the result of high drying temperatures following grain barley harvest.

These results indicate that high nitrogen contributions from leguminous cover crops are unlikely to result in protein levels exceeding malting standards.

4. Discussion

Most malting barley production in North America occurs in the dry Great Plains and West Coast regions with relatively little in the humid Northeast and Midwest [1]. As a result, most malting barley research has focused on different cropping systems than those discussed in this experiment. Many experiments have been performed in much drier conditions [16–20,22,24,39] or with spring planted cultivars [3,16,17,22,24,39]. Furthermore, many studies have found substantial differences in malting quality even in winter malting barley cultivars [10,11,14,18]. Caution should be used when generalizing experimental results given the diversity of growing conditions across North America and the relative novelty of winter malting barley cultivation in the Northeast.

This experiment indicated that even though malting barley is quite sensitive to nitrogen level in grains [13,14,18], leguminous crops may be grown before winter malting barley without damaging the yield or malting quality of the barley crop (Table 5). This result differs somewhat from research on malting barley production following legume cover crops further north in New England. Darby et al. and Surjawan et al. [3,23] reported that sunn hemp cover crop before winter barley reduced the next summer's yield in Vermont [23] and that a pea/oat/vetch cover crop slightly reduced malting quality in spring barley in Maine [3]. That said, malting barley is much more sensitive to nitrogen in the spring than in the fall [13] and Darby et al. [23] did not find effects of either sunn hemp or crimson clover on malting quality.

In the American West, past studies have explored whether growing spring malting barley after legume cash crops could impact the grain yield and malting quality [16,17,39]. Sainju [17] found that pea residue retained soil nitrate better than bare fallow, while Sainju et al. [39] and Turkington et al. [16] found that planting spring barley after peas did not cause negative quality characteristics such as high protein content, which have been commonly seen from excess nitrogen fertilizer [13,14,18]. However, these experiments were done with a spring barley following peas seed harvest, and the residues would not have contained nearly as much nitrogen as a legume cover crop incorporated into the soil, as was performed by Surjawan et al. [3].

The sunn hemp growth and nitrogen content (Table 2, Figures 1–3) in this experiment were similar to those seen after a similar amount of time (45 days) by Clark [25] in New York, although the overall sunn hemp production was lower than that seen by Clark after 60 days or by Mansoer et al. in the American Southeast [27]. Crimson clover (CC) and the summer fallow (NC) produced much lower biomass and nitrogen yield (Table 2, Figure 1; Figure 3) but did not lead to lower barley yields (Table 5). Together these results suggest that the barley was neither in need of nor hurt by the extra nitrogen supplied by the SH and SH + CC treatments. Given that malting barley can have disease issues when grown directly after a grass crop [1], it is agronomically important that winter malting barley be grown after a high nitrogen producer like sunn hemp without negative effects. Additionally, this research supports the idea recommended by Shrestha and Lindsey [1] that winter malting barley could be grown after soybeans (*Glycine max* (L.) Merr.). The growing season in Massachusetts may be too short for this to be a feasible cropping system. However, these results show that farmers further south should not be especially concerned that nitrogen from a previous soybean (or other legume) crop would negatively affect grain quality of winter malting barley. Additionally, longer growing seasons under climate change conditions and the development of shorter season soybeans could make this sort of crop system more attractive in the future in Massachusetts and other New England states.

While the differences seen in this trial based on barley seeding rate were minimal, previous studies have found that an increased seeding rate led to better overall malting quality and lower protein content in particular [20,22,24] in spring planted barley in western North America. The results from this experiment do not substantively contradict these earlier findings. While farmers are unlikely to see large yield gains from higher seeding rates, there may indeed be a reduction in variability from a relatively small investment in seed. Indeed, this research agrees with Darby et al. [23], who also found that similar increases in seeding rate can promote winter survival, although not always affecting final yield or malting quality overall.

The overall malting quality of barley measured in this study was similar to that seen in other winter malting barley in the Northeast. As in other studies, the barley grain did not meet all malting quality standards [11,13,23]. The best practices for malting barley production in the Northeast are still under development and producers should not expect to get malting quality grain every year [2,14]. Given that malting barley in the Northeast can be quite variable and is much more affected by cultivar selection [3,10,11,14], spring fertilizer application [13], and harvest management [2,12], the results of the current experiment indicated that regional farmers can plan their cropping system without worrying that a previous leguminous crop may cause quality issues in following winter malting barley.

5. Conclusions

It remains challenging to achieve superior malting barley in the Northeastern United States. This could lead farmers to form the impression that there is little flexibility if they want to meet the required quality standards. However, while excess nitrogen has been shown to lead to poor malting quality, growing high-nitrogen-producing legumes before planting winter malting barley is unlikely to reduce the quality of the succeeding barley. Specifically, this study demonstrates that following a legume cover crop with winter malting barley does not reduce grain yield or malting quality in terms of protein content, test weight, 1000 kernel weight, germinative energy, falling number, or DON content. These results show that winter malting barley can be integrated into crop rotations with leguminous plants without negative impacts on barley growth, yield, and grain quality.

Author Contributions: Conceptualization, M.H. and H.D.; supervision, M.H.; formal analysis, A.S. (Arthur Siller) and A.S. (Alexandra Smychkovich); writing—original draft preparation, A.S. (Arthur Siller); writing—review and editing, A.S. (Arthur Siller), M.H., H.D. and A.S. (Alexandra Smychkovich). All authors have read and agreed to the published version of the manuscript.

Funding: This study was partially funded by Northeast Sustainable Agriculture Research and Education, project number GNE 13-066.

Data Availability Statement: The data presented in this study are openly available at UMass Scholarworks.

References

1. Shrestha, R.K.; Lindsey, L.E. Agronomic Management of Malting Barley and Research Needs to Meet Demand by the Craft Brew Industry. *Agron. J.* **2019**, *111*, 1570–1580. [CrossRef]
2. Lazor, J. *The Organic Grain Grower: Small-Scale, Holistic Grain Production for the Home and Market Producer*; Chelsea Green Publishing: White River Junction, VT, USA, 2013.
3. Surjawan, I.; Dougherty, M.P.; Camire, M.E.; Jemison, J.J. *Malting Quality of Maine-Grown Barley*; MR433: Onono, ME, USA, 2004; p. 433. Available online: https://digitalcommons.library.umaine.edu/cgi/viewcontent.cgi?article=1009&context=aes_miscreports (accessed on 29 September 2021).
4. Robertson, L.D.; Wesenberg, D.M. Major uses of barley. In *Idaho Spring Barley Production Guide*; Bull 742; Robertson, L.D., Stark, J.C., Eds.; Ag Publications: Moscow, ID, USA, 2003; Available online: http://www.extension.uidaho.edu/publishing/pdf/BUL/BUL0742.pdf (accessed on 29 September 2021).

5.	Brewers Association. *Malting Barley Characteristics for Craft Brewers*; Brewers Publications: Boulder, CO, USA, 2020; Available online: https://cdn.brewersassociation.org/wp-content/uploads/2019/06/Malting_Barley_Characteristics_For_Craft_Brewers.pdf (accessed on 29 September 2021).

6.	Hmielowski, T. From Field to Pint: Malting Barley in the Eastern U.S. *Crop. Soils* **2017**, *50*, 10–13. [CrossRef]

7.	Wrigley, C.; Batey, I.; Miskelly, D. *Cereal Grains: Assessing and Managing Quality*; Woodhead Publishing: Cambridge, UK, 2016.

8.	BMBRI (Brewing and Malting Barley Research Institute). Quality Factors in Malting Barley. 2010. Available online: http://bmbri.ca/wp-content/uploads/2016/10/Quality-Factors-in-Malting-Barley-May-2010.pdf (accessed on 29 September 2021).

9.	ASBC (American Society of Brewing Chemists). *Methods of Analysis*, 14th ed.; The Society: St. Paul, MN, USA, 2009.

10.	Darby, H.; Cummings, E.; Emick, H. *Organic Winter Malt Barley Variety Trial*; UVM Extension: Burlington, VT, USA, 2017. Available online: https://scholarworks.uvm.edu/cgi/viewcontent.cgi?article=1055&context=nwcsp (accessed on 29 September 2021).

11.	Hashemi, M.; Zandvakili, O.R.; Shureshjani, R.A.; Etemadi, F.; Darby, H. Evaluating Winter Barley Cultivar Using Data Envelopment Analysis Models. *J. Autom. Sin.* **2019**, *11*, 22. [CrossRef]

12.	Mallett, J. *Malt: A Practical Guide from Field to Brewhouse*; Brewers Publications: Boulder, CO, USA, 2014.

13.	Siller, A.; Hashemi, M.; Wise, C.; Smychkovich, A.; Darby, H. Date of Planting and Nitrogen Management for Winter Malt Barley Production in the Northeast, USA. *Agronomy* **2021**, *11*, 797. [CrossRef]

14.	Zhong, B.H.W.; Wiersma, J.J.; Sheaffer, C.C.; Steffenson, B.J.; Smith, K.P. Assessment of Winter Barley in Minnesota: Relationships among Cultivar, Fall Seeding Date, Winter Survival, and Grain Yield. *Crop. Forage Turfgrass Manag.* **2019**, *5*, 190055. [CrossRef]

15.	USDA NASS (National Agricultural Statistics Service). Small Grains 2020 Summary 09/30/2020. 2020. Available online: https://downloads.usda.library.cornell.edu/usda-esmis/files/5t34sj573/7p88d6171/bn999x131/smgr0920.pdf (accessed on 29 September 2021).

16.	Turkington, T.K.; O'Donovan, J.T.; Edney, M.J.; Juskiw, P.E.; McKenzie, R.H.; Harker, K.N.; Clayton, G.W.; Xi, K.; Lafond, G.P.; Irvine, R.B.; et al. Effect of Crop Residue, Nitrogen Rate and Fungicide Application on Malting Barley Productivity, Quality, and Foliar Disease Severity. *Can. J. Plant Sci.* **2012**, *92*, 577–588. [CrossRef]

17.	Sainju, U.M. Tillage, Cropping Sequence, and Nitrogen Fertilization Influence Dryland Soil Nitrogen. *Agron. J.* **2013**, *105*, 1253–1263. [CrossRef]

18.	Castro, A.; Petrie, S.; Budde, A.; Corey, A.; Hayes, P.; Kling, J.; Rhinhart, K. Variety and N Management Effects on Grain Yield and Quality of Winter Barley. *Crop Manag.* **2008**, *7*, 1–8. [CrossRef]

19.	Stevens, W.B.; Sainju, U.M.; Caesar-TonThat, T.; Iversen, W.M. Malt Barley Yield and Quality Affected by Irrigation, Tillage, Crop Rotation, and Nitrogen Fertilization. *Agron. J.* **2015**, *107*, 13. [CrossRef]

20.	Edney, M.J.; O'Donovan, J.T.; Turkington, T.K.; Clayton, G.W.; McKenzie, R.; Juskiw, P.; Lafond, G.P.; Brandt, S.; Grant, C.A.; Harker, K.N.; et al. Effects of Seeding Rate, Nitrogen Rate and Cultivar on Barley Malt Quality: Effects of Seeding Rate, Nitrogen Rate and Cultivar on Barley Malt Quality. *J. Sci. Food Agric.* **2012**, *92*, 2672–2678. [CrossRef]

21.	Delogu, G.; Cattivelli, L.; Pecchioni, N.; De Falcis, D.; Maggiore, T.; Stanca, A.M. Uptake and Agronomic Efficiency of Nitrogen in Winter Barley and Winter Wheat. *Eur. J. Agron.* **1998**, *9*, 11–20. [CrossRef]

22.	McKenzie, R.H.; Middleton, A.B.; Bremer, E. Fertilization, Seeding Date, and Seeding Rate for Malting Barley Yield and Quality in Southern Alberta. *Can. J. Plant Sci.* **2005**, *85*, 603–614. [CrossRef]

23.	Darby, D.H.; Emick, H.; Cummings, E. *Winter Barley Seeding Rate, Cover Crop and Variety Trial*; UVM Extension: Burlington, VT, USA, 2016. Available online: https://scholarworks.uvm.edu/cgi/viewcontent.cgi?article=1113&context=nwcsp (accessed on 29 September 2021).

24.	O'Donovan, J.T.; Turkington, T.K.; Edney, M.J.; Clayton, G.W.; McKenzie, R.H.; Juskiw, P.E.; Lafond, G.P.; Grant, C.A.; Brandt, S.; Harker, K.N.; et al. Seeding Rate, Nitrogen Rate, and Cultivar Effects on Malting Barley Production. *Agron. J.* **2011**, *103*, 709–716. [CrossRef]

25.	Clark, S. *Seeding Rates and Harvest Dates on Productivity of 'Tropic Sun' Sunn Hemp in South-Central New York*; Final Study Report: USDA-NRCS, Big Flats Plant Materials Center: Corning, NY, USA, 2016. Available online: https://www.nrcs.usda.gov/Internet/FSE_PLANTMATERIALS/publications/nypmcsr12923.pdf (accessed on 29 September 2021).

26.	Clark, A. *Managing Cover Crops Profitably*, 3rd ed.; Sustainable Agriculture Network: Beltsville, MD, USA, 2008.

27.	Mansoer, Z.; Reeves, D.W.; Wood, C.W. Suitability of Sunn Hemp as an Alternative Late-Summer Legume Cover Crop. *Soil Sci. Soc. of Am. J.* **1997**, *61*, 246–253. [CrossRef]

28.	Hinds, J.; Hooks, C.R.R. Population Dynamics of Arthropods in a Sunn-Hemp Zucchini Interplanting System. *Crop Prot.* **2013**, *53*, 6–12. [CrossRef]

29.	Hooks, C.R.R.; Hinds, J.; Zobel, E.; Patton, T. Impact of Crimson Clover Dying Mulch on Two Eggplant Insect Herbivores: Impact of Crimson Clover on Insects in Eggplant. *J. Appl. Entomol.* **2013**, *137*, 170–180. [CrossRef]

30.	Yang, X.M.; Drury, C.F.; Reynolds, W.D.; Phillips, L.A. Nitrogen Release from Shoots and Roots of Crimson Clover, Hairy Vetch, and Red Clover. *Can. J. Soil. Sci.* **2020**, *100*, 179–188. [CrossRef]

31.	Sechtig, A. *QuickChem Method 12-107-04-1-A. Determination of Nitrate/Nitrite by Flow Injection Analysis*; Lachat Instruments 6645 West Mill Road: Milwaukee WI, USA, 2001.

32.	James, C. *A Manual of Assessment Keys for Plant Diseases*; American Phytopathological Society: St. Paul, MN, USA, 1971.

33.	AACC. *AACC Approved Methods of Analysis*, 11th ed.; Method 56-81.04 Determination of Falling Number; Cereals & Grains Association: St. Paul, MN, USA, 2000.

34. Luo, D.; Ganesh, S.; Koolaard, J. Predictmeans: Calculate Predicted Means for Linear Models. R Package Version 1.0.6. Available online: https://CRAN.R-project.org/package=predictmeans (accessed on 29 September 2021).
35. Bates, D.; Mächler, M.; Bolker, B.; Walker, S. Fitting Linear Mixed-Effects Models Using lme4. *J. Stat. Softw.* **2015**, *67*, 1–48. [CrossRef]
36. R Core Team. *R: A Language and Environment for Statistical Computing*; R Foundation for Statistical Computing: Vienna, Austria, 2017. Available online: https://www.R-project.org/ (accessed on 29 September 2021).
37. Jiwan, S.R.; Singh, W.S. Weed composition and nutrient uptake by weeds in sole and intercrops during rabi season. *Indian J. Ecol.* **2021**, *48*, 300–303.
38. da Conceição de Matos, C.; da Silva Teixeira, R.; da Silva, I.R.; Costa, M.D.; da Silva, A.A. Interspecific competition changes nutrient: Nutrient ratios of weeds and maize. *J. Plant Nutr. Soil Sci.* **2019**, *182*, 286–295. [CrossRef]
39. Sainju, U.M.; Lenssen, A.W.; Barsotti, J.L. Dryland Malt Barley Yield and Quality Affected by Tillage, Cropping Sequence, and Nitrogen Fertilization. *Agron. J.* **2013**, *105*, 329–340. [CrossRef]

Article

Comparative Effectiveness of Four Nitrification Inhibitors for Mitigating Carbon Dioxide and Nitrous Oxide Emissions from Three Different Textured Soils

Yafei Guo, Asif Naeem and Karl H. Mühling *

Institute of Plant Nutrition and Soil Science, Kiel University, Hermann-Rodewald-Strasse 2, 24118 Kiel, Germany; yguo@plantnutrition.uni-kiel.de (Y.G.); anaeem@plantnutrition.uni-kiel.de (A.N.)
* Correspondence: khmuehling@plantnutrition.uni-kiel.de; Tel.: +49-(431)-880-3189

Abstract: Nitrification inhibitors (NIs) can be used to reduce both NO_3^--N leaching and N_2O-N emissions. However, the comparative efficacies of NIs can be strongly affected by soil type. Therefore, the efficacies of four nitrification inhibitors (dicyandiamide (DCD), 3, 4-dimethylpyrazole phosphate (DMPP), nitrogenous mineral fertilizers containing the DMPP ammonium stabilizer (ENTEC) and active ingredients: 3.00–3.25% 1, 2, 4-triazole and 1.50–1.65% 3-methylpyrazole (PIADIN)) were investigated in three different textured N-fertilized (0.5 g NH_4^+-N kg^{-1} soil) soils of Schleswig-Holstein, namely, Marsch (clayey), Östliches Hügelland (loamy) and Geest (sandy) under a controlled environment. Total CO_2-C and N_2O-N emissions were significantly higher from Marsch than Östliches Hügelland and Geest. In Marsch, DMPP showed the highest inhibitory effect on CO_2-C emission (50%), followed by PIADIN (32%) and ENTEC (16%). In Östliches Hügelland, DCD and PIADIN showed the highest and equal inhibitory effect on CO_2-C emission (73%), followed by DMPP (64%) and ENTEC (36%). In Marsch and Östliches Hügelland, DCD showed the stronger inhibitory effect on N_2O-N emission (86% and 47%) than DMPP (56% and 30%) and PIADIN (54% and 16%). In Geest, DMPP was more effective in reducing N_2O-N emission (88%) than PIADIN (70%) and DCD (33%). Thus, it can be concluded that DCD is a better NI for clay and loamy soils, while DMPP and PIADIN are better for sandy soils to inhibit soil nitrification and gaseous emissions.

Keywords: nitrification inhibitors; soil type; CO_2 and N_2O emissions; soil nitrogen dynamic

Citation: Guo, Y.; Naeem, A.; Mühling, K.H. Comparative Effectiveness of Four Nitrification Inhibitors for Mitigating Carbon Dioxide and Nitrous Oxide Emissions from Three Different Textured Soils. *Nitrogen* **2021**, 2, 155–166. https://doi.org/10.3390/nitrogen2020011

Academic Editor: Tida Ge

Received: 23 March 2021
Accepted: 9 April 2021
Published: 13 April 2021

Publisher's Note: MDPI stays neutral with regard to jurisdictional claims in published maps and institutional affiliations.

1. Introduction

Nitrogen is an important element for plant growth in agro-ecosystems [1], but the effectiveness of applied fertilizer N in crops rarely exceeds 40% [2]. Chemical N fertilizer constitutes approximately 75% of the total EU input of reactive N [3], and between 40% and 70% of the fertilizer N applied is lost to the atmosphere or the hydrosphere [4]. The majority of applied N is lost from agriculture through ammonia (NH_3) volatilization, gaseous emissions of nitrous oxide (N_2O) and di-nitrogen (N_2) and nitrate (NO_3^-) leaching [5]. Soil NO_3^--N leaching and N_2O emission are processes responsible for both N losses from agricultural soils as well as environmental pollution [6,7]. Beyond its powerful greenhouse effect, N_2O is also a major ozone-depleting substance involved in the destruction of the protective ozone layer in the stratosphere [8].

The concentration of N_2O, which is about 300 times more reactive a gas than CO_2, has risen from a pre-industrial value of 270 ppb to 319 ppb in 2005 [9], primarily due to agricultural practices and increased use of industrial fertilizers [1,10]. It has been estimated that agricultural soils produce 2.8 (1.7–4.8) Tg N_2O-N $year^{-1}$ and contribute approximately 65% of the atmospheric N_2O loading [8,9]. In Europe, N_2O emissions from agricultural soils contribute about 70% of the total annual N_2O emissions (European Environment Agency, 2015). Application of N to soils as chemical or organic fertilizers stimulates nitrous oxide (N_2O) emissions, mainly through the processes of denitrification and nitrification [11,12]. Nitrification is an aerobic process in which ammonium (NH_4^+) is first oxidized to nitrite (NO_2^-) and then nitrate (NO_3^-) [1], and it plays a key role in the soil N cycle [13]. During the oxidation of NH_4^+ to NO_2^-, N_2O can be produced as an intermediate and liberated into the atmosphere [1]. Denitrification is an anaerobic microbial process in which organic carbon is used as an energy source and NO_3^- is reduced to gaseous N compounds, including N_2 and N_2O [14].

Nitrification inhibitors (NIs) are used to improve the efficiency of N fertilizers through decreasing both NO_3^- leaching and gaseous N emissions [14,15]. NIs can decelerate the rate of soil nitrification by deactivating the enzyme ammonia monooxygenase (AMO) responsible for catalyzing ammonia oxidation, the first and rate-limiting step of nitrification, which is produced by ammonia-oxidizing archaea (AOA) and ammonia-oxidizing bacteria (AOB) [16]. As NO_3^- is the initial required substrate for denitrification, the use of NIs decreases N_2O emissions from both processes (nitrification and denitrification) [1].

The production of N_2O, and in turn CO_2 emission, in soil and its inhibition by NIs are complex processes, which can be influenced by different factors such as physicochemical characteristics of the soil. Among these, texture is a major soil characteristic that governs various soil properties, and hence the relative effectiveness of NIs may be different in different textured soils. Soil texture can influence the effectiveness of NIs by affecting their stability/persistence and absorption in soils. Some NIs such as 3, 4-dimethylpyrazole phosphate (DMPP) have been shown to affect N_2O emission by decreasing soil pH [17], which can also vary with soil texture. Some studies have tested the effectiveness of NIs to reduce N_2O emissions but results varied considerably because the studies were carried out under different soil conditions [1]. Thus, it is difficult to draw any specific conclusions about the N_2O mitigation potential of NIs in different textured soils.

This study investigates the effectiveness of four NIs, namely, dicyandiamide (DCD), 3, 4-dimethylpyrazole phosphate (DMPP), nitrogenous mineral fertilizers containing the DMPP ammonium stabilizer (ENTEC) and active ingredients: 3.00–3.25% 1, 2, 4-triazole and 1.50–1.65% 3-methylpyrazole (PIADIN) under three vastly different textured soils (clayey, loamy and sandy). We hypothesized that (i) clay contents would have a positive effect on soil N_2O emissions; and (ii) DMPP would have a better performance than DCD, PIADIN and ENTEC in reducing soil CO_2 and N_2O emissions and the conversion of NH_4^+-N to NO_3^--N under a range of soils with different textures. The objective of this study was to evaluate the variation in soil CO_2 and N_2O emissions, and NH_4^+-N and NO_3^--N concentrations following the application of the abovementioned NIs in clayey (Marsch), loamy (Östliches Hügelland) and sandy (Geest) soils.

2. Materials and Methods

2.1. Collection, Preparation and Characterization of Soil

Three soils varying in soil texture were collected from the following three geological regions of Schleswig-Holstein, Germany: (1) the sandy outwash region (Geest, the outwash region is dominated by Brunic Arenosols or Cambisols, Podzols and Gleysols, as well as Histosols), (2) the Weichselian glacial region in the east (Östliches Hügelland, the Weichselian glacial deposits contain very fertile Luvisols, Cambisols, Anthrosols derived from colluvic material, Gleysols and Rheic Histosols), and (3) the marshland with alluvial deposits in the west (Marsch, the marshland includes different types of Fluvic Gleysols and Histosols) [18]. The selected soils represent the major soil types or climatic zones of the state of Schleswig-Holstein. The soils belong to three different natural grasslands of the state.

Soils from the upper 20 cm soil horizon were collected. Visible plant residues and stones were removed by passing soil through a 2 mm sieve. A representative subsample of each soil was analyzed for salient characteristics, namely, particle size distribution (clay, sand and silt), pH, total carbon (C) and N, NH_4^+-N and NO_3^--N following standard methods (Table 1).

Table 1. Salient characteristics of the experimental soils.

Characteristic	Marsch	Östliches Hügelland	Geest
Silt (%)	10.6 ± 3.9	22.3 ± 9.1	19.2 ± 7.8
Sand (%)	44.3 ± 5.4	62.3 ± 8.7	75.0 ± 6.6
Clay (%)	45.1 ± 3.8	15.4 ± 4.6	5.8 ± 2.3
Texture	clayey	loamy	sandy
Total C (g kg^{-1})	20.8 ± 0.34	12.0 ± 0.50	13.2 ± 0.19
Total N (g kg^{-1})	1.37 ± 0.01	1.13 ± 0.05	1.11 ± 0.02
NH_4^+-N (mg kg^{-1})	0.86 ± 0.02	1.39 ± 0.06	3.56 ± 0.06
NO_3^--N	10.9 ± 0.32	13.47 ± 0.59	0.08 ± 0.02
pH	6.00 ± 1.72	7.40 ± 2.45	5.50 ± 2.03

Values (mean $\pm$ SE, $n = 4$).

2.2. Incubation Experiment

Four NIs, namely, dicyandiamide (DCD), 3, 4-dimethylpyrazole phosphate (DMPP), nitrogenous mineral fertilizers containing the DMPP ammonium stabilizer (ENTEC) and PIADIN (active ingredients: 3.00–3.25% 1, 2, 4-triazole and 1.50–1.65% 3-methylpyrazole), were tested with the three different soil types (Marsch, Östliches Hügelland and Geest). For comparison, a control without the addition of NI was also included for each soil type. The soils were packed into cylindrical pots (15 cm diameter and 33 cm length, sealed at the bottom) to achieve a bulk density of 1.4 g cm^{-3} with 20 cm depth. The experimental treatments, each having four replications, were arranged in a completely randomized design. All the pots were fertilized with 0.5g NH_4^+-N kg^{-1} soil using ammonium sulfate ((NH_4)$_2SO_4$) salt in solution form. The NIs were applied at 5% of applied NH_4^+-N (i.e., 25 mg kg^{-1} soil in solution form). The control treatment involved the application of N fertilizer and deionized water only. Deionized water was added each day to maintain the moisture content equivalent to the water-holding capacity of the soils. Following treatment application, pots were incubated for a period of 57 days in a climatic chamber adjusted to a consistent temperature (15 °C), soil moisture (80% soil water-holding capacity) and air humidity (50%).

2.3. Collection and Measurement of Emitted CO_2 and N_2O

The rate of CO_2 and N_2O emissions was calculated by measuring the concentration of these gases from each pot at different times. The CO_2 and N_2O samples were collected once a day during the first week, once after two days during the second week, and once after three days during the rest of the incubation period. For sampling, the pots were closed first and then gas samples were collected at 0, 20, 40 and 60 min. A 10-mL syringe with a hypodermic needle was used to collect the gas samples. The gas samples were stored in pre-evacuated Chromacol glass vials with chloro-butyl rubber lids that prevent the leakage of gas samples. Each gas sampling was carried out between 09:00 and 11:00 am. Except for the times when gas samples were collected, the pots were left open.

The concentrations of CO_2 and N_2O in the gas samples were measured by gas chromatograph (Agilent 7890A GC, Agilent, CA, USA). The rate of CO_2 and N_2O emissions from each pot (ppm/min) during lid closure was calculated using headspace volume of the pot and a linear relation between CO_2 and N_2O concentrations and time [19]. The flux of CO_2-C (μg h^{-1} kg^{-1}) and N_2O-N (ng h^{-1} kg^{-1}) was calculated with the following equation:

$$EN_2O\text{-}N = \frac{R \times 60 \times V_{gas} \times AR}{W_{soil} \times V_m} \times 2 \times 1000 \qquad (1)$$

$$ECO_2\text{-}C = \frac{R \times 60 \times V_{gas} \times AR}{W_{soil} \times V_m} \qquad (2)$$

where ECO_2/N_2O is the flux of CO_2-C (μg h^{-1} kg^{-1}) and N_2O-N (ng h^{-1} kg^{-1}), R is the rate of CO_2 and N_2O emissions from each pot (ppm/min), V_{gas} is the gas volume in pot

(L), W_{soil} is the weight of dry soil in pot (kg), AR is the relative atomic mass of C and N, i.e., 12 and 14, respectively, and V_m is the molar volume of gas which is 23.7 L/mol at 15 °C. Total CO_2 and N_2O emissions during the experimental period were calculated from the daily emissions of the gases. The relative lowering of total N_2O emission (%) from NI-treated soils as compared to control was regarded as the efficiency of NIs.

2.4. Analysis of NH_4^+ and NO_3^- in Soil

Soil samples (0–20 cm depth) were collected from the pots on day 1, 15, 29, 43 and 57 of incubation for the measurement of NH_4^+ and NO_3^- concentration. Each soil sample was divided into two subsamples; one was oven-dried at 105 °C for 8 h to calculate water content while the other was used for determination of NH_4^+ and NO_3^-. For the analysis of soil mineral N, 10 g of fresh soil was mixed with 40 mL of 0.0125 M $CaCl_2$ solution (1:4) and shaken for 1 h. After centrifugation for 10 min, the extracts were filtered through Whatman filter paper No. 40 and stored at 4 °C. The extracts were analyzed for NH_4^+ and NO_3^- concentrations using a continuous flow analyzer (San^{++} Automated Wet Chemistry Analyzer—Continuous Flow Analyzer (CFA), Skalar, The Netherlands).

2.5. Statistical Analysis

Soil CO_2-C and N_2O-N total emissions and NH_4^+-N and NO_3^--N concentrations across different incubation times were compared by one-way analysis of variance (ANOVA). Treatment means for total CO_2 and N_2O emissions were compared using two-way ANOVA. The significance of differences between individual means and incubation times was determined using a Tukey's honest significant difference (HSD) test. The statistical analyses were performed by R statistical software (University of Auckland, Oakland, CA, USA) at a confidence level of 95% ($p \leq 0.05$).

3. Results

3.1. Fluxes of CO_2-C and N_2O-N

Marsch had the highest CO_2-C flux, followed by Geest and the lowest in Östliches Hügelland (Figure 1). The flux of CO_2-C progressively decreased with time in all the NI treatments under Marsch and Geest soils (Figure 1). In Marsch, DMPP had the lowest CO_2-C flux during the incubation period (Figure 1, Table 2). CO_2-C flux did not differ among the NI treatments during the incubation period in Geest soil (Figure 1). DCD, DMPP and PIADIN had the lowest CO_2-C flux at 7 days of incubation in Östliches Hügelland soil (Figure 1).

Marsch had the highest N_2O-N flux, followed by Östliches Hügelland and the lowest in Geest soil (Figure 2). Marsch and Östliches Hügelland soils with DCD and ENTEC applied had the lowest and the highest N_2O-N flux, respectively, compared with the other NIs during the incubation period (Figure 2). In Geest, ENTEC showed the highest N_2O-N flux during the incubation period while DMPP and PIADIN showed the lowest N_2O-N flux (Figure 2).

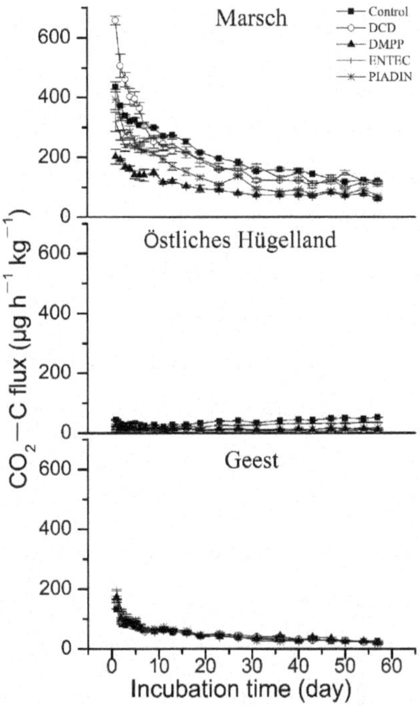

Figure 1. The flux of CO_2-C emission as affected by control, dicyandiamide (DCD), 3, 4-dimethylpyrazole phosphate (DMPP), nitrogenous mineral fertilizers containing the DMPP ammonium stabilizer (ENTEC) and active ingredients: 3.00–3.25% 1, 2, 4-triazole and 1.50–1.65% 3-methylpyrazole (PIADIN) treatments in Marsch, Östliches Hügelland and Geest soils. The data points are means of four independent pot replicates, and error bars represent standard errors of the means ($n = 4$).

Table 2. The F test values for CO_2-C flux, N_2O-N flux, NH_4^+-N and NO_3^--N under different texture soils with four nitrification inhibitors applied.

Soil Type	Treatment	CO_2-C Flux	N_2O-N Flux	NH_4^+-N	NO_3^--N
Marsch	Control	0.001	0.016	0.000	0.000
	DCD	0.003	0.014	0.000	0.000
	DMPP	0.045	0.021	0.093	0.089
	ENTEC	0.015	0.040	0.000	0.000
	PIADIN	0.006	0.032	0.000	0.000
Östliches Hügelland	Control	0.013	0.140	0.000	0.000
	DCD	0.085	0.017	0.008	0.007
	DMPP	0.124	0.003	0.000	0.000
	ENTEC	0.004	0.001	0.000	0.000
	PIADIN	0.045	0.017	0.062	0.000
Geest	Control	0.000	0.004	0.001	0.000
	DCD	0.000	0.004	0.000	0.000
	DMPP	0.001	0.011	0.000	0.000
	ENTEC	0.000	0.000	0.000	0.000
	PIADIN	0.001	0.034	0.000	0.002

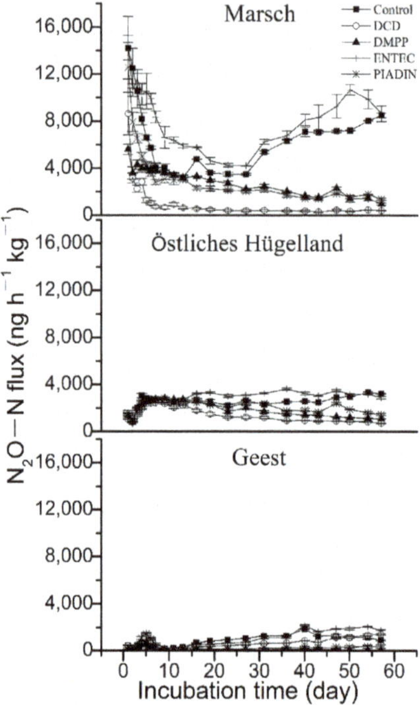

Figure 2. The flux of N_2O-N emission as affected by control, DCD, DMPP, ENTEC and PIADIN treatments in Marsch, Östliches Hügelland and Geest soils. The data points are means of four independent pot replicates, and error bars represent standard errors of the means ($n = 4$).

3.2. Total Emissions of CO_2-C and N_2O-N

Total CO_2-C emission during the incubation period was significantly higher from Marsch soil (274 mg kg^{-1}) than Östliches Hügelland soil (54.6 mg kg^{-1}) and Geest soil (60.3 mg kg^{-1}) (Table 3). In Marsch soil, DMPP showed the highest inhibitory effect on CO_2 emission (50%), followed by PIADIN (32%) and the lowest by ENTEC (16%). DCD did not influence CO_2-C emission from Marsch soil (Table 3). In Östliches Hügelland soil, DCD and PIADIN showed the highest and equal inhibitory effect on CO_2 emission (73%), followed by DMPP (64%) and the lowest by ENTEC (36%). The effect of the NIs on CO_2 emission from Geest soils was nonsignificant.

Table 3. Total CO_2-C emission during the incubation period from Marsch, Östliches Hügelland and Geest soils with control, DCD, DMPP, ENTEC and PIADIN applied under controlled conditions.

	Marsch		Östliches Hügelland		Geest	
NI	**Emission (mg kg^{-1})**	**Decrease (%)**	**Emission (mg kg^{-1})**	**Decrease (%)**	**Emission (mg kg^{-1})**	**Decrease (%)**
Control	274 ± 1.3a	-	54.6 ± 2.3e	-	60.3 ± 0.4e	-
DCD	261 ± 11.5a	5	14.8 ± 1.7g	73	64.4 ± 1.2e	−6.8
DMPP	136 ± 3.8d	50	19.8 ± 0.3g	64	59.6 ± 4.4e	1.2
ENTEC	231 ± 11.3b	16	35.1 ± 0.5f	36	65.8 ± 1.4e	−9.1
PIADIN	187 ± 8.4c	32	14.5 ± 1.0g	74	65.0 ± 3.3e	−7.8

The values (mean ± SE) are means of four independent pot replicates. The values indicated with the same lowercase letter(s) are not significantly different at $p = 0.05$.

Total N_2O-N emission was also the highest from Marsch soil (8051 µg kg^{-1}), followed by Östliches Hügelland soil (3516 µg kg^{-1}) and the lowest from Geest soil (1313 µg kg^{-1}) (Table 4). In Marsch and Östliches Hügelland soils, DCD showed the stronger inhibitory effect on N_2O-N emission (86% and 47%, respectively) compared with DMPP (56% and 30%, respectively) and PIADIN (54% and 16%, respectively). In Geest soil, DMPP was more effective in reducing N_2O-N emission (88%) than PIADIN (70%) and DCD (33%).

Table 4. Total N_2O-N emissions during the incubation period from Marsch, Östliches Hügelland and Geest soils with control, DCD, DMPP, ENTEC and PIADIN applied under controlled conditions.

NI	Marsch		Östliches Hügelland		Geest	
	Emission (mg kg^{-1})	Decrease (%)	Emission (mg kg^{-1})	Decrease (%)	Emission (mg kg^{-1})	Decrease (%)
Control	8051 ± 279b	-	3516 ± 24d		1313 ± 19i	-
DCD	1157 ± 222i	86	1861 ± 102g	47	879 ± 34j	33
DMPP	3533 ± 81d	56	2467 ± 74f	30	157 ± 23k	88
ENTEC	9849 ± 280a	−22	4169 ± 42c	−18	1569 ± 17h	−19
PIADIN	3708 ± 408dc	54	2960 ± 105e	16	387 ± 83k	70

The values (mean ± SE) are means of four independent pot replicates. The values indicated with the same lowercase letter(s) are not significantly different at $p = 0.05$.

3.3. Soil NH_4^+-N and NO_3^--N Concentrations

In Marsch soil, DCD, DMPP and PIADIN slowed down the nitrification process and maintained higher NH_4^+-N concentration and lower NO_3^--N concentration as compared to control (Figure 3; Table 2). ENTEC did not affect NH_4^+-N and NO_3^--N concentrations, which showed the same trend as that of control Marsch soil (Figure 3). NH_4^+-N concentration decreased and NO_3^--N concentration increased progressively during the incubation period in control treatment of Östliches Hügelland soil, whereas DCD, DMPP and PIADIN maintained NH_4^+-N concentration at high levels and kept NO_3^--N at lower levels as compared to control (Figure 3). Similarly, DCD, DMPP, ENTEC and PIADIN slowed down the decrease in NH_4^+-N and the increase in NO_3^--N concentrations as compared to control in Geest soil (Figure 3). DMPP and PIADIN showed lower NO_3^--N concentration as compared to DCD and ENTEC in Geest soil (Figure 3).

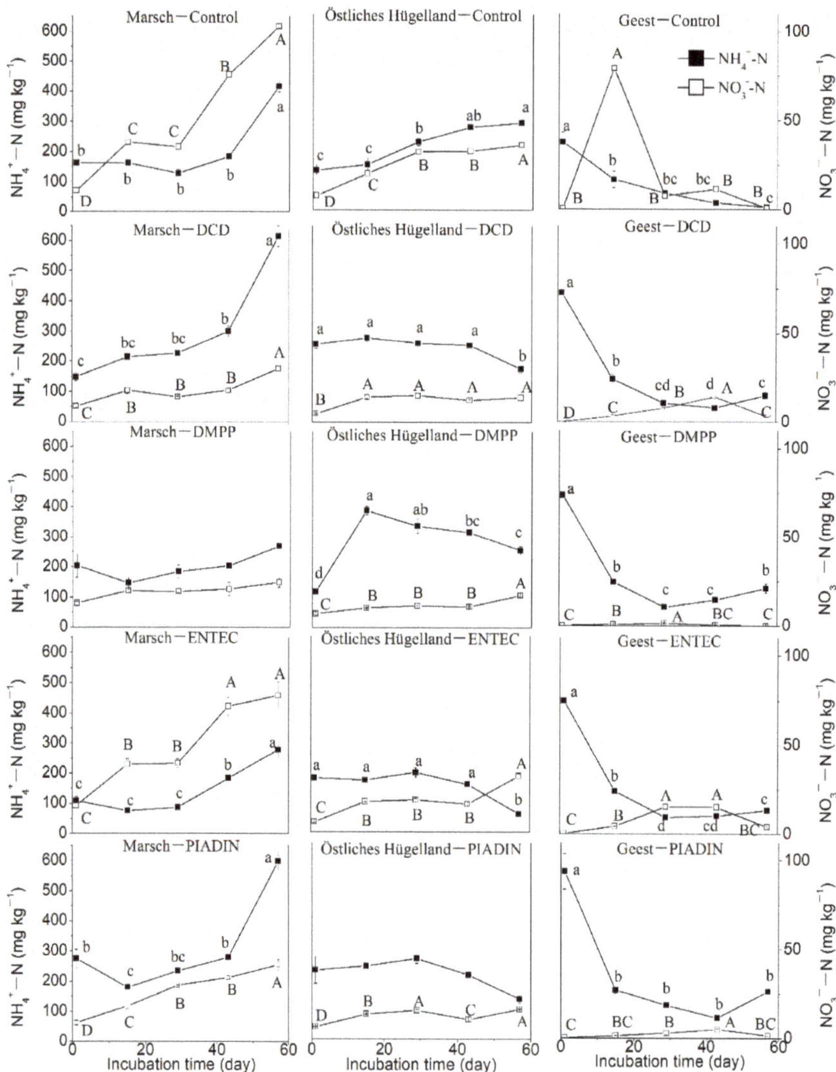

Figure 3. Changes in NH_4^+-N and NO_3^--N concentrations in Marsch, Östliches Hügelland and Geest soil with control, DCD, DMPP, ENTEC and PIADIN applied. The data values are means of four independent pot replicates, and error bars represent standard errors of the means ($n = 4$). Incubation times indicated by the same lowercase (NH_4^+) or uppercase (NO_3^-) letter(s) do not differ significantly at $p \leq 0.05$.

4. Discussion

4.1. Effect of Soil Type

Our results showed that CO_2-C emission from Marsch soil was higher than from Östliches Hügelland and Geest soils (Figure 1, Table 3). This may be explained by the fact that Marsch soil had more organic matter and clay content than the other soils, and CO_2 emission is positively linked to soil organic carbon (SOC) content [14]. Soil organic carbon content is linearly correlated to soil respiration [20]. Thomsen et al. [21] found that the emission of CO_2 from the sandiest soil was lower than from the heavier textured soils due to a low content of potentially mineralizable native SOC. Moreover, CO_2-C emissions from native SOC increased with increasing clay content but this relationship was ascribed to the different mineralogy of the soils [22].

Marsch soil showed the highest N_2O-N emission, followed by Östliches Hügelland soil and the lowest by Geest soil (Table 4). Organic C and clay content of the three soils also followed the same decreasing order (Table 1). Microbial community structure of a soil plays an important role in defining the rate of nitrification and denitrification in soil [23], and a soil with higher microbial activity could result in higher N_2O emissions [24]. Nitrogen mineralization and N_2O emissions are influenced by soil organic matter [25] and microbial population [26]. Since nitrification is primarily an autotrophic process, with heterotrophic nitrification accounting for only 20% [27], C-substrate availability and N_2O emissions from denitrification and nitrification are always positively related to each other [28,29]. Abbasi et al. [30] found that the process of denitrification and production of N_2O were smaller in arable soil deficient in organic carbon compared with grassland soil with plenty of organic carbon. Cébron et al. [31] reported that the presence of organic carbon promoted the population of nitrifying bacteria in a clay-textured soil. Thus, higher organic carbon content in Marsch soil could be one possible reason for higher N_2O emission from this soil. Moreover, higher N_2O emission from Marsch than Geest soil may also be explained by the higher pH of this soil. Fan et al. [32] reported significantly higher N_2O emission rates from three alkaline soils (pH 7.6–8.2) as compared to an acidic soil with a pH of 5.6.

Nitrification is also directly influenced by soil texture [25]. A fine-textured soil can retain more water and create more frequent and longer anaerobic conditions than a coarse- or medium-textured soil, and thus may result in faster denitrification and N_2O emission [33]. Overall, the cumulative N_2O-N emission from a clay soil was significantly higher than those from a loamy soil [33] and therefore N_2O emissions increase with increasing clay content.

4.2. Effect of NIs

Except for CO_2-C emission from Geest soil and N_2O-N emission from ENTEC-treated soil, NIs significantly decreased CO_2-C and N_2O-N emissions in all other cases, but the magnitude of the inhibitory effect did not show any specific consistent trend with the soil texture (Tables 3 and 4). Different results have been reported by previous researchers regarding the relative efficacy of NIs in varying textured soils. Barth et al. [18] found that NIs are more efficient in light soils than heavy soils, whereas Akiyama et al. [34] found that the effectiveness of NIs was relatively consistent across the various soil types. According to Fisk et al. [35], the efficacy of DMPP and DCD diminished with the addition of soil organic matter, and Marsden et al. [36] reported decreased efficacy with higher clay content. Volpi et al. [14] found that NIs were effective only in the soil with the highest nitrification activity and the lowest clay content.

In our study, we found that the comparative effectiveness of the NIs in inhibiting CO_2-C and N_2O-N emissions depended on soil type (Tables 3 and 4). DCD had the highest inhibitory effect on soil nitrification in clay soil of Marsch, followed by loam soil of Östliches Hügelland and lowest in sandy soil of Geest (Table 4). Ernfors et al. [37] reported that the efficacy of DCD was soil-specific, whereas Wakelin et al. [38] found that the efficacy of DCD was not related to the soil type. The lowest efficacy of DCD in sandy soil as found in our study may be explained by the lowest decomposition of this NI in sandy soil as compared to heavy textured soils [1]. Moreover, McGeough et al. [7] found that DCD may not be an effective NI under heterotrophic nitrification, which is proportionally the dominant form of nitrification in sandy soil as compared to a heavy textured soil.

DMPP was reported to be highly specific and more effective in inhibiting nitrification as compared to DCD [39]. We found that the inhibitory effect of DMPP on N_2O emission was more pronounced in the sandy soil of Geest than in the loamy and clay soils (Table 4). In short-term incubation experiments, Barth et al. [40] found that decreasing sand content reduced the efficacy of DMPP in retarding NH_4^+ oxidation. They found that the adsorption capacity of DMPP is positively correlated with clay content, and that a lower effectiveness of DMPP may be due to the adsorption of the NI on silt and clay particles. In their later study, the same group of researchers reported that DMPP had more pronounced inhibition of N_2O emission in sandy than in loam soil [41]. The same explanation as given above for the effect of DMPP stands true for PIADIN, which showed the same trend of effectiveness in different textured soils as that of DMPP. DMPP has relatively low mobility [39–43], mineralizes slowly and thus has a longer-lasting inhibitory effect on nitrification than DCD [44,45].

Soil texture may affect organic carbon turnover by adsorption of organic carbon onto surfaces of clay or organic complexes [46]. Thus, the effect of different NIs on organic carbon decomposition could also be affected by soil texture. Our results showed that DCD did not influence the soil organic carbon decomposition in Marsch soil, and thus CO_2-C emission from DCD-treated soil was the same as that from control soil (Table 3). DMPP had the best inhibitory effect on the organic carbon decomposition in Marsch soil. The NIs DCD, DMPP and PIADIN effectively inhibited the decomposition of organic carbon in Östliches Hügelland soil and resulted in the lowest CO_2-C emissions (Table 3). ENTEC had the smallest inhibitory effect on the soil organic carbon decomposition in Östliches Hügelland soil.

5. Conclusions

CO_2-C and N_2O-N emissions were higher from N-fertilized clayey soils than from lighter textured loamy and sandy soils. DCD was proven the most effective NI in decreasing CO_2-C and N_2O-N emissions and inhibiting nitrification in the clayey and loamy soils. On the other hand, DMPP and PIADIN could decrease CO_2-C and N_2O-N emissions more effectively in the sandy soil. ENTEC remained the least effective in inhibiting CO_2-C emission, whereas it did not inhibit N_2O-N emission from the studied soils.

It is concluded that clayey soil has more gaseous emissions and NIs perform differently depending on the soil texture to inhibit soil nitrification and gaseous emissions. DCD is a better NI for clay and loamy soils while DMPP and PIADIN are better for sandy soil to inhibit soil nitrification and gaseous emissions. ENTEC is ineffective in all soil textures.

Author Contributions: Conceptualization, K.H.M.; methodology, software, writing—original draft preparation, Y.G.; writing—review and editing, A.N. and K.H.M.; supervision, K.H.M.; funding acquisition, K.H.M. All authors have read and agreed to the published version of the manuscript.

Funding: This research was funded by Ph.D. project funding from the Society of Energy and Climate (EKSH) of Schleswig-Holstein, grant number 14/12-24, and by a George Forster Post-Doctorate Fellowship grant from the Alexander von Humboldt Foundation.

Data Availability Statement: The data presented in this study are available on request from the corresponding author. The data are not publicly available due to restrictions.

Acknowledgments: K.H.M. and Y.G. acknowledge Ph.D. project funding from the Society of Energy and Climate (EKSH) of Schleswig-Holstein (14/12-24). A.N. is thankful to the Alexander von Humboldt Foundation for a George Forster Post-Doctorate Fellowship grant.

Conflicts of Interest: The authors declare no conflict of interest.

References

1. Gilsanz, C.; Báez, D.; Misselbrook, T.H.; Dhanoa, M.S.; Cárdenas, L.M. Development of emission factors and efficiency of two nitrification inhibitors, DCD and DMPP. *Agric. Ecosyst. Environ.* **2016**, *216*, 1–8. [CrossRef]
2. Chen, D.; Suter, H.; Islam, A.; Edis, R.; Freney, J.R.; Walker, C.N. Prospects of improving efficiency of fertiliser nitrogen in Australian agriculture: A review of enhanced efficiency fertilisers. *Soil Res.* **2008**, *46*, 289–301. [CrossRef]
3. Van Grinsven, H.J.M.; Spiertz, J.H.J.; Westhoek, H.J.; Bouwman, A.F.; Erisman, J.W. Nitrogen use and food production in European regions from a global perspective. *J. Agric. Sci.* **2014**, *152*, 9–19. [CrossRef]
4. Sutton, M.A.; Howard, C.M.; Erisman, J.W.; Billen, G.; Bleeker, A.; Grennfelt, P.; van Grinsven, H.; Grizetti, B. *The European Nitrogen Assessment: Sources, Effects and Policy Perspectives*; Cambridge University Press: Cambridge, NY, USA, 2011.

5. Liu, R.; Hayden, H.L.; Hu, H.; He, J.; Suter, H.; Chen, D. Effects of the nitrification inhibitor acetylene on nitrous oxide emissions and ammonia-oxidizing microorganisms of different agricultural soils under laboratory incubation conditions. *Appl. Soil Ecol.* **2017**, *119*, 80–90. [CrossRef]
6. Harrison, R.; Webb, J. A review of the effect of N fertilizer type on gaseous emissions. In *Advances in Agronomy*; Sparks, D.L., Ed.; Elsevier Academic Press Inc.: San Diego, CA, USA, 2001; Volume 73, pp. 65–108.
7. McGeough, K.; Watson, C.; Müller, C.; Laughlin, R.; Chadwick, D. Evidence that the efficacy of the nitrification inhibitor dicyandiamide (DCD) is affected by soil properties in UK soils. *Soil Biol. Biochem.* **2016**, *94*, 222–232. [CrossRef]
8. Ravishankara, A.R.; Daniel, J.S.; Portmann, R.W. Nitrous Oxide (N2O): The Dominant Ozone-Depleting Substance Emitted in the 21st Century. *Science* **2009**, *326*, 123–125. [CrossRef]
9. IPCC. Climate change 2007: The physical science basis. In *Contribution of Working Group I to the Fourth Assessment Report of the Intergovernmental Panel on Climate Change*; Solomon, S., Qin, D., Manning, M., Eds.; Cambridge University Press: Cambridge, UK, 2007.
10. Hénault, C.; Grossel, A.; Mary, B.; Roussel, M.; Léonard, J. Nitrous oxide emission by agricultural soils: A review of spatial andtemporal variability for mitigation. *Pedosphere* **2021**, *22*, 426–433. [CrossRef]
11. Hu, H.-W.; Chen, D.; He, J.-Z. Microbial regulation of terrestrial nitrous oxide formation: Understanding the biological pathways for prediction of emission rates. *FEMS Microbiol. Rev.* **2015**, *39*, 729–749. [CrossRef] [PubMed]
12. Zhang, J.; Müller, C.; Cai, Z. Heterotrophic nitrification of organic N and its contribution to nitrous oxide emissions in soils. *Soil Biol. Biochem.* **2015**, *84*, 199–209. [CrossRef]
13. Khalil, M.I.; Buegger, F.; Schraml, M.; Gutser, R.; Richards, K.G.; Schmidhalter, U. Gaseous Nitrogen Losses from a Cambisol Cropped to Spring Wheat with Urea Sizes and Placement Depths. *Soil Sci. Soc. Am. J.* **2009**, *73*, 1335–1344. [CrossRef]
14. Volpi, I.; Laville, P.; Bonari, E.; o di Nasso, N.N.; Bosco, S. Improving the management of mineral fertilizers for nitrous oxide mitigation: The effect of nitrogen fertilizer type, urease and nitrification inhibitors in two different textured soils. *Geoderma* **2017**, *307*, 181–188. [CrossRef]
15. Boeckx, P.; Xu, X.; Van Cleemput, O. Mitigation of N2O and CH4 Emission from Rice and Wheat Cropping Systems Using Dicyandiamide and Hydroquinone. *Nutr. Cycl. Agroecosyst.* **2005**, *72*, 41–49. [CrossRef]
16. Zhang, L.-M.; Hu, H.-W.; Shen, J.-P.; He, J.-Z. Ammonia-oxidizing archaea have more important role than ammonia-oxidizing bacteria in ammonia oxidation of strongly acidic soils. *ISME J.* **2011**, *6*, 1032–1045. [CrossRef]
17. Barth, G.; Von Tucher, S.; Schmidhalter, U.; Otto, R.; Motavalli, P.; Ferraz-Almeida, R.; Sattolo, T.M.S.; Cantarella, H.; Vitti, G.C. Performance of nitrification inhibitors with different nitrogen fertilizers and soil textures. *J. Plant Nutr. Soil Sci.* **2019**, *182*, 694–700. [CrossRef]
18. Horn, R.; Mordhorst, A.; Fleige, H.; Zimmermann, I.; Burbaum, B.; Filipinski, M.; Cordsen, E. Soil type and land use effects on tensorial properties of saturated hydraulic conductivity in northern Germany. *Eur. J. Soil Sci.* **2020**, *71*, 179–189. [CrossRef]
19. Venterea, R.T.; Petersen, S.O.; De Klein, C.A.; Pedersen, A.R.; Noble, A.D.L.; Rees, R.M.; Gamble, J.D.; Parkin, T.B. Global Research Alliance N2O chamber methodology guidelines: Flux calculations. *J. Environ. Qual.* **2020**, *49*, 1141–1155. [CrossRef]
20. Lou, Y.; Zhou, L. *Soil Respiration and the Environment*; Academic Press: New York, NY, USA, 2006; p. 33.
21. Thomsen, I.K.; Schjønning, P.; Olesen, J.E.; Christensen, B.T. C and N turnover in structurally intact soils of different texture. *Soil Biol. Biochem.* **2003**, *35*, 765–774. [CrossRef]
22. Parfitt, R.L.; Salt, G.J.; Saggar, S. Effect of leaching and clay content on carbon and nitrogen mineralisation in maize and pasture soils. *Soil Res.* **2001**, *39*, 535. [CrossRef]
23. Butterbach-Bahl, K.; Baggs, E.M.; Dannenmann, M.; Kiese, R.; Zechmeister-Boltenstern, S. Nitrous oxide emissions from soils: How well do we understand the processes and their controls? *Philos. Trans. R. Soc. B Biol. Sci.* **2013**, *368*, 20130122. [CrossRef]
24. Schaufler, G.; Kitzler, B.; Schindlbacher, A.; Skiba, U.; Sutton, M.A.; Zechmeister-Boltenstern, S. Greenhouse gas emissions from European soils under different land use: Effects of soil moisture and temperature. *Eur. J. Soil Sci.* **2010**, *61*, 683–696. [CrossRef]
25. Stehfest, E.; Bouwman, L. N2O and NO emission from agricultural fields and soils under natural vegetation: Summarizing available measurement data and modeling of global annual emissions. *Nutr. Cycl. Agroecosyst.* **2006**, *74*, 207–228. [CrossRef]
26. Hart, S.C.; Nason, G.E.; Myrold, D.D.; Perry, D.A. Dynamics of Gross Nitrogen Transformations in an Old-Growth Forest: The Carbon Connection. *Ecology* **1994**, *75*, 880–891. [CrossRef]
27. Liu, R.; Suter, H.; Hayden, H.; He, J.; Chen, D. Nitrate production is mainly heterotrophic in an acid dairy soil with high organic content in Australia. *Biol. Fertil. Soils* **2015**, *51*, 891–896. [CrossRef]
28. Li, C.; Frolking, S.; Butterbach-Bahl, K. Carbon Sequestration in Arable Soils is Likely to Increase Nitrous Oxide Emissions, Offsetting Reductions in Climate Radiative Forcing. *Clim. Chang.* **2005**, *72*, 321–338. [CrossRef]
29. Wan, Y.; Ju, X.; Ingwersen, J.; Schwarz, U.; Stange, C.F.; Zhang, F.; Streck, T. Gross Nitrogen Transformations and Related Nitrous Oxide Emissions in an Intensively Used Calcareous Soil. *Soil Sci. Soc. Am. J.* **2009**, *73*, 102–112. [CrossRef]
30. Abbasi, M.K.; Hina, M.; Tahir, M.M. Effect of Azadirachta indica (neem), sodium thiosulphate and calcium chloride on changes in nitrogen transformations and inhibition of nitrification in soil incubated under laboratory conditions. *Chemosphere* **2011**, *82*, 1629–1635. [CrossRef]
31. Cébron, A.; Berthe, T.; Garnier, J. Nitrification and Nitrifying Bacteria in the Lower Seine River and Estuary (France). *Appl. Environ. Microbiol.* **2003**, *69*, 7091–7100. [CrossRef]

32. Fan, C.; Li, B.; Xiong, Z. Nitrification inhibitors mitigated reactive gaseous nitrogen intensity in intensive vegetable soils from China. *Sci. Total Environ.* **2018**, *612*, 480–489. [CrossRef]

33. Köster, J.R.; Well, R.; Dittert, K.; Giesemann, A.; Lewicka-Szczebak, D.; Mühling, K.H.; Herrmann, A.; Lammel, J.; Senbayram, M. Soil denitrification potential and its influence on N_2O reduction and N_2O isotopomer ratios. *Rapid Commun. Mass Spectrom.* **2013**, *27*, 2363–2373. [CrossRef]

34. Akiyama, H.; Yan, X.; Yagi, K. Evaluation of effectiveness of enhanced-efficiency fertilizers as mitigation options for N_2O and NO emissions from agricultural soils: Meta-analysis. *Glob. Chang. Biol.* **2009**, *16*, 1837–1846. [CrossRef]

35. Fisk, L.; Maccarone, L.; Barton, L.; Murphy, D. Nitrapyrin decreased nitrification of nitrogen released from soil organic matter but not amoA gene abundance at high soil temperature. *Soil Biol. Biochem.* **2015**, *88*, 214–223. [CrossRef]

36. Marsden, K.A.; Marín-Martínez, A.J.; Vallejo, A.; Hill, P.W.; Jones, D.L.; Chadwick, D.R. The mobility of nitrification inhibitors under simulated ruminant urine deposition and rainfall: A comparison between DCD and DMPP. *Biol. Fertil. Soils* **2016**, *52*, 491–503. [CrossRef]

37. Ernfors, M.; Brennan, F.P.; Richards, K.G.; McGeough, K.L.; Griffiths, B.S.; Laughlin, R.J.; Watson, C.J.; Philippot, L.; Grant, J.; Minet, E.P.; et al. The nitrification inhibitor dicyandiamide increases mineralization–immobilization turnover in slurry-amended grassland soil. *J. Agric. Sci.* **2014**, *152*, 137–149. [CrossRef]

38. Wakelin, S.; Williams, E.; O'Sullivan, C.A.; Cameron, K.C.; Di, H.J.; Cave, V.; O'Callaghan, M. Predicting the efficacy of the nitrification inhibitor dicyandiamide in pastoral soils. *Plant Soil* **2014**, *381*, 35–43. [CrossRef]

39. Dittert, K.; Bol, R.; King, R.; Chadwick, D.; Hatch, D. Use of a novel nitrification inhibitor to reduce nitrous oxide emission from15N-labelled dairy slurry injected into soil. *Rapid Commun. Mass Spectrom.* **2001**, *15*, 1291–1296. [CrossRef]

40. Barth, G.; Von Tucher, S.; Schmidhalter, U. Influence of soil parameters on the effect of 3,4-dimethylpyrazole-phosphate as a nitrification inhibitor. *Biol. Fertil. Soils* **2001**, *34*, 98–102. [CrossRef]

41. Barth, G.; Von Tucher, S.; Schmidhalter, U. Effectiveness of 3,4-Dimethylpyrazole Phosphate as Nitriflcation Inhibitor in Soil as Influenced by Inhibitor Concentration, Application Form, and Soil Matric Potential. *Pedosphere* **2008**, *18*, 378–385. [CrossRef]

42. Di, H.J.; Cameron, K.C.; Shen, J.-P.; Winefield, C.S.; O'Callaghan, M.; Bowatte, S.; He, J.-Z. Ammonia-oxidizing bacteria and archaea grow under contrasting soil nitrogen conditions. *FEMS Microbiol. Ecol.* **2010**, *72*, 386–394. [CrossRef]

43. Di, H.J.; Cameron, K.C.; Sherlock, R.R. Comparison of the effectiveness of a nitrification inhibitor, dicyandiamide, in reducing nitrous oxide emissions in four different soils under different climatic and management conditions. *Soil Use Manag.* **2007**, *23*, 1–9. [CrossRef]

44. Weiske, A.; Benckiser, G.; Herbert, T.; Ottow, J. Influence of the nitrification inhibitor 3,4-dimethylpyrazole phosphate (DMPP) in comparison to dicyandiamide (DCD) on nitrous oxide emissions, carbon dioxide fluxes and methane oxidation during 3 years of repeated application in field experiments. *Biol. Fertil. Soils* **2001**, *34*, 109–117. [CrossRef]

45. Irigoyen, I.; Muro, J.; Azpilikueta, M.; Aparicio-Tejo, P.; Lamsfus, A.C. Ammonium oxidation kinetics in the presence of nitrification inhibitors DCD and DMPP at various temperatures. *Soil Res.* **2003**, *41*, 1177–1183. [CrossRef]

46. Oades, J.M. The retention of organic matter in soils. *Biogeochemistry* **1988**, *5*, 35–70. [CrossRef]

Nitrogen

Article

Short-Term Effect of Nitrogen Fertilization on Carbon Mineralization during Corn Residue Decomposition in Soil

Tanjila Jesmin *, Dakota T. Mitchell and Richard L. Mulvaney

Department of Natural Resources and Environmental Sciences, University of Illinois, Turner Hall, 1102 S. Goodwin Avenue, Urbana, IL 61801, USA; mitchelldako@gmail.com (D.T.M.); mulvaney@illinois.edu (R.L.M.)
* Correspondence: jesmin2@illinois.edu

Abstract: The effect of N fertilization on residue decomposition has been studied extensively; however, contrasting results reflect differences in residue quality, the form of N applied, and the type of soil studied. A 60 d laboratory incubation experiment was conducted to ascertain the effect of synthetic N addition on the decomposition of two corn (*Zea mays* L.) stover mixtures differing in C:N ratio by continuous monitoring of CO_2 emissions and periodic measurement of microbial biomass and enzyme activities involved in C and N cycling. Cumulative CO_2 production was greater for the high than low N residue treatment, and was significantly increased by the addition of exogenous N. The latter effect was prominent during the first month of incubation, whereas N-treated soils produced less CO_2 in the second month, as would be expected due to more rapid substrate depletion from microbial C utilization previously enhanced by greater N availability. The stimulatory effect of exogenous N was verified with respect to active biomass, microbial biomass C and N, and cellulase and protease activities, all of which were significantly correlated with cumulative CO_2 production. Intensive N fertilization in modern corn production increases the input of residues but is not conducive to soil C sequestration.

Keywords: aerobic incubation; CO_2 production; microbial metabolism; enzyme activities; active biomass; gross mineralization/immobilization

Citation: Jesmin, T.; Mitchell, D.T.; Mulvaney, R.L. Short-Term Effect of Nitrogen Fertilization on Carbon Mineralization during Corn Residue Decomposition in Soil. *Nitrogen* **2021**, *2*, 444–460. https://doi.org/10.3390/nitrogen2040030

Academic Editor: Jacynthe Dessureault-Rompré

Received: 17 September 2021
Accepted: 25 October 2021
Published: 27 October 2021

Publisher's Note: MDPI stays neutral with regard to jurisdictional claims in published maps and institutional affiliations.

1. Introduction

Modern cereal production relies on intensive N fertilization to increase grain yield and thereby enhances the input of crop residues. The latter effect is important for agricultural sustainability, as residues protect the surface soil against erosion loss, serve as a source of nutrients for plants and soil microbes, and sustain the soil microflora by supplying C as an energy source. Moreover, crop residues are essential to the formation of soil organic matter, although the efficiency of this conversion is necessarily reduced by liberation of CO_2 during microbial decomposition, which depends not only on environmental factors (temperature and moisture) but also N availability and residue quality (i.e., chemical composition and C:N ratio) [1].

The effect of exogenous N addition on residue decomposition has been the subject of numerous investigations, but contrasting results have been reported. A positive effect has been found in some studies [2,3], whereas in others, N addition has reduced C mineralization [4–6] or has had no net effect [7–9]. The disparities can be attributed in part to the type of residue, which affects the proportions of cellulose, lignin, and other constituents that differ in their ease of decomposition [10], and further interactions arise in relation to the size and placement of residues. Moreover, discrepancies can occur because of variation in the fertilizer N rate relative to the soil's N supplying capacity and can also reflect differences in the form of N applied, as microbial N utilization is greater with NH_4^+ than NO_3^- [11,12] but can be inhibited by acidifying N sources such as $(NH_4)_2SO_4$ or NH_4Cl. Another source of inconsistencies in the effect of exogenous N on residue decomposition is the method of incubation, which can be compromised if aerobic conditions are not maintained throughout a study period with continuous collection of CO_2 [13].

The effect of N availability on C mineralization during residue decomposition can be clarified by investigating the impact on microbial biomass production and/or extracellular enzyme activities involved in microbial C and N cycling. Previous studies have shown that incorporation of N with residue or other carbonaceous substrates increases activities for cellulase and protease, two of the major enzymes responsible for C and N mineralization [14–17]; however, a negative effect is also possible when cellulase activity is limited by a low substrate concentration in ligneous materials [18] or when protease is repressed by a substantial concentration of NH_4^+ as the end product [16]. Microbial biomass content tends to follow changes in enzymatic activities and CO_2 production during residue decomposition [19] and can either be increased [20] or decreased [21] by the addition of N. These changes would necessarily affect the dynamics of soil and residue N through mineralization and immobilization [21–23].

Despite a massive input of residues when corn (*Zea mays* L.) is repeatedly grown with synthetic N fertilization in long-term cropping experiments, the usual trend over time is a decline in profile storage of soil organic C (SOC) [24–26]. Such findings motivated the laboratory incubation study reported herein, which utilized continuous CO_2 monitoring and relevant microbial indicators to test the null hypotheses that (1) exogenous or endogenous N increases C and N mineralization during corn residue decomposition, (2) the increased mineralization is due to stimulation of microbial biomass production and enzyme activities, and (3) the effectiveness of N for enhancing mineralization will be reduced by declining substrate availability.

2. Materials and Methods

2.1. Soil Studied

For use in comparing the decomposition of different residues, a bulk sample of surface (0–20 cm) soil was collected in early May 2019 from a Mollisol mapped as the Ipava series [Fine, smectitic, mesic Aquic Argiudolls (Chernozem)] near Farmer City (40°15′12.6″ N 88°34′59.4″ W) in central Illinois, USA. The sampling site had been cropped to a corn−soybean (*Glycine max* L. Merr) rotation for more than 40 years, during which the fertilizer N rate for corn was 180 kg ha^{-1}. The soil sample, collected in a 38 L polyethylene tote box, was sieved (2 mm screen) in the field-moist condition with removal of macro residues from the 2018 corn crop, thoroughly homogenized, and then returned to the tote box for no more than 2 weeks of storage in a refrigerator at 4 °C. A subsample was air-dried for triplicate analyses to determine the properties reported in Table 1.

Table 1. Physicochemical properties of the soil studied.

Property	Value	Reference
pH (soil:water ratio, 1:1)	6.6	[27]
Organic C (g kg^{-1})	21.5	[28]
Total N (g kg^{-1})	1.8	[29,30]
C:N ratio	12.2	
Potentially mineralizable N (mg kg^{-1})	258	[31]
Bioavailable P (mg kg^{-1})	22	[32]
Sand (g kg^{-1})	158	[33]
Silt (g kg^{-1})	595	[33]
Clay (g kg^{-1})	247	[33]
Water-holding capacity (WHC, mL kg^{-1})	635	[34]

2.2. Residues

Corn stover was collected within one day after harvest in late September from subplots of contrasting fertility under continuous corn and a corn−soybean rotation at the historic Morrow Plots in Urbana, Illinois, USA. The stover samples, consisting of leaves, stalks, husks, and cobs, were allowed to dry for one week in a forced-air oven at 50 °C, subsequently ground to <2 mm using a Model 4 Wiley mill (Thomas Scientific, Swedesboro, NJ, USA), and then transferred to air-tight Mason jars for storage at room temperature. Before use, a composite mixture was prepared for each rotation by combining the four different residues according to the proportions given by Pordesimo et al. [35] for aboveground corn biomass. The two mixtures were characterized (Table 2) for organic C, total N, C:N ratio, and major organic fractions by proximate analysis to estimate the water-soluble fraction, lignin, and cellulose + hemicellulose [36].

Table 2. Characterization of corn residue mixtures used in incubation study. Data reported as a mean of duplicate or triplicate determinations. Values for total N and water-soluble fraction differ significantly at $p < 0.001$. Crude protein calculated as total N $\times$ 6.25 [37].

Residue Mixture	Cropping System	Organic C (g kg^{-1})	Total N (g kg^{-1})	Crude Protein (g kg^{-1})	C:N	Proximate Analysis (g kg^{-1})		
						Water-Soluble Fraction	Lignin	Cellulose + Hemicellulose
High N	Corn−corn (fertilized)	407	6.7	42	61	28.0	14.5	53.7
Low N	Corn−soybean (unfertilized)	404	4.3	27	94	18.0	13.8	52.9

2.3. Incubation Procedure

For preincubation, 252 samples of field-moist soil (50 g dry weight equivalent) were weighed into 120 mL polypropylene specimen containers and adjusted to 40% WHC by the addition of deionized water using a Metrohm 665 Dosimat (Metrohm, Herisau, Switzerland). To maintain moisture content and aerobic conditions, the containers were each sealed with Parafilm (Alcan Packaging, Neenah, WI, USA) that was punctured by forming 8–10 holes with a syringe needle. Preincubation was carried out for 7 d under dark conditions in a constant-temperature room maintained at 25 °C.

After preincubation, the following treatments were randomized among nine sets of replicate soil samples: (1) no amendment (control), (2) high N residue (HNR), (3) HNR + KNO$_3$, (4) HNR + (NH$_4$)$_2$SO$_4$, (5) low N residue (LNR), (6) LNR + KNO$_3$, (7) LNR + (NH$_4$)$_2$SO$_4$, (8) KNO$_3$, and (9) (NH$_4$)$_2$SO$_4$. Residues were added at the rate of 10 mg dry weight g^{-1} soil, which was selected to roughly represent modern corn production in the Midwestern USA, followed by thorough mixing for uniform incorporation. Nitrogen as NO$_3^-$ or NH$_4^+$ was applied as uniformly as possible using the Dosimat to dispense a solution that supplied 0.1 mg N g^{-1} soil. All soil samples were treated with sufficient deionized water to bring the moisture content to 50% WHC by using the Dosimat, three specimen containers from each treatment (total of 27) were weighed for monitoring soil moisture content, and every specimen container was transferred to a 1.9 L wide-mouth Mason jar. Three jars from each treatment-specific set of samples, and three additional jars with no soil, were sealed using lids equipped with a pair of ball valves for atmospheric sampling, and the remainder were sealed using standard jar lids. All jars were returned to the constant-temperature room for 60 d of incubation in the dark at 25 °C.

After incubation for 7, 14, 30, 45, and 60 d, five moist soil samples per treatment were thoroughly mixed, and the composite sample was analyzed for soil pH, active biomass, microbial biomass C and N, cellulase and protease activities, and gross N mineralization and immobilization.

2.4. Atmospheric Analyses

At 1 d intervals during the first week of incubation, and at 2 or 3 d intervals thereafter, the Mason jars equipped with gas sampling lids (Figure 1) were each connected to a sampling tube and circulating pump for atmospheric sample collection following the technique described by Horgan et al. [38]. After sampling, lids were removed from all jars in use for incubation and also from the three used for background atmospheric sampling, and the jars were left open for one hour of aeration. If necessary, deionized water was added to replace evaporative losses before reattaching jar lids, and incubation was resumed in darkness at 25 °C.

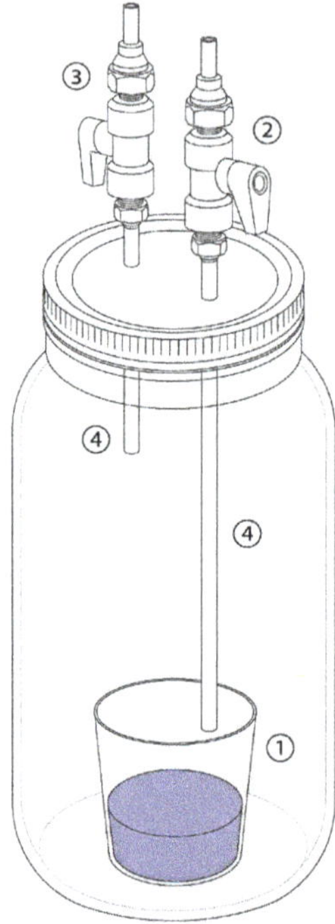

Figure 1. Unit used for incubation with atmospheric sampling by the system of Horgan et al. [38], consisting of a specimen container with soil (1) in a 1.9 L Mason jar equipped with a lid having inlet (2) and outlet (3) ball valves (item # 38EF92, Grainger, Lake Forest, IL, USA) connected to 6.4 mm O.D. brass tubing (4).

Analyses for CO_2 and O_2 were performed using a Hewlett-Packard Model 5790A gas chromatograph (GC) (Agilent Technologies, Santa Clara, CA, USA) equipped with an eight-port sampling valve (Valco Instruments Co., Houston, TX, USA) employing dual 0.5 mL sample loops, a Tracor U-90 ultrasonic detector (Tracor, Austin, TX, USA), and a Hewlett-Packard Model 3390A reporting integrator. This instrument used ultra-high purity He as the carrier gas, Porapak Q for separation of CO_2 at 50 °C, and molecular sieve 5A for separation of O_2 + Ar at 25 °C. Calibration was carried out for every set of analyses using certified mixtures of CO_2 in He (Matheson, Joliet, IL, USA) and of O_2 in N_2 (Airgas, Radnor, PA, USA).

2.5. Soil pH

Duplicate 5 g (dry weight) samples from each treatment were mixed with sufficient deionized water to obtain a 1:1 soil:water suspension, and pH was measured using a glass electrode [27].

2.6. Microbial Biomass Parameters

The active component of microbial biomass was estimated by a modified version of the biokinetics method described by Van de Werf and Verstraete [39]. In the modified method,

5 g (dry weight) of soil was adjusted to 60% WHC with or without the addition of glucose medium and then incubated (25 °C, 6.67 h) in a 250 mL straight-sided glass jar equipped with a gas-tight lid having a ball valve for CO_2 analysis using the GC system previously described. Active biomass was calculated using the equation given by Van de Werf and Verstraete [39] from the increase in CO_2 production observed for glucose-treated samples.

The chloroform fumigation/extraction procedure described by Vance et al. [40] was utilized to measure soil microbial biomass C (MBC) and N (MBN). Following K_2SO_4 extraction, organic C was determined by dichromate oxidation [28] for calculation of MBC (B_c) using the equation proposed by Vance et al. [40], $B_c = F_c/0.45$, where F_c = [(organic C extracted by K_2SO_4 from fumigated soil) − (organic C extracted by K_2SO_4 from non-fumigated soil)], and the value in the denominator represents the proportion of biomass C mineralized to CO_2.

Microbial biomass N was estimated from the difference between fumigated and unfumigated extracts when analyzed for total N by Kjeldahl digestion [29] and diffusion [30]. A value of 0.54 was assumed as the correction factor for calculating biomass N (B_N) by an equation that follows the same form as the one given in the previous paragraph for biomass C.

2.7. Enzyme Activities

Two of the major enzymes involved in C and N mineralization—cellulase and protease—were assayed to determine how their activities changed over the incubation period. Cellulase activity was determined by a *para*-nitrophenol (*p*NP) method [41] using *p*NP-β-D-cellobioside substrate prepared in modified universal buffer (MUB, pH 6.5). For this purpose, triplicate 1 g soil samples (dry weight equivalent) were treated with 5 mL of 12 mM substrate solution, and the mixture was briefly vortexed and then incubated for 2 h at 37 °C. Hydrolysis was terminated immediately following incubation by adding 4 mL of 0.1 M tris buffer (pH 12) and 1 mL of 2 M $CaCl_2$, a 1 mL aliquot of the supernatant was centrifuged to remove sediment, and *p*NP was quantified from absorbance measurements at 405 nm.

Protease assays were performed using the casein-based technique of Ladd and Butler [42] with modifications described by Jesmin et al. [43]. Briefly, 1 g of soil (dry weight equivalent) was treated with 2.5 mL of tris buffer (pH 8) and 2.5 mL of sodium caseinate solution (50 mg casein g^{-1} soil) and incubated for 2 h at 50 °C on a shaking sand bath. After incubation, trichloroacetic acid was added to terminate hydrolytic activity, the supernatant obtained by centrifugation was treated with Na_2CO_3 solution and Folin reagent, and the tyrosine released was determined spectrophotometrically at 650 nm.

2.8. Gross N Mineralization and Immobilization

For comparing treatment effects on N mineralization and immobilization, pool-dilution measurements were performed using quadruplicate 10 g (dry weight) samples of soil from each composite mixture. Two of these samples were treated with 100 mL of 2 M KCl followed by 0.64 mL of deionized water containing 40 μg of N as $(^{15}NH_4)_2SO_4$ (30 atom % ^{15}N), and the remaining samples were incubated at 25 °C for 3 d after the same addition of $(^{15}NH_4)_2SO_4$. In both cases, mineral N was extracted by shaking soil samples with 2 M KCl for 1 h and filtering the resulting soil suspension through Whatman no. 42 filter paper (GE Healthcare, Maidstone, UK) in a Büchner funnel under vacuum. The extracts were analyzed for exchangeable NH_4^+-N by accelerated diffusion methods of Khan et al. [44], followed by ^{15}N analysis using an automated Rittenberg system [45]. Gross rates of mineralization and immobilization were calculated using the zero-order equations derived by Kirkham and Bartholomew [46].

2.9. Statistical Analyses

Replicate data were characterized by computing means and standard deviations. PAST version 3.22 [47] was used to test for homogeneity (Levene's test) and normality (Shapiro−Wilk's test), to evaluate the significance of treatment effects by a one-way analysis of variance (ANOVA), and to carry out mean comparisons by Tukey's procedure ($p < 0.05$ to 0.001).

3. Results and Discussion

3.1. Residue Quality

Besides increasing grain yields and the quantity of above- and below-ground residue inputs, N fertilization has an intrinsic effect on residue quality by increasing biomass content of total N or crude protein [48–50], thereby lowering the C:N ratio [51,52]. These changes are apparent from Table 2, which shows that total N was significantly greater for the high N residue (HNR) collected after using fertilizer N, which caused a substantial decline in C:N ratio. Table 2 also shows that fertilization led to a significant increase in the water-soluble fraction obtained by proximate analysis. This fraction, consisting largely of monomeric sugars (predominately glucose and fructose) along with related alditols, aliphatic acids, and inorganic ions [53], enriches the soil in labile organic C when leached from residue and is readily utilized during microbial decomposition [54–56]. The proximate analyses reported by Table 1 are consistent with previous reports that N fertilization is more effective for increasing soluble than structural carbohydrates in corn residues [57–59].

3.2. C Mineralization

The effects of endogenous and exogenous N on microbial decomposition during the 60 d incubation period are documented by temporal (Figure 2) and cumulative (Figure 3) data for CO_2-C production. As expected, the amount of CO_2-C collected was always greater for incubations with rather than without residue (Figure 2), the cumulative effect being at least a four-fold difference (Figure 3) with O_2 concentrations that always exceeded 0.20 kPa O_2 kPa^{-1} to ensure aerobic conditions. Examination of the inset panel in Figure 2A reveals that CO_2 production without residue was significantly ($p < 0.01$) reduced by the presence of KNO_3 (PN) or $(NH_4)_2SO_4$ (AS), although neither effect was significant when evaluated using the entire dataset (Figure 3). Such findings can be explained by 'microbial N mining', whereby N limitation stimulates microbial attack on indigenous organic matter [60–62], the result being greater C mineralization than would occur when microbial N demand is alleviated by the input of mineral N. Moreover, the high salt index of KNO_3 [63] would have limited microbial activity and biomass due to plasmolysis caused by osmotic stress [64–66], whereas the acidifying effect of $(NH_4)_2SO_4$ [67], which is documented by Figure 4, would have been more important than salinization for inhibiting heterotrophic C oxidation [68–70]. The findings in Figures 2 and 3 are consistent with previous reports that, in the absence of residue inputs, N fertilization usually decreases soil respiration [22,71,72].

Given that the two residue mixtures studied differed considerably in their N contents, an increase in CO_2 production was expected in comparing the HNR with the LNR treatment. This was indeed observed, as collection of CO_2 was significantly greater for HNR than LNR during the first 10 d of incubation but not thereafter (Figure 2B,C). Due to the initial enhancement, a significant increase also occurred in the cumulative emission of CO_2 (Figure 3), which in terms of the C applied was equivalent to 66% for HNR and 50% for LNR. The latter finding is in line with previous studies showing that a higher N content promotes microbial decomposition when crop residues incubate following their incorporation in soil [73–76], but also reflects the fact that HNR was substantially greater in the water-soluble fraction (Table 2). According to Shi and Marschner [55], this fraction serves as a key source of energy to support active growth by the heterotrophic microflora, which promotes residue decomposition during the early stage of incubation.

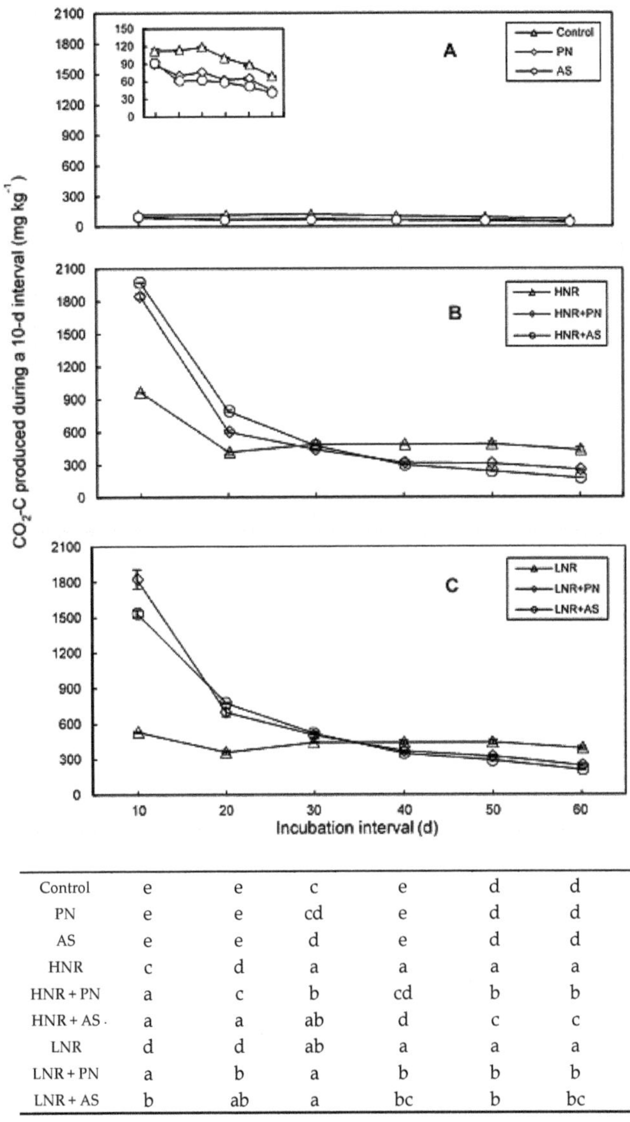

Control	e	e	c	e	d	d
PN	e	e	cd	e	d	d
AS	e	e	d	e	d	d
HNR	c	d	a	a	a	a
HNR + PN	a	c	b	cd	b	b
HNR + AS	a	a	ab	d	c	c
LNR	d	d	ab	a	a	a
LNR + PN	a	b	a	b	b	b
LNR + AS	b	ab	a	bc	b	bc

Figure 2. Total quantity of CO_2-C produced by soil in 10 d intervals during a 60 d aerobic incubation involving the following nine treatments: (**A**) unamended control, potassium nitrate (PN), ammonium sulfate (AS); (**B**) high N residue (HNR) with or without PN (HNR + PN) or AS (HNR + AS); and (**C**) low N residue (LNR) with or without PN (LNR + PN) or AS (LNR + AS). Data shown as a mean from triplicate incubations with standard error bars and a table for mean comparisons. Within a given incubation interval, treatments followed by the same letter do not differ significantly at $p < 0.05$. When compared at a smaller scale (**A**), CO_2-C was significantly greater ($p < 0.01$) for the control than for the PN or AS treatment.

Regardless of which residue mixture was incorporated prior to incubation, cumulative CO_2 production, ranging from 82 to 88% of the residue C applied, was significantly increased by the presence of exogenous NH_4^+ or NO_3^- relative to the HNR and LNR treatments. This is evident from Figure 3, which also shows that the two N sources did not differ in their effects on cumulative CO_2 production, presumably because NO_3^- utilization is promoted by the presence of carbonaceous residues. The finding that addition of mineral N promoted liberation of CO_2 during decomposition of corn residue is consistent with results previously obtained in many relevant incubation studies [2,61,73,76–79] and can presumably be attributed to microbial N utilization for cellular synthesis and metabolism. In some cases, exogenous N has had no significant effect on soil respiration in the presence of corn residue [9,21,80], which may reflect variations in incubation procedure, the type of soil studied, and/or the relative rates of residue and N addition. Reports that decomposition is unaffected or even inhibited by the addition of mineral N are more common from studies with more ligneous plant materials such as wheat (*Triticum aestivum* L.) or rice (*Oryza sativa* L.) straw, tree bark, or sawdust [4,81–83].

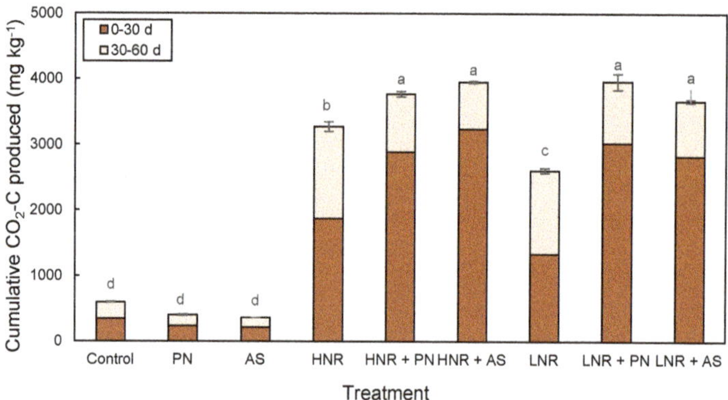

Figure 3. Cumulative CO_2-C produced by soil during half or all of a 60 d aerobic incubation involving an unamended control and the following eight treatments: potassium nitrate (PN), ammonium sulfate (AS), high N residue (HNR) with or without PN (HNR + PN) or AS (HNR + AS), and low N residue (LNR) with or without PN (LNR + PN) or AS (LNR + AS). Data shown as a mean from triplicate incubations with standard error bars obtained for the total amount of CO_2 collected. Treatments do not differ significantly ($p < 0.05$) for the entire 60 d incubation period when bars are accompanied by the same letter.

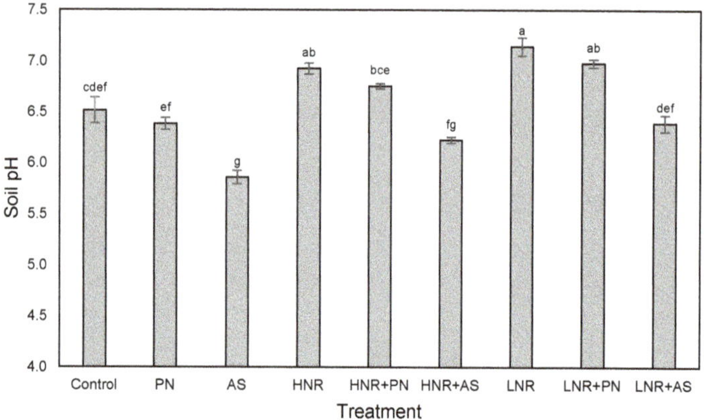

Figure 4. Soil pH during a 60 d aerobic incubation involving an unamended control and the following eight treatments: potassium nitrate (PN), ammonium sulfate (AS), high N residue (HNR) with or without PN (HNR + PN) or AS (HNR + AS), and low N residue (LNR) with or without PN (LNR + PN) or AS (LNR + AS). Data shown as a mean from duplicate incubations for 7 and 60 d with standard error bars. Treatments do not differ significantly ($p < 0.05$) when bars are accompanied by the same letter.

Besides increasing the cumulative production of CO_2 from residue-treated soil, exogenous N shifted the temporal pattern of decomposition, such that 76 to 82% of the CO_2 collected was liberated in the first month of incubation, as compared to 51% for the LNR and 57% for the HNR treatment (Figure 3). A substantial decline subsequently occurred for the fertilized but not the unfertilized treatments with residue, and the difference was usually significant (Figure 2). This shift has previously been observed in numerous incubation studies [2,4,76,80] and can be explained by microbial utilization and subsequent depletion of labile constituents released by residue decomposition.

3.3. Microbial Parameters

3.3.1. Active Biomass

The response of microbial communities to a readily available supply of substrate can be estimated by the assay of active biomass, which in the present project involved a short-term measurement of respiration following the addition of glucose. The results (Figure 5) show that, when averaged over the entire incubation period, active biomass for residue-treated soil was higher with than without exogenous N, although the difference was not necessarily significant due to limited replication. The importance of an active microbial fraction to soil respiration is also evident from the finding that active biomass was highly correlated ($r = 0.94$, $p < 0.001$) with cumulative CO_2 production in the incubation study, as in previous work by Alvarez and Alvarez [84]. A decline in active biomass was expected with the loss of residue C over time, and this was indeed observed when a 2 to 34% decrease occurred during the second month of incubations with residue, as opposed to a corresponding increase of 19 to 119% for treatments without residue (Figure 5). The latter increase, which was most pronounced for the AS treatment, reflects a drastic stimulation of glucose-responsive microbial growth and respiration following an intensifying level of starvation as substrate depletion was exacerbated by aerobic incubation. Under such conditions, the introduction of substrate triggered a microbial shift from the potentially active fraction to an active physiological state, thereby increasing the active biomass pool [85,86] that represented 13 to 35% of the total biomass C.

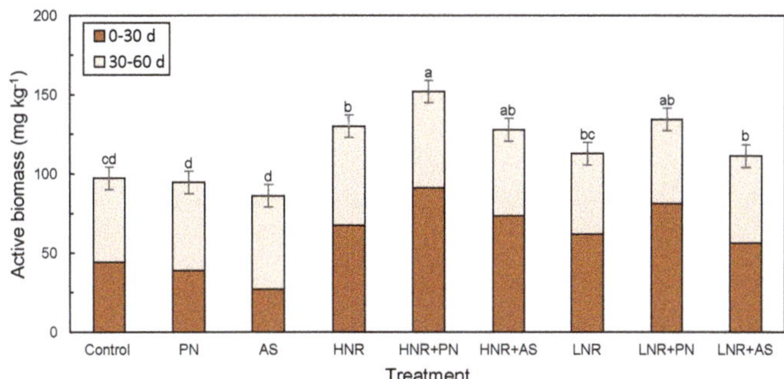

Figure 5. Active biomass measured at five intervals (7, 14, 30, 45, and 60 d) during a 60 d aerobic incubation involving an unamended control and the following eight treatments: potassium nitrate (PN), ammonium sulfate (AS), high N residue (HNR) with or without PN (HNR + PN) or AS (HNR + AS), and low N residue (LNR) with or without PN (LNR + PN) or AS (LNR + AS). Data for each incubation interval shown as a mean from duplicate subsamples with standard error bars. Statistical analyses were performed after averaging data for all five intervals, and treatments do not differ significantly ($p < 0.05$) when bars are accompanied by the same letter.

3.3.2. Microbial Biomass C and N

Unlike active biomass, MBC was unaffected by the presence or absence of exogenous N, with significant differences only being observed in comparing some but not all averages for treatments with and without residue (Figure 6A). Similar findings have previously been reported in relevant incubation studies by Muhammad et al. [21], Song et al. [87], and Li et al. [88], and reflect the contrasting effects of residue and mineral N inputs on microbial growth and activity. Organic substrates supply C essential to microbial biomass synthesis, while inorganic N predominantly contributes to extracellular enzyme activities [89]. The view that N can be more important for promoting microbial activities than growth is relevant to the present study because residue treatments with and without inorganic N differed significantly in active biomass and cumulative CO_2 production (Figures 3 and 5) but not in MBC (Figure 6A), despite a significant correlation of the latter two parameters ($r = 0.81$, $p < 0.01$).

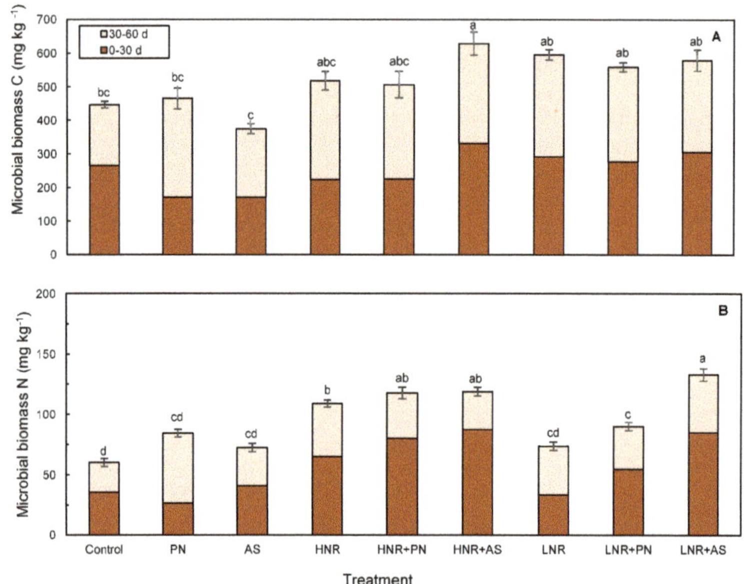

Figure 6. Microbial biomass C (**A**) and microbial biomass N (**B**) measured at five intervals (7, 14, 30, 45, and 60 d) during a 60 d aerobic incubation involving an unamended control and the following eight treatments: potassium nitrate (PN), ammonium sulfate (AS), high N residue (HNR) with or without PN (HNR + PN) or AS (HNR + AS), and low N residue (LNR) with or without PN (LNR + PN) or AS (LNR + AS). Data for each incubation interval shown as a mean from duplicate subsamples with standard error bars. Statistical analyses were performed after averaging data for all five intervals, and treatments do not differ significantly ($p < 0.05$) when bars are accompanied by the same letter.

To better clarify the interacting effects of C and N on total microbial biomass, the MBN fraction was measured at the same incubation intervals utilized in characterizing MBC. The results (Figure 6B) reveal that MBN was significantly greater for the HNR, HNR + PN, and HNR + AS treatments as compared to soils incubated without residue, and with the exception of LNR + AS, significantly lower for the LNR than the HNR treatments. Both findings can be explained by differences in the supply of assimilable N, following similar reports in several previous studies [90–92]. A significant correlation was obtained between MBN and cumulative CO_2 production ($r = 0.78$, $p < 0.05$), indicating that organic C mineralization released C and N substrates for microbial biomass synthesis [93]. For the HNR + PN and LNR + PN treatments that supplied ample C and N, MBC was 2 to 25% greater in the second than in the first month of incubation, whereas the corresponding change for the HNR + AS and LNR + AS treatments was a 10% decrease due to soil acidification. Relative to the same temporal periods, a 35 to 64% decrease in MBN was observed for these four treatments, reflecting a decline in microbial activity due to substrate limitation.

3.3.3. Enzymes Involved in C and N Mineralization

Given the fundamental role of C-degrading enzymes in residue decomposition [16], assays of cellulase activity were performed at periodic intervals throughout the current study. As shown by Figure 7A, cellulase activities were significantly greater for treatments with rather than without residue regardless of whether LNR or HNR was used, and significant increases were also observed upon incubating residue-treated soils with exogenous N. A positive effect of residue addition would be expected for an inducible enzyme such as cellulase [94], while a stimulatory effect of mineral N is documented for cellulosic substrates by several previous studies [15,17,94–96], presumably because increased N availability enhances the activities of the wide variety of cellulolytic soil fungi and bacteria. In line with previous reports by Geisseler and Horwath [16] and Luo et al. [17], cellulase activity was strongly correlated ($r = 0.90$, $p < 0.001$) with cumulative CO_2 production.

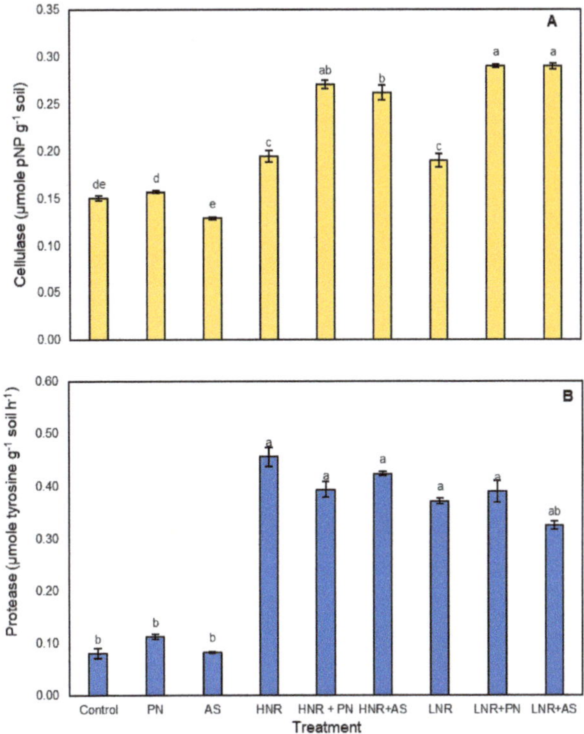

Figure 7. Mean cellulase (**A**) and protease (**B**) activities measured at five intervals (7, 14, 30, 45, and 60 d) during a 60 d aerobic incubation involving an unamended control and the following eight treatments: potassium nitrate (PN), ammonium sulfate (AS), high N residue (HNR) with or without PN (HNR + PN) or AS (HNR + AS), and low N residue (LNR) with or without PN (LNR + PN) or AS (LNR + AS). Data shown as a mean from triplicate incubations with standard error bars. Treatments do not differ significantly ($p < 0.05$) when bars are accompanied by the same letter.

To further evaluate the effect of substrate availability on the microbial turnover of C and N, assays were carried out concurrently for protease as well as cellulase. Proteolytic enzymes play a key role in soil N mineralization, comprising a group of extracellular hydrolases that convert proteins to peptides and/or amino acids [97]. Figure 7B reveals that soil protease activities were much lower when residues were absent with the control, PN, and AS treatments, owing to the lack of C sources for heterotrophic microbial metabolism and extracellular enzyme synthesis. Protease activities increased markedly for the remaining treatments that all involved residue addition. Such increases were induced in response to ample availability of organic C and N, but there is no evidence of any effect due to the input of mineral N. Similar findings from previous research concerning soil protease activities have been linked to substrate C:N ratios and the repressive effect of NH_4^+ or NO_3^- [16,98–100]. In the present study, protease activity was significantly correlated with cumulative CO_2 ($r = 0.94$, $p < 0.001$), active biomass ($r = 0.89$, $p < 0.001$), MBC ($r = 0.76$, $p < 0.05$), and cellulase activity ($r = 0.71$, $p < 0.05$), which is consistent with previous findings by Geisseler and Horwath [16,98], Geisseler et al. [99], and Mishra et al. [101].

3.4. N Mineralization

Because soil C and N cycling is closely coupled, ^{15}N pool dilution was used to evaluate the treatments under investigation with respect to gross N mineralization and immobilization. The results are summarized by Figure 8A, which shows that immobilization always exceeded mineralization, even when there was no residue addition with the control, PN, and AS treatments. Both findings can be attributed to the use of a soil that contained carbonaceous residues when collected from a field previously cropped to corn [102]. Despite limited replication that minimized significant differences, net immobilization (Figure 8B) tended to be higher for treatments involving the addition of NH_4^+ than NO_3^-, as would be expected from the microbial preference previously documented for N utilization [11,12,22]. Gross mineralization and immobilization were highly correlated ($r = 0.91$, $p < 0.001$), as would be expected given their concurrent occurrence as the central processes in soil N cycling [22].

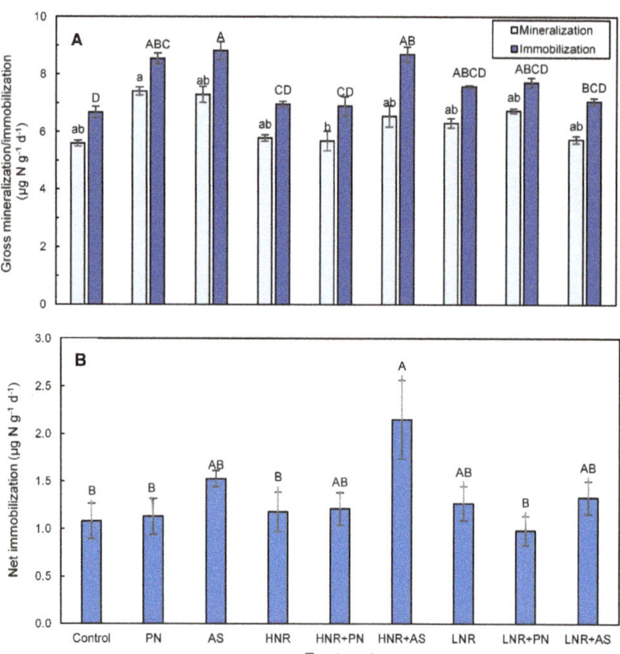

Figure 8. Gross mineralization/immobilization (**A**) and net immobilization (**B**) measured at five intervals (7, 14, 30, 45, and 60 d) during a 60 d aerobic incubation involving an unamended control and the following eight treatments: potassium nitrate (PN), ammonium sulfate (AS), high N residue (HNR) with or without PN (HNR + PN) or AS (HNR + AS), and low N residue (LNR) with or without PN (LNR + PN) or AS (HNR + AS). Values reported as a mean with standard error bars representing duplicate data collected before and after a 3 d incubation with (^{15}NH$_4$)$_2$SO$_4$. Treatments do not differ significantly ($p < 0.05$) when bars are accompanied by the same lowercase (mineralization) or uppercase (immobilization) letter.

4. Conclusions

The soil incubation study reported herein supports the view that N availability is an important factor in the decomposition of corn stover, as cumulative CO_2 production after 60 d was directly affected by residue N content and a substantial increase occurred from the application of mineral N, although there was no significant impact on gross N mineralization. A stimulatory effect of exogenous N was observed for active biomass and cellulase activity but not for MBC, MBN, or protease. With respect to C mineralization, this effect was much more pronounced in the first than in the second month of incubation, as would be expected due to a greater decline in substrate availability.

These findings have practical implications for modern corn production that relies on high planting rates and intensive N fertilization to increase grain yield and is often assumed to promote SOC storage by enhancing residue inputs. On the contrary, five decades of synthetic N fertilization led to a net decline in profile storage of SOC, relative to treatment-specific baseline data collected for the historic Morrow Plots [26], which is consistent with similar evidence from numerous other long-term cropping experiments throughout the world. Such findings would be expected if fertilizer and residue N promote heterotrophic C utilization during microbial decomposition, as demonstrated by the short-term incubation approach adopted for the present project.

Author Contributions: Conceptualization, T.J. and R.L.M.; methodology, T.J. and R.L.M.; software, T.J.; validation, T.J. and R.L.M.; formal analysis, T.J. and R.L.M.; investigation, T.J., D.T.M. and R.L.M.; resources, R.L.M.; data curation, T.J.; writing—original draft preparation, T.J. and R.L.M.; writing—review and editing, T.J. and R.L.M.; visualization, T.J. and R.L.M.; supervision, R.L.M.; project administration, R.L.M.; funding acquisition, R.L.M. All authors have read and agreed to the published version of the manuscript.

Funding: This research received no external funding.

Institutional Review Board Statement: Not applicable.

Informed Consent Statement: Not applicable.

Acknowledgments: Appreciation is expressed to Michelle Wander, Department of NRES, and Lynn Connor, USDA Agricultural Research Service, for providing access to some of the laboratory equipment used in our work. We also thank Andrew Margenot, Department of Crop Sciences, for providing expertise regarding enzyme assay techniques; Vander Nunes, Department of NRES, for assistance with electronic technical illustration; Tim Smith, Cropsmith, for assisting with soil sample collection; Wenyan Li, South China Agricultural University, for constructive suggestions regarding experimental design; and Kelsey Griesheim, Lucas Goldschmidt, Jill Allen, and Jenna Mattes for technical support with various phases of the project.

Conflicts of Interest: The authors declare no conflict of interest.

Sample Availability: Samples of the compounds are not available from the authors.

References

1. Grzyb, A.; Wolna-Maruwka, A.; Niewiadomska, A. Environmental factors affecting the mineralization of crop residues. *Agronomy* **2020**, *10*, 1951. [CrossRef]
2. Green, C.J.; Blackmer, A.M.; Horton, R. Nitrogen effects on conservation of carbon during corn residue decomposition in soil. *Soil Sci. Soc. Am. J.* **1995**, *59*, 453–459. [CrossRef]
3. Moran, K.K.; Six, J.; Horwath, W.R.; van Kessel, C. Role of mineral-nitrogen in residue decomposition and stable soil organic matter formation. *Soil Sci. Soc. Am. J.* **2005**, *69*, 1730–1736. [CrossRef]
4. Lueken, H.; Hutcheon, W.L.; Paul, E.A. The influence of nitrogen on the decomposition of crop residues in the soil. *Can. J. Soil Sci.* **1962**, *42*, 276–288. [CrossRef]
5. Henriksen, T.M.; Breland, T.A. Nitrogen availability effects on carbon mineralization, fungal and bacterial growth, and enzyme activities during decomposition of wheat straw in soil. *Soil Biol. Biochem.* **1999**, *31*, 1121–1134. [CrossRef]
6. Wang, W.J.; Baldock, J.A.; Dalal, R.C.; Moody, P.W. Decomposition dynamics of plant materials in relation to nitrogen availability and biochemistry determined by NMR and wet-chemical analysis. *Soil Biol. Biochem.* **2004**, *36*, 2045–2058. [CrossRef]
7. Recous, S.; Robin, D.; Darwis, D.; Mary, B. Soil inorganic N availability: Effect on maize residue decomposition. *Soil Biol. Biochem.* **1995**, *27*, 1529–1538. [CrossRef]

8. Foereid, B.; de Neergaard, A.; Høgh-Jensen, H. Turnover of organic matter in a *Miscanthus* field: Effect of time in *Miscanthus* cultivation and inorganic nitrogen supply. *Soil Biol. Biochem.* **2004**, *36*, 1075–1085. [CrossRef]

9. Al-Kaisi, M.M.; Kwaw-Mensah, D.; Ci, E. Effect of nitrogen fertilizer application on corn residue decomposition in Iowa. *Agron. J.* **2017**, *109*, 2415–2427. [CrossRef]

10. Alexander, M. *Introduction to Soil Microbiology*, 2nd ed.; Wiley: New York, NY, USA, 1977.

11. Rice, C.W.; Tiedje, J.M. Regulation of nitrate assimilation by ammonium in soils and in isolated soil microorganisms. *Soil Biol. Biochem.* **1989**, *21*, 597–602. [CrossRef]

12. Ma, Q.; Zheng, J.; Watanabe, T.; Funakawa, S. Microbial immobilization of ammonium and nitrate fertilizers induced by starch and cellulose in an agricultural soil. *Soil Sci. Plant Nutr.* **2021**, *67*, 89–96. [CrossRef]

13. Gao, H.; Chen, X.; Wei, J.; Zhang, Y.; Zhang, L.; Chang, J.; Thompson, M.L. Decomposition dynamics and changes in chemical composition of wheat straw residue under anaerobic and aerobic conditions. *PLoS ONE* **2016**, *11*, e0158172. [CrossRef]

14. Asmar, F.; Eiland, F.; Nielsen, N.E. Effect of extracellular-enzyme activities on solubilization rate of soil organic nitrogen. *Biol. Fertil. Soils* **1994**, *17*, 32–38. [CrossRef]

15. Carreiro, M.M.; Sinsabaugh, R.L.; Repert, D.A.; Parkhurst, D.F. Microbial enzyme shifts explain litter decay responses to simulated nitrogen deposition. *Ecology* **2000**, *81*, 2359–2365. [CrossRef]

16. Geisseler, D.; Horwath, W.R. Relationship between carbon and nitrogen availability and extracellular enzyme activities in soil. *Pedobiologia* **2009**, *53*, 87–98. [CrossRef]

17. Luo, R.; Fan, J.; Wang, W.; Luo, J.; Kuzyakov, Y.; He, J.S.; Chu, H.; Ding, W. Nitrogen and phosphorus enrichment accelerates soil organic carbon loss in alpine grassland on the Qinghai-Tibetan Plateau. *Sci. Total Environ.* **2019**, *650*, 303–312. [CrossRef] [PubMed]

18. Dimarogona, M.; Topakas, E.; Christakopoulos, P. Cellulose degradation by oxidative enzymes. *Comput. Struct. Biotechnol. J.* **2012**, *2*, e201209015. [CrossRef] [PubMed]

19. Ajwa, H.A.; Dell, C.J.; Rice, C.W. Changes in enzyme activities and microbial biomass of tallgrass prairie soil as related to burning and nitrogen fertilization. *Soil Biol. Biochem.* **1999**, *31*, 769–777. [CrossRef]

20. Singh, H.; Singh, K.P. Effect of residue placement and chemical fertilizer on soil microbial biomass under tropical dryland cultivation. *Biol. Fertil. Soils* **1993**, *16*, 275–281. [CrossRef]

21. Muhammad, W.; Vaughan, S.M.; Dalal, R.C.; Menzies, N.W. Crop residues and fertilizer nitrogen influence residue decomposition and nitrous oxide emission from a Vertisol. *Biol. Fertil. Soils* **2011**, *47*, 15–23. [CrossRef]

22. Jansson, S.L. Tracer studies on nitrogen transformations in soil with special attention to mineralization-immobilization relationships. *Ann. R. Agric. Coll. Swed.* **1958**, *24*, 101–306.

23. Zaman, M.D.; Di, H.J.; Cameron, K.C.; Frampton, C.M. Gross nitrogen mineralization and nitrification rates and their relationships to enzyme activities and the soil microbial biomass in soils treated with dairy shed effluent and ammonium fertilizer at different water potentials. *Biol. Fertil. Soils* **1999**, *29*, 178–186. [CrossRef]

24. Brye, K.R.; Gower, S.T.; Norman, J.M.; Bundy, L.G. Carbon budgets for a prairie and agroecosystems: Effects of land use and interannual variability. *Ecol. Appl.* **2002**, *12*, 962–979. [CrossRef]

25. Wilts, A.R.; Reicosky, D.C.; Allmaras, R.R.; Clapp, C.E. Long-term corn residue effects: Harvest alternatives, soil carbon turnover, and root-derived carbon. *Soil Sci. Soc. Am. J.* **2004**, *68*, 1342–1351. [CrossRef]

26. Khan, S.A.; Mulvaney, R.L.; Ellsworth, T.R.; Boast, C.W. The myth of nitrogen fertilization for soil carbon sequestration. *J. Environ. Qual.* **2007**, *36*, 1821–1832. [CrossRef]

27. Peters, J.B.; Nathan, M.V.; Laboski, C.A.M. pH and lime requirement. In *Recommended Chemical Soil Test Procedures for the North Central Region*; Nathan, M., Gelderman, R., Eds.; University of Missouri: Columbia, MO, USA, 2015; pp. 4.1–4.7. Available online: https://extension.missouri.edu/media/wysiwyg/Extensiondata/Pub/pdf/specialb/sb1001.pdf (accessed on 3 March 2021).

28. Mebius, L.J. A rapid method for the determination of organic carbon in soil. *Anal. Chim. Acta* **1960**, *22*, 120–124. [CrossRef]

29. Bremner, J.M. Nitrogen-total. In *Methods of Soil Analysis Part 3-Chemical Methods*; Sparks, D.L., Ed.; Soil Science Society of America: Madison, WI, USA, 1996; pp. 1085–1121. [CrossRef]

30. Stevens, W.B.; Mulvaney, R.L.; Khan, S.A.; Hoeft, R.G. Improved diffusion methods for nitrogen and ^{15}nitrogen analysis of Kjeldahl digests. *J. AOAC Int.* **2000**, *83*, 1039–1046. [CrossRef] [PubMed]

31. Khan, S.A.; Mulvaney, R.L.; Hoeft, R.G. A simple soil test for detecting sites that are nonresponsive to nitrogen fertilization. *Soil Sci. Soc. Am. J.* **2001**, *65*, 1751–1760. [CrossRef]

32. Bray, R.H.; Kurtz, L.T. Determination of total, organic, and available forms of phosphorus in soils. *Soil Sci.* **1945**, *59*, 39–46. [CrossRef]

33. Bouyoucos, G.J. Hydrometer method improved for making particle size analyses of soils. *Agron. J.* **1962**, *54*, 464. [CrossRef]

34. Bremner, J.M.; Shaw, K. Denitrification in soil. I. Methods of investigation. *J. Agric. Sci.* **1958**, *51*, 40–52. [CrossRef]

35. Pordesimo, L.O.; Edens, W.C.; Sokhansanj, S. Distribution of aboveground biomass in corn stover. *Biomass Bioenerg.* **2004**, *26*, 337–343. [CrossRef]

36. Harper, S.H.T.; Lynch, J.M. The chemical components and decomposition of wheat straw leaves, internodes and nodes. *J. Sci. Food Agric.* **1981**, *32*, 1057–1062. [CrossRef]

37. Jones, D.B. *Factors for Converting Percentages of Nitrogen in Foods and Feeds into Percentages of Protein*; Circular No. 183; USDA: Washington, DC, USA, 1931.

38. Horgan, B.P.; Mulvaney, R.L.; Branham, B.E. Determination of atmospheric volume for direct field measurement of denitrification in soil cores. *Soil Sci. Soc. Am. J.* **2001**, *65*, 511–516. [CrossRef]

39. Van de Werf, H.; Verstraete, W. Estimation of active soil microbial biomass by mathematical analysis of respiration curves: Calibration of the test procedure. *Soil Biol. Biochem.* **1987**, *19*, 261–265. [CrossRef]

40. Vance, E.D.; Brookes, P.C.; Jenkinson, D.S. An extraction method for measuring soil microbial biomass C. *Soil Biol. Biochem.* **1987**, *19*, 703–707. [CrossRef]

41. Margenot, A.J.; Nakayama, Y.; Parikh, S.J. Methodological recommendations for optimizing assays of enzyme activities in soil samples. *Soil Biol. Biochem.* **2018**, *125*, 350–360. [CrossRef]

42. Ladd, J.N.; Butler, J.H.A. Short-term assays of soil proteolytic enzyme activities using proteins and dipeptide derivatives as substrates. *Soil Biol. Biochem.* **1972**, *4*, 19–30. [CrossRef]

43. Jesmin, T.; Margenot, A.J.; Mulvaney, R.L. A comprehensive method for casein-based assay of soil protease activity. *Commun. Soil Sci. Plant Anal.* **2021**, in press.

44. Khan, S.A.; Mulvaney, R.L.; Mulvaney, C.S. Accelerated diffusion methods for inorganic-nitrogen analysis of soil extracts and water. *Soil Sci. Soc. Am. J.* **1997**, *61*, 936–942. [CrossRef]

45. Mulvaney, R.L.; Fohringer, C.L.; Bojan, V.J.; Michlik, M.M.; Herzog, L.F. A commercial system for automated nitrogen isotope-ratio analysis by the Rittenberg technique. *Rev. Sci. Instrum.* **1990**, *61*, 897–903. [CrossRef]

46. Kirkham, D.; Bartholomew, W.V. Equations for following nutrient transformations in soil, utilizing tracer data. *Soil Sci. Soc. Am. Proc.* **1954**, *18*, 33–34. [CrossRef]

47. Paleontological Statistics Software Package for Education and Data Analysis (PAST). Available online: https://past.en.lo4d.com/windows (accessed on 27 July 2021).

48. Terman, G.L.; Noggle, J.C. Nutrient concentration changes in corn as affected by dry matter accumulation with age and response to applied nutrients. *Agron. J.* **1973**, *65*, 941–945. [CrossRef]

49. Perry, L.J., Jr.; Olson, R.A. Yield and quality of corn and grain sorghum grain and residues as influenced by N fertilization. *Agron. J.* **1975**, *67*, 816–818. [CrossRef]

50. Li, H.; Li, L.; Wegenast, T.; Longin, C.F.; Xu, X.; Melchinger, A.E.; Chen, S. Effect of N supply on stalk quality in maize hybrids. *Field Crops Res.* **2010**, *118*, 208–214. [CrossRef]

51. Russell, A.E.; Cambardella, C.A.; Laird, D.A.; Jaynes, D.B.; Meek, D.W. Nitrogen fertilizer effects on soil carbon balances in Midwestern U.S. agricultural systems. *Ecol. Appl.* **2009**, *19*, 1102–1113. [CrossRef]

52. Jantalia, C.P.; Halvorson, A.D. Nitrogen fertilizer effects on irrigated conventional tillage corn yields and soil carbon and nitrogen pools. *Agron. J.* **2011**, *103*, 871–878. [CrossRef]

53. Chen, S.F.; Mowery, R.A.; Scarlata, C.J.; Chambliss, C.K. Compositional analysis of water-soluble materials in corn stover. *J. Agric. Food Chem.* **2007**, *55*, 5912–5918. [CrossRef]

54. Kalbitz, K.; Solinger, S.; Park, J.H.; Michalzik, B.; Matzner, E. Controls on the dynamics of dissolved organic matter in soils: A review. *Soil Sci.* **2000**, *165*, 277–304. [CrossRef]

55. Shi, A.D.; Marschner, P. Soil respiration and microbial biomass after residue addition are influenced by the extent by which water-extractable organic C was removed from the residues. *Eur. J. Soil Biol.* **2014**, *63*, 28–32. [CrossRef]

56. Li, Z.Q.; Zhao, B.Z.; Zhang, J.B. Effects of maize residue quality and soil water content on soil labile organic carbon fractions and microbial properties. *Pedosphere* **2016**, *26*, 829–838. [CrossRef]

57. Schulze, W. Stickstoffdüngung und Zuckerbildung bei Mais. *Albrecht-Thaer-Arch* **1964**, *8*, 711–723. [CrossRef]

58. Baghdadi, A.; Balazadeh, M.; Kashani, A.; Golzardi, F.; Gholamhoseini, M.; Mehrnia, M. Effect of pre-sowing and nitrogen application on forage quality of silage corn. *Agron. Res.* **2017**, *15*, 11–23.

59. Gallagher, M.E.; Hockaday, W.C.; Masiello, C.A.; Snapp, S.; McSwiney, C.P.; Baldock, J.A. Biochemical suitability of crop residues for cellulosic ethanol: Disincentives to nitrogen fertilization in corn agriculture. *Environ. Sci. Technol.* **2011**, *45*, 2013–2020. [CrossRef] [PubMed]

60. Moorhead, D.L.; Sinsabaugh, R.L. A theoretical model of litter decay and microbial interaction. *Ecol. Monogr.* **2006**, *76*, 151–174. [CrossRef]

61. Chen, R.; Senbayram, M.; Blagodatsky, S.; Myachina, O.; Dittert, K.; Lin, X.; Blagodatskaya, E.; Kuzyakov, Y. Soil C and N availability determine the priming effect: Microbial N mining and stoichiometric decomposition theories. *Glob. Chang. Biol.* **2014**, *20*, 2356–2367. [CrossRef]

62. Meyer, N.; Welp, G.; Bornemann, L.; Amelung, W. Microbial nitrogen mining affects spatio-temporal patterns of substrate-induced respiration during seven years of bare fallow. *Soil Biol. Biochem.* **2017**, *104*, 175–184. [CrossRef]

63. Rader, L.F.; White, L.M.; Whittaker, C.W. The salt index-A measure of the effect of fertilizers on the concentration of the soil solution. *Soil Sci.* **1943**, *55*, 201–218. [CrossRef]

64. Pathak, H.; Rao, D.L.N. Carbon and nitrogen mineralization from added organic matter in saline and alkali soils. *Soil Biol. Biochem.* **1998**, *30*, 695–702. [CrossRef]

65. Adviento-Borbe, M.A.A.; Doran, J.W.; Drijber, R.A.; Dobermann, A. Soil electrical conductivity and water content affect nitrous oxide and carbon dioxide emissions in intensively managed soils. *J. Environ. Qual.* **2006**, *35*, 1999–2010. [CrossRef]

66. Rousk, J.; Elyaagubi, F.K.; Jones, D.L.; Godbold, D.L. Bacterial salt tolerance is unrelated to soil salinity across an arid agroecosystem salinity gradient. *Soil Biol. Biochem.* **2011**, *43*, 1881–1887. [CrossRef]

67. Pierre, W.H. Nitrogenous fertilizers and soil acidity: I. Effect of various nitrogenous fertilizers on soil reaction. *J. Am. Soc. Agron.* **1928**, *20*, 254–269. [CrossRef]

68. McAndrew, D.W.; Malhi, S.S. Long-term N fertilization of a solonetzic soil: Effects on chemical and biological properties. *Soil Biol. Biochem.* **1992**, *24*, 619–623. [CrossRef]

69. Malhi, S.S.; Nyborg, M.; Harapiak, J.T.; Heier, K.; Flore, N.A. Increasing organic C and N in soil under bromegrass with long-term N fertilization. *Nutr. Cycl. Agroecosyst.* **1997**, *49*, 255–260. [CrossRef]

70. Blagodatskaya, E.V.; Anderson, T.H. Adaptive responses of soil microbial communities under experimental acid stress in controlled laboratory studies. *Appl. Soil Ecol.* **1999**, *11*, 207–216. [CrossRef]

71. Chandra, P.; Bollen, W.B. Effect of wheat straw, nitrogenous fertilizers, and carbon-to-nitrogen ratio on organic decomposition in a sub-humid soil. *J. Agric. Food Chem.* **1960**, *8*, 19–24. [CrossRef]

72. Ramirez, K.S.; Craine, J.M.; Fierer, N. Nitrogen fertilization inhibits soil microbial respiration regardless of the form of nitrogen applied. *Soil Biol. Biochem.* **2010**, *42*, 2336–2338. [CrossRef]

73. Schmidt, U.; Schmidt, G. Zur Frage der Mineralisation von Ernterückständen in Abhängigkeit vom Stickstoffgehalt. *Z. Pflernähr. Düng. Bodenk.* **1963**, *101*, 99–109. [CrossRef]

74. Reinertsen, S.A.; Elliott, L.F.; Cochran, V.L.; Campbell, G.S. Role of available carbon and nitrogen in determining the rate of wheat straw decomposition. *Soil Biol. Biochem.* **1984**, *16*, 459–464. [CrossRef]

75. Janzen, H.H.; Kucey, R.M.N. C, N, and S mineralization of crop residues as influenced by crop species and nutrient regime. *Plant Soil* **1988**, *106*, 35–41. [CrossRef]

76. Sharma, M.P.; Habib, A. Decomposition rate of crop residue and their nutrients release pattern under different nitrogen levels in Inceptisol. *Ann. Agric. Res. New Ser.* **2005**, *26*, 441–446.

77. Sakala, W.D.; Cadisch, G.; Giller, K.E. Interactions between residues of maize and pigeonpea and mineral N fertilizers during decomposition and N mineralization. *Soil Biol. Biochem.* **2000**, *32*, 679–688. [CrossRef]

78. Conde, E.; Cardenas, M.; Ponce-Mendoza, A.; Luna-Guido, M.L.; Cruz-Mondragón, C.; Dendooven, L. The impacts of inorganic nitrogen application on mineralization of ^{14}C-labelled maize and glucose, and on priming effect in saline alkaline soil. *Soil Biol. Biochem.* **2005**, *37*, 681–691. [CrossRef]

79. Chen, H.; Billen, N.; Stahr, K.; Kuzyakov, Y. Effects of nitrogen and intensive mixing on decomposition of ^{14}C-labelled maize (*Zea mays* L.) residue in soils of different land use types. *Soil Till. Res.* **2007**, *96*, 114–123. [CrossRef]

80. Potthoff, M.; Dyckmans, J.; Flessa, H.; Muhs, A.; Beese, F.; Joergensen, R.G. Dynamics of maize (*Zea mays* L.) leaf straw mineralization as affected by the presence of soil and the availability of nitrogen. *Soil Biol. Biochem.* **2005**, *37*, 1259–1266. [CrossRef]

81. Allison, F.E.; Murphy, R.M. Comparative rates of decomposition in soil of wood and bark particles of several hardwood species. *Soil Sci. Soc. Am. Proc.* **1962**, *26*, 463–466. [CrossRef]

82. Sain, P.; Broadbent, F.E. Decomposition of rice straw in soils as affected by some management factors. *J. Environ. Qual.* **1977**, *6*, 96–100. [CrossRef]

83. Fog, K. The effect of added nitrogen on the rate of decomposition of organic matter. *Biol. Rev.* **1988**, *63*, 433–462. [CrossRef]

84. Alvarez, C.R.; Alvarez, R. Short-term effects of tillage systems on active soil microbial biomass. *Biol. Fertil. Soils* **2000**, *31*, 157–161. [CrossRef]

85. De Nobili, M.; Contin, M.; Mondini, C.; Brookes, P.C. Soil microbial biomass is triggered into activity by trace amounts of substrate. *Soil Biol. Biochem.* **2001**, *33*, 1163–1170. [CrossRef]

86. Blagodatskaya, E.; Kuzyakov, Y. Active microorganisms in soil: Critical review of estimation criteria and approaches. *Soil Biol. Biochem.* **2013**, *67*, 192–211. [CrossRef]

87. Song, Y.; Song, C.; Tao, B.; Wang, J.; Zhu, X.; Wang, X. Short-term responses of soil enzyme activities and carbon mineralization to added nitrogen and litter in a freshwater marsh of Northeast China. *Eur. J. Soil Biol.* **2014**, *61*, 72–79. [CrossRef]

88. Li, L.J.; Zhu-Barker, X.; Ye, R.; Doane, T.A.; Horwath, W.R. Soil microbial biomass size and soil carbon influence the priming effect from carbon inputs depending on nitrogen availability. *Soil Biol. Biochem.* **2018**, *119*, 41–49. [CrossRef]

89. Bach, E.M.; Hofmockel, K.S. Coupled carbon and nitrogen inputs increase microbial biomass and activity in prairie bioenergy systems. *Ecosystems* **2015**, *18*, 417–427. [CrossRef]

90. Paul, G.C.; Solaiman, A.R. Changes of microbial biomass carbon and nitrogen in upland sugarcane soil amended with different organic materials. *Commun. Soil Sci. Plant Anal.* **2005**, *35*, 2433–2447. [CrossRef]

91. Treseder, K.K. Nitrogen additions and microbial biomass: A meta-analysis of ecosystem studies. *Ecol. Lett.* **2008**, *11*, 1111–1120. [CrossRef] [PubMed]

92. Abbasi, M.K.; Khizar, A. Microbial biomass carbon and nitrogen transformations in a loam soil amended with organic–inorganic N sources and their effect on growth and N-uptake in maize. *Ecol. Eng.* **2012**, *39*, 123–132. [CrossRef]

93. Mgelwa, A.S.; Hu, Y.L.; Xu, W.B.; Ge, Z.Q.; Yu, T.W. Soil carbon and nitrogen availability are key determinants of soil microbial biomass and respiration in forests along urbanized rivers of southern China. *Urban For. Urban Green.* **2019**, *43*, 126351. [CrossRef]

94. Suto, M.; Tomita, F. Induction and catabolite repression mechanisms of cellulase in fungi. *J. Biosci. Bioeng.* **2001**, *92*, 305–311. [CrossRef]

95. Saiya-Cork, K.R.; Sinsabaugh, R.L.; Zak, D.R. The effects of long term nitrogen deposition on extracellular enzyme activity in an *Acer saccharum* forest soil. *Soil Biol. Biochem.* **2002**, *32*, 1309–1315. [CrossRef]

96. Keeler, B.L.; Hobbie, S.E.; Kellogg, L.E. Effects of long-term nitrogen addition on microbial enzyme activity in eight forested and grassland sites: Implications for litter and soil organic matter decomposition. *Ecosystems* **2009**, *12*, 1–15. [CrossRef]

97. Landi, L.; Renella, G.; Giagnoni, L.; Nannipieri, P. Activities of proteolytic enzymes. In *Methods of Soil Enzymology*; Dick, R.P., Ed.; Soil Science Society of America: Madison, WI, USA, 2011; pp. 247–260. [CrossRef]

98. Geisseler, D.; Horwath, W.R. Regulation of extracellular protease activity in soil in response to different sources and concentrations of nitrogen and carbon. *Soil Biol. Biochem.* **2008**, *40*, 3040–3048. [CrossRef]

99. Geisseler, D.; Joergensen, R.G.; Ludwig, B. Potential soil enzyme activities are decoupled from microbial activity in dry residue-amended soil. *Pedobiologia* **2012**, *55*, 253–261. [CrossRef]

100. Chen, H.; Li, D.; Zhao, J.; Xiao, K.; Wang, K. Effects of nitrogen addition on activities of soil nitrogen acquisition enzymes: A meta-analysis. *Agric. Ecosyst. Environ.* **2018**, *252*, 126–131. [CrossRef]

101. Mishra, S.; Di, H.J.; Cameron, K.C.; Monaghan, R.; Carran, A. Gross nitrogen mineralisation rates in pastural soils and their relationships with organic nitrogen fractions, microbial biomass and protease activity under glasshouse conditions. *Biol. Fertil. Soils* **2005**, *42*, 45–53. [CrossRef]

102. Green, C.J.; Blackmer, A.M. Residue decomposition effects on nitrogen availability to corn following corn or soybean. *Soil Sci. Soc. Am. J.* **1995**, *59*, 1065–1070. [CrossRef]

Nitrogen

MDPI

Biochar Applied with Inorganic Nitrogen Improves Soil Carbon, Nitrate and Ammonium Content of a Sandy Loam Temperate Soil

Peter Omara [1,2,*], Lawrence Aula [1], Fred Otim [2], Alfred Obia [2], Joao Luis Bigatao Souza [1] and Daryl Brain Arnall [1]

[1]   Department of Plant and Soil Sciences, Oklahoma State University, Stillwater, OK 74078, USA; aula@okstate.edu (L.A.); bigatao@okstate.edu (J.L.B.S.); b.arnall@okstate.edu (D.B.A.)
[2]   Department of Agronomy, Gulu University, Gulu P.O. Box 166, Uganda; otimakio@gmail.com (F.O.); a.obia@gu.ac.ug (A.O.)
*   Correspondence: peter.omara@okstate.edu

Abstract: Biochar is suggested to improve soil properties. However, its combination with inorganic nitrogen (N) fertilizer in temperate soils is not well understood. This study compared the effect of fertilizer N-biochar-combinations (NBC) and fertilizer-N (FN) on total soil N (TSN), soil organic carbon (SOC), soil nitrate (NO_3^-–N), and ammonium (NH_4^+–N). Soil samples were taken from experiments at Efaw and Lake Carl Blackwell (LCB), Oklahoma, USA with ten treatments consisting of three N rates (50, 100, and 150 kg N ha^{-1}) and three biochar rates (5, 10, and 15 t ha^{-1}). Results at Efaw showed greater TSN and SOC under NBC compared to FN by 3 and 21%, respectively. No percentage difference was observed for NH_4^+–N while NO_3^-–N was lower by 7%. At LCB, TSN, SOC, NO_3^-–N, and NH_4^+–N were higher under NBC by 5, 18, 24, and 10%, respectively, compared to FN. Whereas application of biochar improved SOC at both sites, NO_3^-–N and NH_4^+–N were only significant at LCB site with a sandy loam soil but not at Efaw with silty clay loam. Therefore, biochar applied in combination with inorganic N can improve N availability with potential to increase crop N uptake on coarse textured soils.

Keywords: biochar; total nitrogen; nitrate; ammonium; soil organic carbon

Citation: Omara, P.; Aula, L.; Otim, F.; Obia, A.; Souza, J.L.B.; Arnall, D.B. Biochar Applied with Inorganic Nitrogen Improves Soil Carbon, Nitrate and Ammonium Content of a Sandy Loam Temperate Soil. *Nitrogen* **2022**, *3*, 90–100. https://doi.org/10.3390/nitrogen3010007

Academic Editor: Jacynthe Dessureault-Rompré

Received: 27 December 2021
Accepted: 18 February 2022
Published: 23 February 2022

Publisher's Note: MDPI stays neutral with regard to jurisdictional claims in published maps and institutional affiliations.

1. Introduction

Biochar is a stable carbon (C) rich material formed through pyrolysis of organic materials [1,2]. Application of biochar to the soil is suggested to improve soil properties in addition to C sequestration [3–6]. It is reported to be beneficial in improving soil physical, biological, and chemical properties which include, among others, soil organic carbon (SOC), water retention capacity, cation exchange capacity (CEC), total soil nitrogen (TSN), and soil pH, hence contributing to soil fertility [7,8].

Soil organic C, one of the most important biological properties that determines quality of soil, is believed to be improved through application of biochar. Some research reports have documented the contribution of fertilizer-N (FN) in increasing SOC stock. They argue that FN increases quantity of crop residues added to the soil as a result of improved biomass production. Generally, high rates of FN inconsistently affect SOC where increases are observed in some cases while manure application more frequently increases surface soil SOC [9]. Biochar application with >90% of C in recalcitrant forms more consistently increases SOC. However, there are contradictory conclusions on the role of biochar in enhancing SOC storage. Some researchers have reported negative priming effect of biochar to the native SOC as a result of increasing the rate of evolution of carbon dioxide hence less storage [10,11]. This could be due to short term oxidation of the labile biochar compounds [12]. If the soil is inherently poor in SOC, application of biochar will reduce the evolution of CO_2 while the opposite would be observed in soils rich in organic C [13].

Besides, C loss is always very small relative to the amount of C stored within the biochar itself [14]. In contrast, Cross and Sohi [12] reported that application of biochar did not, for the most part, indicate negative priming of the native SOC and that application of biochar could stabilize native SOC in grassland soils. Applying a combination of N and biochar could contribute to the increase in the SOC storage.

With evidence of increased SOC following biochar application, soil N is likely to increase. Soil N is present mostly in organic compounds which consist of both particulate organic N and dissolved organic N. The particulate organic N include the N in living organisms and detritus. On the other hand, dissolved organic N consists of a wide range of organic substances, such as free amino acids, and proteins, among others [15]. Biochar soil incorporation is suggested to increase the buildup of organic N. Prommer et al. [16] reported that application of inorganic N in combination with biochar had a synergistic effect by activating the belowground build-up of soil organic N. They explained that biochar reduces the transformation rates of the native soil organic N as plants and microbes draw from the inorganic fertilizer N. Bai et al. [17] added that changes in microbial processes and activities on soil organic N following biochar soil application are mediated primarily by abiotic factors such as rainfall and temperature. Therefore, biochar has a great potential in building soil organic N.

Plants take up N in the inorganic form; NO_3^- and NH_4^+ which are susceptible to losses such as volatilization, denitrification, runoff, and leaching [18–20]. Biochar application may improve inorganic N retention through alteration of CEC and anion exchange capacity (AEC) with the greatest benefit on sandy soils and this has been demonstrated by many studies [21–25]. The increased AEC of biochar reduces leaching of NO_3^-–N while the CEC increases the adsorption of NH_4^+–N. Therefore, the application of inorganic N with biochar may reduce loss and increase uptake of both NO_3^-–N and NH_4^+–N. The objective of this study was to compare the effect of fertilizer N-biochar-combinations (NBC) and FN on soil NO_3^-–N, NH_4^+–N, SOC, and TSN. We hypothesized greater soil N content and improved SOC under NBC compared to NF following harvest of maize.

2. Materials and Methods

2.1. Experimental Sites and Design

Field trials were conducted for two years in the summer cropping season of 2018 and 2019 at Efaw Agronomy Research Station (36°08'12.6" N 97°06'25.8" W) and Lake Carl Blackwell research farm (36°08'58.0" N 97°17'19.3" W), near Stillwater, OK, USA. Efaw Agronomy Research Station had Ashport silty clay loam (fine-silty, mixed, superactive, thermic Fluventic Haplustoll) soil. Lake Carl Blackwell had Pulaski fine-sandy loam (coarse/loamy, mixed nonacid, thermic Udic Ustifluvent) soil [26]. The treatments included; 0, 50, 100, and 150 kg ha^{-1} of FN with no biochar; and 5, 10 and 15 t ha^{-1} of biochar with no FN. The three NBC treatments were 50 kg N plus 5 t ha^{-1} biochar, 100 kg N plus 10 t ha^{-1} biochar, and 150 kg N plus 15 t ha^{-1} biochar. In the second year, treatments were applied to the same exact plots used in the first year. Biochar was obtained from Wakefield Agricultural Carbon (Columbia, MO, USA), a USDA certified biochar producing company. Physical and chemical properties of Southern Yellow Pine biochar pyrolyzed at 500 °C, and the initial soil conditions are included in Table 1. All the N and biochar treatments were applied prior to planting of maize. Total rainfall and average air temperature (April to September) in 2018 and 2019 at Stillwater, OK, USA were obtained from Oklahoma Mesonet (Figure 1).

Table 1. Physical and chemical properties of soft wood (Southern Yellow Pine) biochar supplied by Wakefield Biochar, Columbia, Missouri; the initial soil chemical properties at Lake Carl Blackwell (LCB) and Efaw research sites, Stillwater, OK, USA.

Biochar/ Site	pH	K	Ca	Mg	Mn	Fe	BD	TP	TN	TOC
Unit		mg kg^{-1}	mg kg^{-1}	mg kg^{-1}	mg kg^{-1}	mg kg^{-1}	g cm^{-1}	mg kg^{-1}	g kg^{-1}	g kg^{-1}
Biochar	7.4	612	4128	1225	234	595	0.48	4.53	5.9	876.7
LCB	5.7	349	804	207	x	x	x	12	0.8	9.1
Efaw	5.6	153	1466	354	x	x	x	13	0.7	6.8

TP, total phosphate; TN, total nitrogen; TOC, total organic carbon; BD, bulk density; x, values not determined. Initial soil properties were determined before the first year of biochar application.

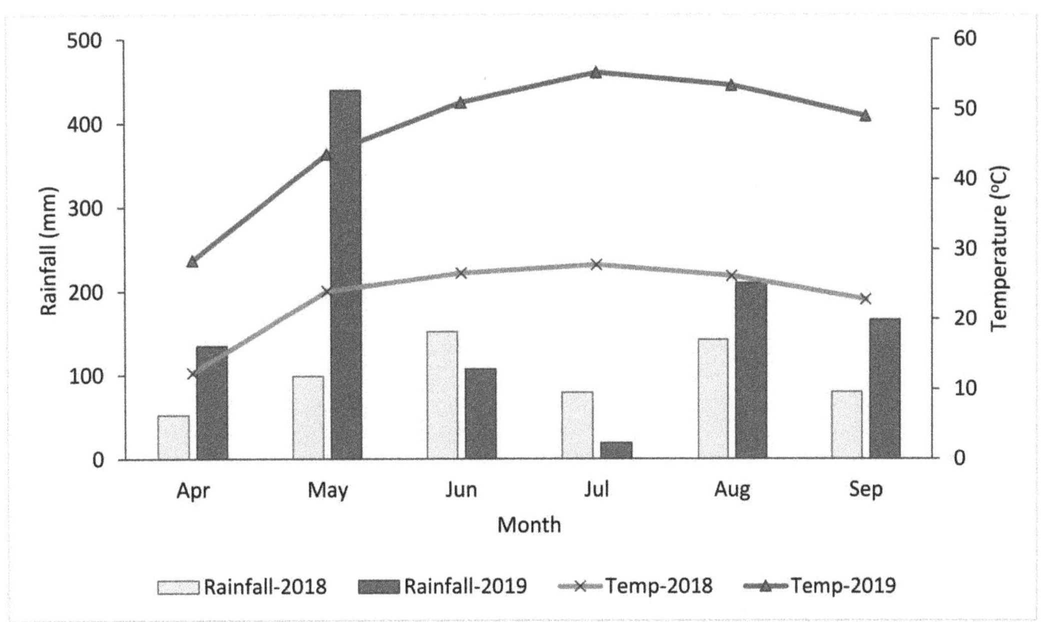

Figure 1. Total rainfall and average air temperature (April to September) in 2018 and 2019 at Stillwater, OK, USA.

Fertilizer-N was applied as urea ammonium nitrate—UAN (28:0:0). Fertilizer-N, biochar and NBC treatments were surface applied. Biochar was broadcast and incorporated at a 15 cm soil depth using a 2720 John Deere Disk Ripper (John Deere, Moline, IL, USA). This incorporation ensured an in-depth mixing of the biochar-N fertilizer complex with soil materials for the respective treatments.

2.2. Data Collection and Analysis

Composite soil samples, 15–20 cores per plot at 0–15 cm, were collected five months after biochar application following harvest of maize in 2018 and 2019. Soil samples were oven-dried for 48 h at 65 °C, and ground to pass through a 1 mm sieve size to remove larger aggregates and plant roots. The extraction of inorganic N (NO$_3^-$–N and NH$_4^+$–N) was carried out from 5 g of soil with 25 mL 1 M KCl after shaking for 30 min on a rotary shaker at 200 rpm. The extracts were filtered with 0.45 µm Whatman filter paper and then analyzed using automated Lachat QuickChem 8500 Series 2 Flow Injection Analyzer (Hach Co., Loveland, CO, USA). The SOC and TSN contents were determined from 200 mg of soil using dry combustion [27] at 950 °C with LECO Truspec CN dry combustion analyzer LECO CN628 (LECO Inc., St. Joseph, MI, USA).

2.3. Statistical Analysis

In this study, the independent variables were contrasts, treatments, and replications while the dependent variables were NO_3^--N, NH_4^+-N, TSN, and SOC. Data were analyzed separately for each year and separated by location. The GLM procedure of the SAS statistical package was used in the analysis of variance (ANOVA) [28]. The combined ANOVA tested for the effect of the independent variables as well as key interactions on the response variables. For all the response variables, the difference between treatment means from NBC and FN were compared using single-degree-of-freedom orthogonal contrasts [29,30]. In addition to the level of statistical significance from ANOVA, the standard error (S.E) of means for each treatment and the coefficient of variation (CV) were used to indicate the precision of measurement and the extent of variability within and between groups, respectively. For each response variable, treatment means and the corresponding S.E were presented in a table that combined experimental sites and years. Additionally, contrasts that compared specific treatments of interests were presented in the bottom half of the table with corresponding F and *p*-values for each site and year.

3. Results

3.1. Soil Nitrate N

The results at Efaw location in 2018 did not show any significant difference ($p = 0.0534$) in soil NO_3^--N content between treatments (Table 2). At each fertilizer rate, soil NO_3^--N was higher under NBC compared to FN by 5 and 7% at 50 and 100 kg N ha^{-1}, respectively. At 150 kg N ha^{-1}, orthogonal contrast showed that NBC was significantly ($p = 0.0259$) lower than FN by 31%. The highest soil NO_3^--N of 6.4 mg kg^{-1} was observed under FN at 150 kg N ha^{-1} while the lowest (3.8 mg kg^{-1}) was observed at 10 t ha^{-1} biochar with no N applied. In 2019, results showed significant differences in soil NO_3^--N content ($p < 0.0001$) between treatments (Table 2). However, contrasts between NBC and FN did not show significant difference in soil NO_3^--N at all N rates. Soil NO_3^--N was lower under NBC than observed FN by 5, 9, and 5% at 50, 100, and 150 kg N ha^{-1}, respectively. The highest soil NO_3^--N (5.9 mg kg^{-1}) was observed under FN at 150 kg N ha^{-1} while the lowest (4.1 mg kg^{-1}) was observed under 5 t ha^{-1} of biochar with no N applied.

At LCB, the 2018 ANOVA results showed significant differences in soil NO_3^--N ($p < 0.0001$) between treatments (Table 2). For each fertilizer rate, contrasts between NBC and FN did not show significant difference in soil NO_3^--N at 50 kg N ha^{-1} ($p = 0.1702$). However, significant differences were seen at 100 kg N ha^{-1} ($p = 0.0003$) and 150 kg N ha^{-1} ($p < 0.0001$). Nitrate under NBC was higher than that observed under FN by 11, 29, and 40% at 50, 100, and 150 kg N ha^{-1}, respectively. The highest NO_3^--N (3.9 mg kg^{-1}) was seen under NBC at 100 kg N ha^{-1} while the lowest (2.0 mg kg^{-1}) was observed at 150 kg N ha^{-1} under FN. In 2019, results were similar to that of 2018 where significant differences in NO_3^--N ($p = 0.001$) were observed between treatments (Table 2). Contrasts between NBC and FN did not show a significant difference in NO_3^--N at 50 kg N ha^{-1} ($p = 0.3134$) and 100 kg N ha^{-1} ($p = 0.0891$), while significant difference was observed at 150 kg N ha^{-1} ($p = 0.02$). The observed differences showed higher NO_3^--N under NBC than FN by 16, 23, and 27% at 50, 100, and 150 kg N ha^{-1}, respectively. The highest soil NO_3^--N (7.0 mg kg^{-1}) was observed under NBC at 150 kg N ha^{-1} while the lowest (3.7 mg kg^{-1}) was observed at the check plot.

Table 2. Mean nitrate N for treatments plus the associated contrasts between N fertilizer and biochar-N combinations at Efaw and Lake Carl Blackwell, Stillwater, Oklahoma. 2018 and 2019.

Treatment	N Rate	Biochar	NO_3^-–N at Efaw				NO_3^-–N at LCB			
	kg ha^{-1}	t ha^{-1}	mg kg^{-1}				mg kg^{-1}			
			2018		2019		2018		2019	
			mean	±S.E	Mean	±S.E	mean	±S.E	mean	±S.E
1	0	0	4.24	0.42	4.05	0.24	2.13	0.24	3.71	0.71
2	50	0	4.76	0.14	4.81	0.54	2.66	0.17	3.94	0.21
3	100	0	4.94	0.23	5.33	0.11	2.76	0.18	4.39	0.17
4	150	0	6.38	0.22	5.88	0.18	1.96	0.06	5.11	0.28
5	0	5	4.45	0.70	4.05	0.02	2.15	0.17	3.95	0.09
6	0	10	3.80	0.39	4.22	0.03	2.14	0.06	4.00	0.11
7	0	15	4.02	0.63	4.18	0.12	2.24	0.26	4.09	0.07
8	50	5	5.01	0.35	4.58	0.04	2.98	0.03	4.67	0.70
9	100	10	5.29	0.32	4.88	0.17	3.87	0.26	5.68	0.21
10	150	15	4.86	0.85	5.58	0.01	3.27	0.12	6.98	0.88
Pr > F			0.0534		<0.0001		<0.0001		0.001	
C.V, %			17.3		7.7		11.5		16.6	
Contrasts			F	Pr > F	F	Pr > F	F	Pr > F	F	Pr > F
2 vs. 8			0.17	0.6840	0.43	0.5222	2.13	0.1702	1.11	0.3134
3 vs. 9			0.35	0.5640	1.63	0.2262	25.02	0.0003	3.42	0.0891
4 vs. 10			6.45	0.0259	0.74	0.4067	34.60	<0.0001	7.18	0.0200
2, 3 & 4 vs. 8, 9 & 10			0.78	0.3944	2.60	0.1326	50.79	<0.0001	10.39	0.0073

C.V, coefficient of variation between treatments; S.E, standard error for replicated means (±SE, $n = 3$). Nitrogen fertilizer was applied as urea ammonium nitrate—UAN (28:0:0). Biochar was applied immediately following UAN and incorporated to a depth of 15 cm.

3.2. Soil Ammonium N

The ANOVA at the Efaw location in 2018 did not show significant difference in soil NH_4^+–N content ($p = 0.9268$) between treatments (Table 3). At each fertilizer rate, soil NH_4^+–N content was higher under NBC compared to FN by 8 and 9% at 50 and 100 kg N ha^{-1}, respectively, while a decrease under NBC by 3% was observed at 150 kg N ha^{-1} compared to FN. The highest NH_4^+–N of 21.2 mg kg^{-1} was seen under NBC at 100 kg N ha^{-1} while the lowest (17.6 mg kg^{-1}) was observed at the check plot. In 2019, results showed significant differences in NH_4^+–N content ($p = 0.0009$) between treatments (Table 3). However, contrasts between NBC and FN did not show significant difference in NH_4^+–N at all contrasted N rates. Soil NH_4^+–N content decreased under NBC by 13% at 150 kg N ha^{-1} compared to FN. The highest NH_4^+–N (5.5 mg kg^{-1}) was observed under FN at 150 kg N ha^{-1} while the lowest (3.9 mg kg^{-1}) was observed under 5 t ha^{-1} of biochar with no N applied.

At LCB, the 2018 results showed significant differences in soil NH_4^+–N ($p = 0.016$) between treatments (Table 3). For each fertilizer rate, contrasts between NBC and FN did not show significant difference in soil NH4$^+$–N at 50 kg N ha^{-1} ($p = 0.3546$) while significant differences were seen at 100 kg N ha^{-1} ($p = 0.026$) and 150 kg N ha^{-1} ($p = 0.0182$). Soil NH_4^+–N was higher under NBC than FN by 6, 14, and 14% at 50, 100, and 150 kg N ha^{-1}, respectively. The highest NH_4^+–N (31 mg kg^{-1}) was observed at 150 kg N ha^{-1} under NBC while the lowest (23 mg kg^{-1}) was observed at 10 t ha^{-1} of biochar with no N fertilizer applied. In 2019, results were similar to that of 2018 where significant difference in NH_4^+–N ($p < 0.0001$) was observed between treatments (Table 3). Contrasts between NBC and FN did not show any significant difference in soil NH_4^+–N at 50 kg N ha^{-1} ($p = 0.8881$) and 100 kg N ha^{-1} ($p = 0.1078$) while significant differences were seen at 150 kg N ha^{-1} ($p = 0.0026$). Soil NH_4^+–N was higher under NBC than FN by 1, 8, and 15% at 50, 100, and 150 kg N ha^{-1}, respectively. The highest soil NH_4^+–N (5.2 mg kg^{-1}) was observed under NBC at 150 kg N ha^{-1} while the lowest (4.1 mg kg^{-1}) was observed at the check plot.

Table 3. Mean ammonium N for treatments plus the associated contrasts between N fertilizer and biochar-N combinations at Efaw and Lake Carl Blackwell, Stillwater, Oklahoma. 2018 and 2019.

Treatment	N Rate	Biochar	NH_4^+–N at Efaw				NH_4^+–N at LCB			
	kg ha^{-1}	t ha^{-1}	mg kg^{-1}				mg kg^{-1}			
			2018		2019		2018		2019	
			mean	±S.E	Mean	±S.E	mean	±S.E	mean	±S.E
1	0	0	17.63	0.59	4.21	0.07	24.43	1.23	4.06	0.03
2	50	0	18.53	1.23	5.01	0.27	24.42	1.13	4.19	0.16
3	100	0	19.33	1.22	5.15	0.29	24.25	0.78	4.24	0.14
4	150	0	20.72	1.25	5.51	0.38	26.32	0.86	4.43	0.09
5	0	5	19.70	2.47	3.87	0.32	24.76	2.27	4.10	0.01
6	0	10	19.03	0.81	4.26	0.10	23.60	0.61	4.11	0.05
7	0	15	19.26	3.53	4.27	0.06	23.98	1.11	4.09	0.06
8	50	5	20.17	1.53	4.40	0.11	25.90	1.44	4.22	0.16
9	100	10	21.15	2.08	4.49	0.15	28.15	0.67	4.59	0.17
10	150	15	20.07	2.54	4.82	0.18	30.52	1.38	5.21	0.15
Pr > F			0.9268		0.0009		0.016		<0.0001	
C.V, %			14.9		8.3		8.4		4.6	
Contrasts			F	Pr > F	F	Pr > F	F	Pr > F	F	Pr > F
2 vs. 8			0.46	0.5121	3.09	0.1044	0.93	0.3546	0.02	0.8881
3 vs. 9			0.56	0.4676	3.51	0.0854	6.45	0.0260	3.02	0.1078
4 vs. 10			0.07	0.7953	3.95	0.0702	7.47	0.0182	14.31	0.0026
2, 3 & 4 vs. 8, 9 & 10			0.45	0.5156	10.52	0.0070	12.96	0.0036	10.70	0.0067

C.V, coefficient of variation between treatments; S.E, standard error for replicated means (±SE, $n = 3$). Nitrogen fertilizer was applied as urea ammonium nitrate—UAN (28:0:0). Biochar was applied immediately following UAN and incorporated to a depth of 15 cm.

3.3. Soil Organic C

At Efaw, the 2018 results indicated an overall significant difference ($p = 0.0016$) in SOC between treatments (Table 4). Contrast comparing NBC and FN at 50 kg N ha^{-1} did not indicate significance difference ($p = 0.6542$) while significance differences in SOC were seen at 100 kg N ha^{-1} ($p = 0.0064$) and 150 kg N ha^{-1} ($p = 0.0018$). Higher SOC observed under NBC than FN correspond to 5, 27, and 31% at 50, 100, and 150 kg N ha^{-1}, respectively. The highest SOC of 9.6 g kg^{-1} was observed under NBC at 150 kg N ha^{-1} while the lowest of 6.6 g kg^{-1} was obtained at 150 kg N ha^{-1} under FN. Similar observations were made in 2019 where significant differences ($p = 0.0007$) in SOC were seen between treatments (Table 4). Contrasts revealed significant differences between NBC and FN at 100 kg N ha^{-1} ($p = 0.018$) and 150 kg N ha^{-1} ($p = 0.0007$). The NBC registered higher SOC than FN by 22 and 35% at 100 and 150 kg N ha^{-1}, respectively. The highest SOC of 11 g kg^{-1} was observed under NBC at 150 kg N ha^{-1} while the lowest of 6.86 g kg^{-1} was obtained at the control plot with no biochar and N applied.

Results at LCB for 2018 did not show an overall significant difference ($p = 0.0758$) in SOC between treatments (Table 4). For each fertilizer rate, contrast comparing NBC with FN did not show significant difference at 50 kg N ha^{-1} ($p = 0.0858$) but differences were seen at 100 kg N ha^{-1} ($p = 0.0058$) and 150 kg N ha^{-1} ($p = 0.0006$). The SOC under NBC was higher than that observed under FN by 17, 21, and 28% at 50, 100, and 150 kg N ha^{-1}, respectively. The highest SOC of 12 g kg^{-1} was observed at 150 kg N ha^{-1} under NBC while the lowest of 8.2 g kg^{-1} was observed at 50 kg N ha^{-1} under FN. In 2019, overall ANOVA showed significant differences ($p = 0.0015$) in SOC between treatments (Table 4). Contrasts between NBC and FN showed significant differences at 50 kg N ha^{-1} ($p = 0.0415$), 100 kg N ha^{-1} ($p = 0.0241$) and 150 kg N ha^{-1} ($p = 0.0335$). The observed differences showed higher SOC under NBC more than under FN by 14, 15, and 12% at 50, 100, and 150 kg N ha^{-1}, respectively. The highest SOC of 13 g kg^{-1} was observed under NBC at 150 kg N ha^{-1} while the lowest of 8.2 g kg^{-1} was seen at 50 kg N ha^{-1} under FN.

Table 4. Mean soil organic C for treatments plus the associated contrasts between N fertilizer and biochar-N combinations at Efaw and Lake Carl Blackwell, Stillwater, Oklahoma. 2018 and 2019.

Treatment	N Rate	Biochar	Soil Organic C at Efaw				Soil Organic C at LCB			
	kg ha^{-1}	t ha^{-1}	g kg^{-1}				g kg^{-1}			
			2018		2019		2018		2019	
			mean	±S.E	Mean	±S.E	Mean	±S.E	mean	±S.E
1	0	0	6.76	0.35	6.86	0.43	9.14	0.98	8.53	0.19
2	50	0	7.05	0.10	7.25	0.44	8.24	0.21	8.23	0.21
3	100	0	6.76	0.37	7.13	0.13	8.40	0.16	8.58	0.16
4	150	0	6.61	0.25	7.34	0.31	8.76	0.12	10.39	0.37
5	0	5	7.79	0.06	8.48	1.13	10.23	1.60	10.48	0.46
6	0	10	8.69	1.03	10.09	0.55	9.79	0.19	9.32	0.73
7	0	15	9.37	0.20	7.03	0.36	10.46	1.38	11.18	1.40
8	50	5	7.39	0.38	7.55	0.63	9.64	0.09	9.58	0.57
9	100	10	9.22	0.96	9.37	0.58	10.91	0.71	10.11	0.25
10	150	15	9.58	0.64	11.01	0.99	12.23	1.04	12.72	0.72
Pr > F			0.0016		0.0007		0.0758		0.0015	
C.V, %			11.8		13.2		14.9		10.9	
Contrasts			F	Pr > F	F	Pr > F	F	Pr > F	F	Pr > F
2 vs. 8			0.21	0.6542	0.14	0.7147	3.50	0.0858	5.21	0.0415
3 vs. 9			10.87	0.0064	7.50	0.0180	11.22	0.0058	6.65	0.0241
4 vs. 10			15.79	0.0018	20.11	0.0007	21.51	0.0006	5.76	0.0335
2, 3 & 4 vs. 8, 9 & 10			19.92	0.0008	19.24	0.0009	28.27	0.0002	17.58	0.0012

C.V, coefficient of variation between treatments; S.E, standard error for replicated means (±SE, n = 3). Nitrogen fertilizer was applied as urea ammonium nitrate—UAN (28:0:0). Biochar was applied immediately following UAN and incorporated to a depth of 15 cm.

3.4. Total Soil N

The ANOVA at Efaw, did not show any significant difference (p = 0.3316) in TSN between treatments in 2018 (Table 5). Total soil N was 14% lower under NBC than observed under FN at 50 kg N ha^{-1}. At 100 and 150 kg N ha^{-1}, TSN was higher under NBC than FN by 6 and 5%, respectively. In 2019, similar observations were made where no significant difference (p = 0.6854) in TSN among treatments (Table 5). At 50 kg N ha^{-1}, TSN was higher under NBC than FN by 10%.

Table 5. Mean total soil N for treatments plus the associated contrasts between N fertilizer and biochar-N combinations at Efaw and Lake Carl Blackwell, Stillwater, Oklahoma. 2018 and 2019.

Treatment	N Rate	Biochar	Total Soil N at Efaw				Total Soil N at LCB			
	kg ha^{-1}	t ha^{-1}	g kg^{-1}				g kg^{-1}			
			2018		2019		2018		2019	
			mean	±S.E	Mean	±S.E	mean	±S.E	mean	±S.E
1	0	0	0.71	0.01	0.79	0.04	0.80	0.03	0.79	0.03
2	50	0	0.82	0.06	0.74	0.05	0.76	0.04	0.77	0.04
3	100	0	0.70	0.05	0.80	0.06	0.75	0.08	0.75	0.02
4	150	0	0.68	0.01	0.73	0.04	0.78	0.00	0.81	0.06
5	0	5	0.78	0.02	0.79	0.04	0.73	0.03	0.85	0.03
6	0	10	0.72	0.04	0.81	0.04	0.84	0.04	0.77	0.01
7	0	15	0.73	0.01	0.72	0.03	0.78	0.04	0.87	0.00
8	50	5	0.71	0.02	0.83	0.04	0.82	0.05	0.81	0.05
9	100	10	0.75	0.04	0.77	0.04	0.82	0.03	0.84	0.03
10	150	15	0.71	0.06	0.80	0.05	0.82	0.05	0.78	0.02
Pr > F			0.3316		0.6854		0.6466		0.2424	

Table 5. *Cont.*

Treatment	N Rate	Biochar	Total Soil N at Efaw				Total Soil N at LCB			
	kg ha^{-1}	t ha^{-1}	g kg^{-1}				g kg^{-1}			
			2018		2019		2018		2019	
C.V, %			8.8		9.7		9		7.2	
Contrasts			F	Pr > F	F	Pr > F	F	Pr > F	F	Pr > F
2 vs. 8			2.77	0.1221	1.62	0.2268	0.85	0.3740	0.39	0.5462
3 vs. 9			0.49	0.4954	0.16	0.6990	1.39	0.2609	2.63	0.1309
4 vs. 10			0.30	0.5921	1.32	0.2730	0.37	0.5533	0.38	0.5512
2, 3 & 4 vs. 8, 9 & 10			0.06	0.8170	1.37	0.2647	2.45	0.1432	0.88	0.3654

C.V, coefficient of variation between treatments; S.E, standard error for replicated means ($\pm$SE, $n = 3$). Nitrogen fertilizer was applied as urea ammonium nitrate—UAN (28:0:0). Biochar was applied immediately following UAN and incorporated to a depth of 15 cm.

At LCB, results for 2018 did not show any significant difference ($p = 0.64$) in TSN between treatments (Table 5). At each fertilizer rate, single degree of freedom contrast did not indicate significant difference in TSN between NBC and FN ($p = 0.143$). Total soil N averaged 0.76, 0.75, and 0.78 g kg^{-1} for FN and 0.82, 0.82, and 0.82 g kg^{-1} for NBC at 50, 100, and 150 kg ha^{-1}, respectively. Similar observations were made in 2019 where ANOVA did not indicate significant difference ($p = 0.2424$) in TSN between treatments (Table 5). Total soil N averaged 0.77, 0.75, and 0.81 g kg^{-1} for FN and 0.81, 0.84, and 0.78 g kg^{-1} for NBC at 50, 100, and 150 kg ha^{-1}, respectively.

4. Discussion

4.1. Soil Inorganic N

The soil NO_3^-–N and NH_4^+–N content were both improved with biochar application. Differences in both soil NO_3^-–N and NH_4^+–N content were seen between experimental years. In 2018, significantly higher soil NO_3^-–N and NH_4^+–N content were observed than in 2019. This could be due to high amount of rainfall received in 2019 that could have caused substantial leaching of inorganic N compared to 2018. Overall, NO_3^-–N across sites and years increased by 8.8% while NH_4^+–N increased by 4.8% with biochar application. Similar observations were made by Yao et al. [31] who reported significant increase in NO_3^-–N (34%) and NH_4^+–N (35%) following biochar application. In addition, Singh et al. [7] observed up to 94% more soil NH_4^+–N under biochar amendment than in the untreated plot. It is important to note that most of the studies that report high proportion of retained inorganic N were soil column leaching experiment compared to the current study that was conducted under field conditions as demonstrated by Libutti et al. [32]. In the current study, positive effects of biochar application in increasing the availability of NO_3^-–N and NH_4^+–N were observed at 10 and 15 t ha^{-1}. With 20 t ha^{-1} of wood biochar, Gao et al. [33] observed NO_3^-–N and NH_4^+–N recovery of 33 and 53%, respectively, under field conditions. In an attempt to offer explanations, Zheng et al. [34] indicated the increase in soil water holding capacity, NH_4^+–N adsorption, and enhanced N immobilization as the main reasons for the increase in recovery of fertilizer-N following biochar soil application. Indeed, increasing the capacity of the soil to hold water increases chances of retaining both NO_3^-–N and NH_4^+–N within soil solution [35]. The enhanced adsorption of NH_4^+–N has been attributed to increase in CEC as a result of the oxidation of aromatic C and formation of carboxyl groups [36]. Lawrinenko and Laird [25] reported an increase in the anion exchange capacity (AEC) of biochar, which reduces leaching of anionic nutrients. They explained that the increased AEC is due to the formation of oxonium functional group ($-O^+$) and non-specific proton adsorption by condensed aromatic rings. Therefore, the rate at which biochar increases the availability of NO_3^-–N and NH_4^+–N is largely dependent on specific biochar production conditions.

4.2. Soil Organic C

The SOC was significantly increased with biochar application. The general ANOVA did not show significant difference in SOC content between experimental years. However, differences in SOC content were observed between experimental sites. Higher SOC content was observed at LCB with a coarse textured sandy loam soil compared to Efaw with silty clay loam. Results averaged across experimental sites and years indicate a 19.3% increase in SOC under biochar soil amendment. The significant impact of biochar on SOC have been well documented [21,25,37]. For instance, Liu et al. [38] observed as high as 40% increase in SOC under biochar treatment. At just 8 t ha^{-1} of biochar derived from wheat straw, Zhang et al. [39] observed 34–80% increase in SOC. Similarly, Gao et al. [33] reported that wood biochar application increased SOC by 33% at 20 t ha^{-1} under tropical soils. Soil organic C increased at all biochar rates used in this study. The apparent and perhaps obvious reason for the increased SOC under biochar soil amendment is the fact that biochar contains high proportion of C by weight compared to other elements. In this study, the pine wood biochar used contained 87% organic C. Indeed, application of material with such high organic C content will certainly increase the SOC of the amended soil. With other factors constant, this implies that the increase in SOC following biochar application is dependent on the rate of biochar application. This notion is consistent with observation in the current study where SOC increased with biochar application rate. In addition, biochar is also known to persist in soil for a long period of time. In the latter case, some researchers have presented evidence on the stability of biochar in the soil and suggested its application as a strategy for soil C sequestration [40,41]. Therefore, application of biochar in agricultural soil are important both from the agronomic and environmental perspectives.

4.3. Total Soil N

Overall, TSN content was not significantly improved following biochar application. However, significant difference was observed between experimental years. The TSN content in 2018 across treatments was lower than observed in 2019, probably due to the cumulative effect since treatments were applied to the same exact plots as in 2018. The TSN content observed at LCB was higher than that at Efaw. This is due to differences in soil type. Biochar effect was significant at the LCB site with a sandy loam soil but not at Efaw with silty clay loam. Across sites and years, an overall observed increase in TSN under biochar soil amendment was 3.7%. This finding is similar to the observations by Agegnehu et al. [25], using waste willow wood (*Salix* spp.) as biochar feedstock. Significant differences between TSN of FN treatment and NBC were not seen. The non-significant response of TSN to biochar application in the above scenarios is probably attributed to limited N in biochar from woody sources, and that was insufficient to support TSN increase within experimental periods and rates used in these studies. Total soil N is a quantity that builds up in soil over a period of time. To illustrate this viewpoint, Omara et al. [42] observed a significant trend in buildup of TSN in a long-term experiment where fertilizer N was applied on a yearly basis. Therefore, the element of time and rate of application, alongside N content of biochar, is paramount in explaining the behavior of TSN following biochar application. Contrary to these findings, Uzoma et al. [21] observed significant increase in TSN using dry cow manure biochar at similar rates as in the current study. Using dry cow manure biochar could have resulted to this significant difference in TSN buildup as compared to biochar from woody sources within the rates used in this study. With hard-wood biochar, Prommer et al. [16] reported significant increase in TSN following biochar application in combination with inorganic N compared to inorganic N alone. They could have seen the positive results due to high rate of biochar applied with up to 72 t ha^{-1} compared to only 15 t ha^{-1} as the maximum application rate of biochar used in the current study. The authors alluded that application of inorganic fertilizer-N in combination with biochar compensate for the reduction in organic N mineralization. Therefore, application rate and nature of biochar seems to play an important role in determining the rate of TSN increase following biochar application.

5. Conclusions

The study used postharvest soil samples taken from fields following maize crop five months after biochar application to compare the effect of FN and NBC on soil NO_3^--N, NH_4^+-N, SOC, and TSN. It was hypothesized that greater recovery of inorganic N and SOC would be observed under NBC compared to NF following harvest of maize. Generally, results averaged over sites and years showed some advantages of applying NBC as opposed to FN where NO_3^--N, NH_4^+-N, SOC, and TSN increased under the combination by 4%, 20%, 9%, and 5%, respectively. Positive impact of the combination was realized with 10 and 15 t ha^{-1} of biochar. The SOC, NO_3^--N, and NH_4^+-N increased with increase in fertilizer rates. Whereas biochar improved SOC at both sites, significant response of NO_3^--N and NH_4^+-N availability to biochar application were only observed at LCB site with sandy loam soil but not at Efaw with silty clay loam. Thus, the application of biochar in combination with fertilizer-N improves N availability with the potential to increase crop N uptake on coarse textured soils compared to soils with fine texture. Therefore, given the pivotal role of C and N in soil quality, this study shows that application of a combination of biochar and inorganic N could be important in the future management of soils which are inherently poor with texture-related limitations. Since the current study did not specifically evaluate the retentive capacity of biochar, future study could estimate the actual adsorption capacity by analyzing the CEC, AEC, and base saturation of soils treated with biochar.

Author Contributions: Conceptualization, P.O.; methodology, P.O.; formal analysis, P.O. and L.A.; investigation, P.O.; resources, D.B.A.; writing—original draft preparation, P.O.; writing—review and editing, L.A., F.O., A.O., J.L.B.S. and D.B.A. All authors have read and agreed to the published version of the manuscript.

Funding: This research received no external funding.

Institutional Review Board Statement: Not applicable.

Informed Consent Statement: Not applicable.

Data Availability Statement: Not applicable.

Acknowledgments: The authors would like to acknowledge and sincerely thank Tyler Lynch and Robert Lemings for their assistance with field work.

Conflicts of Interest: The authors declare no conflict of interest.

References

1. Chan, K.Y.; Zwieten, L.V.; Meszaros, I.; Downie, A.; Joseph, S. Agronomic Values of Green Waste Biochar as a Soil Amendment. *Soil Res.* **2007**, *45*, 629–634. [CrossRef]
2. Woolf, D.; Amonette, J.E.; Street-Perrott, F.A.; Lehmann, J.; Joseph, S. Sustainable biochar to mitigate global climate change. *Nat. Commun.* **2010**, *1*, 56. [CrossRef] [PubMed]
3. Crombie, K.; Mašek, O.; Sohi, S.P.; Brownsort, P.; Cross, A. The Effect of Pyrolysis Conditions on Biochar Stability as Determined by Three Methods. *GCB Bioenergy* **2013**, *5*, 122–131. [CrossRef]
4. Spigarelli, B.P.; Kawatra, S.K. Opportunies and Challenges in Carbon Dioxide Capture. *J. CO₂ Util.* **2013**, *1*, 69–87. [CrossRef]
5. Jindo, K.; Mizumoto, H.; Sawada, Y.; Sonoki, T. Physical and Chemical Characterization of Biochars Derived from Different Agricultural Residues. *Biogeosciences* **2014**, *11*, 6613–6621. [CrossRef]
6. Stavins, R.N. The Costs of Carbon Sequestration: A Revealed-Preference Approach. *Am. Econ. Rev.* **2017**, *89*, 994–1009. [CrossRef]
7. Singh, B.P.; Hatton, B.J.; Singh, B.; Cowie, A.L.; Kathuria, A. Influence of Biochars on Nitrous Oxide Emission and Nitrogen Leaching from Two Contrasting Soils. *J. Environ. Qual.* **2010**, *39*, 1224. [CrossRef]
8. Atkinson, C.J.; Fitzgerald, J.D.; Hipps, N.A. Potential Mechanisms for Achieving Agricultural Benefits from Biochar Application to Temperate Soils: A Review. *Plant Soil* **2010**, *337*, 1–18. [CrossRef]
9. Aula, L.; Macnack, N.; Omara, P.; Mullock, J.; Raun, W.R. Effect of Fertilizer Nitrogen on Soil Organic Carbon, Total Nitrogen, and Soil pH in Long-Term Continuous Winter Wheat (*Triticum aestivum* L.). *Commun. Soil Sci. Plant Anal.* **2016**, *47*, 863–874. [CrossRef]
10. Wardle, D.A.; Nilsson, M.C.; Zackrisson, O. Response to Comment on Fire-Derived Charcoal Causes Loss of Forest Humus. *Science* **2008**, *321*, 1295. [CrossRef]
11. Jones, D.L.; Murphy, D.V.; Khalid, M.; Ahmad, W.; Edwards-Jones, G.; DeLuca, T.H. Short-Term Biochar-Induced Increase in Soil CO2 Release Is Both Biotically and Abiotically Mediated. *Soil Biol. Biochem.* **2011**, *43*, 1723–1731. [CrossRef]

12. Cross, A.; Sohi, S.P. The priming potential of biochar products in relation to labile carbon contents and soil organic matter status. *Soil Boil. Biochem.* **2011**, *43*, 2127–2134. [CrossRef]

13. Kimetu, J.M.; Lehmann, J. Stability and Stabilization of Biochar and Green Manure in Soil with Different Organic Carbon Contents. *Soil Res.* **2010**, *48*, 577–585. [CrossRef]

14. Jones, C.M.; Kammen, D.M. Quantifying Carbon Footprint Reduction Opportunities for U.S. Households and Communities. *Environ. Sci. Technol.* **2011**, *45*, 4088–4095. [CrossRef] [PubMed]

15. Howarth, R.W. Nitrogen in Freshwater Systems and Estuaries Nitrogen. In *Reference Module in Earth Systems and Environmental Sciences*; Elsevier: Amsterdam, The Netherlands, 2014. [CrossRef]

16. Prommer, J.; Wanek, W.; Hofhansl, F.; Trojan, D.; Offre, P.; Urich, T.; Schleper, C.; Sassmann, S.; Kitzler, B.; Soja, G.; et al. Biochar decelerates soil organic nitrogen cycling but stimulates soil nitrification in a temperate arable field trial. *PLoS ONE* **2014**, *9*, 86388. [CrossRef] [PubMed]

17. Bai, S.H.; Reverchon, F.; Xu, C.Y.; Xu, Z.; Blumfield, T.J.; Zhao, H.; Van Zwieten, L.; Wallace, H.M. Wood biochar increases nitrogen retention in field settings mainly through abiotic processes. *Soil Biol. Biochem.* **2015**, *90*, 232–240. [CrossRef]

18. Raun, W.R.; Johnson, G.V. Improving Nitrogen Use Efficiency for Cereal Production. *Agron. J.* **1999**, *91*, 357–363. [CrossRef]

19. Fageria, N.K.; Baligar, V.C. Enhancing nitrogen use efficiency in crop plants. *Adv. Agron.* **2005**, *88*, 97–185. [CrossRef]

20. Omara, P.; Aula, L.; Oyebiyi, F.B.; Raun, W.R. World Cereal Nitrogen Use Efficiency Trends: Review and Current Knowledge. *Agrosyst. Geosci. Environ.* **2019**, *2*, 1–8. [CrossRef]

21. Uzoma, K.C.; Inoue, M.; Andry, H.; Fujimaki, H.; Zahoor, A.; Nishihara, E. Effect of cow manure biochar on maize productivity under sandy soil condition. *Soil Use Manag.* **2011**, *27*, 205–212. [CrossRef]

22. Jiang, T.Y.; Jiang, J.; Xu, R.K.; Li, Z. Adsorption of Pb (II) on variable charge soils amended with rice-straw derived biochar. *Chemosphere* **2012**, *89*, 249–256. [CrossRef] [PubMed]

23. Bruun, E.W.; Petersen, C.T.; Hansen, E.; Holm, J.K.; Hauggaard-Nielsen, H. Biochar amendment to coarse sandy subsoil improves root growth and increases water retention. *Soil Use Manag.* **2014**, *30*, 109–118. [CrossRef]

24. Lawrinenko, M.; Laird, D.A. Anion exchange capacity of biochar. *Green Chem.* **2015**, *17*, 4628–4636. [CrossRef]

25. Agegnehu, G.; Bass, A.M.; Nelson, P.N.; Bird, M.I. Benefits of biochar, compost and biochar–compost for soil quality, maize yield and greenhouse gas emissions in a tropical agricultural soil. *Sci. Total Environ.* **2016**, *543*, 295–306. [CrossRef] [PubMed]

26. Soil Survey Staff. Natural Resources Conservation Service, United States Department of Agriculture. Web Soil Survey. Available online: http://websoilsurvey.sc.egov.usda.gov/ (accessed on 24 June 2020).

27. Schepers, J.S.; Francis, D.D.; Thompson, M.T. Simultaneous determination of total C, total N and 15N on soil and plant material. *Commun. Soil Sci. Plant Anal.* **1989**, *20*, 949–959. [CrossRef]

28. SAS Institute Inc. *Base SAS 9.4 Procedures Guide: Statistical Procedures*, 2nd ed.; SAS Institute Inc.: Cary, NC, USA, 2013.

29. Nogueira, M.C.S. Orthogonal contrasts: Definitions and concepts. *Sci. Agric.* **2004**, *61*, 118–124. [CrossRef]

30. Abdi, H.; Williams, L.J. Contrast analysis. *Encycl. Res. Des.* **2010**, *1*, 243–251.

31. Yao, Y.; Gao, B.; Zhang, M.; Inyang, M.; Zimmerman, A.R. Effect of Biochar Amendment on Sorption and Leaching of Nitrate, Ammonium, and Phosphate in a Sandy Soil. *Chemosphere* **2012**, *89*, 1467–1471. [CrossRef]

32. Libutti, A.; Mucci, M.; Francavilla, M.; Monteleone, M. Effect of biochar amendment on nitrate recovery in a silty clay loam soil. *Ital. J. Agron.* **2016**, *11*, 273–276. [CrossRef]

33. Gao, S.; Hoffman-Krull, K.; Bidwell, A.L.; DeLuca, T.H. Locally produced wood biochar increases nutrient retention and availability in agricultural soils of the San Juan Islands, USA. *Agric. Ecosyst. Environ.* **2016**, *233*, 43–54. [CrossRef]

34. Zheng, H.; Wang, Z.; Deng, X.; Herbert, S.; Xing, B. Impacts of adding biochar on nitrogen retention and bioavailability in agricultural soil. *Geoderma* **2013**, *206*, 32–39. [CrossRef]

35. Libutti, A.; Cammerino, A.R.B.; Francavilla, M.; Monteleone, M. Soil amendment with biochar affects water drainage and nutrient losses by leaching: Experimental evidence under field-grown conditions. *Agronomy* **2019**, *9*, 758. [CrossRef]

36. Liang, B.; Lehmann, J.; Solomon, D.; Kinyangi, J.; Grossman, J.; O'Neill, B.; Skjemstad, J.O.; Thies, J.; Luizão, F.J.; Petersen, J.; et al. Black carbon increases cation exchange capacity in soils. *Soil Sci. Soc. Am. J.* **2006**, *70*, 1719–1730. [CrossRef]

37. Laird, D.A.; Fleming, P.; Davis, D.D.; Horton, R.; Wang, B.; Karlen, D.L. Impact of Biochar Amendments on the Quality of a Typical Midwestern Agricultural Soil. *Geoderma* **2010**, *158*, 443–449. [CrossRef]

38. Liu, S.; Zhang, Y.; Zong, Y.; Hu, Z.; Wu, S.; Zhou, J.; Jin, Y.; Zou, J. Response of soil carbon dioxide fluxes, soil organic carbon and microbial biomass carbon to biochar amendment: A meta-analysis. *GCB Bioenergy* **2016**, *8*, 392–406. [CrossRef]

39. Zhang, M.; Cheng, G.; Feng, H.; Sun, B.; Zhao, Y.; Chen, H.; Chen, J.; Dyck, M.; Wang, X.; Zhang, J.; et al. Effects of straw and biochar amendments on aggregate stability, soil organic carbon, and enzyme activities in the Loess Plateau, China. *Environ. Sci. Pollut. Res.* **2017**, *24*, 10108–10120. [CrossRef]

40. Lorenz, K.; Lal, R. Biochar application to soil for climate change mitigation by soil organic carbon sequestration. *J. Plant Nutri. Soil Sci.* **2014**, *177*, 651–670. [CrossRef]

41. Wang, J.; Xiong, Z.; Kuzyakov, Y. Biochar stability in soil: Meta-analysis of decomposition and priming effects. *GCB Bioenergy* **2016**, *8*, 512–523. [CrossRef]

42. Omara, P.; Aula, L.; Eickhoff, E.M.; Dhillon, J.S.; Lynch, T.; Wehmeyer, G.B.; Raun, W.R. Influence of No-Tillage on Soil Organic Carbon, Total Soil Nitrogen, and Winter Wheat (*Triticum aestivum* L.) Grain Yield. *Int. J. Agron.* **2019**, *2019*, 9632969. [CrossRef]

Article

Nutrient Availability for *Lactuca sativa* Cultivated in an Amended Peatland: An Ionic Exchange Study

Jacynthe Dessureault-Rompré *, Alexis Gloutney and Jean Caron

Soil and Agri-Food Engineering Department, Laval University, 2480 Hochelaga Blvd.,
Quebec City, QC G1V 0A6, Canada; alexis.gloutney.1@ulaval.ca (A.G.); jean.caron@fsaa.ulaval.ca (J.C.)
* Correspondence: jacynthe.dessureault-rompre@fsaa.ulaval.ca

Abstract: Few conservation strategies have been applied to cultivated peatland. This field study over one growth cycle of *Lactuca sativa* examined the effect of plant-based, high-C/N-ratio amendments in a real farming situation on peatland. Plant Root Simulator (PRS®) probes were used directly in the field to assess the impacts of incorporating *Miscanthus x giganteus* straw and *Salix miyabeana* chips on nutrient availability for lettuce. The results showed that lettuce yield decreased by 35% in the miscanthus straw treatment and by 14% in the willow chip treatment. In addition, the nitrogen flux rate was severely reduced during crop growth (75% reduction) and the plant N uptake index was much lower in the amended treatments than in the control. The phosphorus supply rate was also significantly lower (24% reduction) in the willow treatment. The influence of sampling zone was significant as well, with most macro-nutrients being depleted in the root zone and most micro-nutrients being mobilized. Additional work is needed to optimize the proposed conservation strategy and investigate the effects of consecutive years of soil amendment on different vegetable crops and in different types of cultivated peatlands to confirm and generalize the findings of this study. Future field studies should also explore the long-term carbon dynamics under plant-based, high-C/N-ratio amendments to determine if they can offset annual C losses.

Keywords: nitrogen; miscanthus; willow; field experiment; lettuce; plant-based amendment; rhizosphere

Citation: Dessureault-Rompré, J.; Gloutney, A.; Caron, J. Nutrient Availability for *Lactuca sativa* Cultivated in an Amended Peatland: An Ionic Exchange Study. *Nitrogen* **2022**, 3, 26–42. https://doi.org/ 10.3390/nitrogen3010002

Academic Editor: Marouane Baslam

Received: 11 December 2021
Accepted: 10 January 2022
Published: 14 January 2022

Publisher's Note: MDPI stays neutral with regard to jurisdictional claims in published maps and institutional affiliations.

1. Introduction

Peatlands serve a number of essential purposes, including the production of substrates and food [1]. Thanks to their high organic matter content, drained peatlands can be highly productive and suitable for agriculture, grasslands, and forestry. However, the drainage required to make organic soils suitable for agricultural activities creates conditions conducive to rapid degradation and compaction [2]. Wind and water erosion, subsidence, and soil organic matter mineralization leading to CO_2 losses are the main causes of the degraded soil conditions observed in cultivated peatlands [3–5]. Solutions proposed thus far have included restoring cultivated peatlands to their natural state [6–8] or changing land use drastically, to paludiculture, for example [9,10].

More recently, the use of chopped biomass crops as a soil amendment has been proposed as a way to make peatland cultivation more sustainable [11,12]. Implemented in conjunction with optimal water table management and erosion mitigation, this practice may lead to an integrated conservation management approach that would improve the long-term sustainability of cultivated histosols [12–14].

Cultivated peatland in Canada constitutes an important part of the agricultural economy. In the plain of Montreal, in southwestern Quebec, close to 12,000 ha of land is covered in deep organic soils [15]. Although small, the area plays an essential role in the production of high-value vegetable crops, such as *Daucus carota* (carrots), *Lactuca sativa* (lettuce), and *Allium cepa* (onion), supplying fruits and vegetables to Canada and the northeastern United States [16].

In 2016, a large-scale project was set-up in collaboration with local growers to develop conservation strategies adapted to their soils. As Bourdon et al. (2021) explained [11]: "In addition to controlling wind and, to a lesser extent, water erosion, a combination of other strategies is currently being investigated to extend the lifespan of these highly fertile lands including the application of plant biomass with a high carbon (C)/nitrogen (N) ratio (woodchip and grass straw)." Such strategies could potentially compensate for soil losses based on recent C modeling [12,17] and restore physical properties [18].

Miscanthus (*Miscanthus x giganteus*) straw and willow (*Salix miyabeana*) chips have been found to be good candidates for these conservation strategies [12,18,19], despite some anticipated side-effects such as N and P immobilization [11,20]. Miscanthus and short-rotation willow can be grown as biomass crops in degraded land [21,22] with little fertilization and practically no weed control [23,24], making them excellent choices for on-farm soil amendment production. Unlike manure and sewage sludge, for example, these plant-based amendments comply with food safety norms regarding vegetable cultivation and are not a source of pollution for rivers and bodies of water near the amended sites [25,26].

Approximately 14–20% of peatlands are used globally for agriculture, and when drained and cultivated, they represent some of the world's most productive agricultural soils [27]. Hence, the need for the development of sustainable management practice is urgent for these precious soils. The proposed conservation strategy is new and original as it aims to maintain long-term vegetable production on these productive lands while ensuring a massive return of organic matter annually, through biomass crop production on-site, or a biomass supply locally grown on more degraded surfaces.

The general objective of this field study was to assess the impacts of incorporating plant-based amendments, specifically *Miscanthus x giganteus* straw and *Salix miyabeana* chips, on nutrient availability for the roots of *Lactuca sativa* crops cultivated on peatland. We hypothesized that (1) both amendments similarly reduced the availability of nutrients, mainly nitrogen; that (2) the intensity of the reduction observed decreased over time; and that (3) rhizosphere strategies compensate for the effect of amendment.

2. Materials and Methods

2.1. Study Site

The soil at the research site (45°11′ N, 73°20′ W) is a moderately decomposed Haplosaprist [28], with a pH of 5.7 and a carbon and nitrogen content of 46% and 2.0%, respectively.

The average annual precipitation (30-year period) at the experimental site is 961 mm, and the annual average temperature is 6.6 °C. The annual frost-free period is 146 days, with 3289 degree-days above 0 °C (http://climate.weather.gc.ca/climate_normals/).

Although the field experiment was conducted over only one crop growth cycle, the experimental design was rigorously developed and applied, thus assuring the representativeness of the data, which are characteristic of this site and year, specifically.

2.2. Soil Amendment Description

Two biomass crops were selected in this experiment to amend the histosol: *Miscanthus x giganteus* (miscanthus) and *Salix miyabeana* (willow). These amendments were chosen based on the findings of Dessureault-Rompré et al. [12], who observed miscanthus and willow biomass material to be more resistant to degradation than the annual plant sorghum (*Sorghum bicolor*). The harvested biomass crops were bought from nearby producers. The characteristics of the two soil amendments are presented here in Table 1.

Table 1. Characteristics of the chopped plant material.

Characteristics	Miscanthus	Willow
C (%)	45.3	46.4
N (%)	0.198	0.343
P (%)	0.048	0.031
C/N	228	135
C/P	242	381
Hemicellulose (%)	29.8	10.9
Cellulose (%)	23.0	32.2
Lignin (%)	34.8	35.5
Lignin/N	176	103

2.3. Experimental Design

A 1 ha area of a 7 ha field was divided into three sections: (1) control, (2) miscanthus-amended, and (3) willow-amended. Each experimental section covered an area of about 330 m^2. Chopped miscanthus and willow material was added to the soil at a rate of 15 t ha^{-1} (Figure 1). This rate is thought to be adequate to compensate for the average annual C losses observed in this type of soil (Dessureault-Rompré et al., 2020).

Figure 1. Experimental sections (miscanthus on the **left** and willow on the **right**) pictured before the chopped biomass was incorporated into the soil.

On 12 June 2019, the baled miscanthus was chopped and spread using a FP240 (New Holland Agriculture, New Holland, PA, USA) pull-type forage harvester equipped with a hay pickup. The bales were loosened and laid by hand in windrows on the ground to be picked up by the forage harvester and blown from the chute onto the plot. As the willow was already chipped when received, a lime spreader was used to spread it over the plot. The chopped plant material was then incorporated to a depth of 15 cm using a chisel.

2.4. Ionic Exchange Resin

The use of Plant Root Simulator (PRS®) ion exchange resins is an economical and rapid way to quantify the concentrations of a range of nutrients and contaminants in soils by simulating their uptake by plant roots [29–31]. In this study, the soil nutrient supply was assessed using PRS® cationic and anionic exchange resin probes. At six locations in each experimental plot, two cationic exchange resin probes and two anionic exchange resin probes were inserted at a depth of 15 cm in the root zone of two lettuces and into two root exclusion cylinders (bulk soil) measuring about 15 cm in height and 10 cm in diameter (Figure 2). The resin probes remained in the soil for seven days, and they were then replaced with fresh ones (Table 2). The removed resin probes were cleaned with a toothbrush and rinsed with demineralized water to remove any soil particles that could continue to exchange ions with the resins. Although laborious, this cleaning procedure is essential to ensure that no soil particles are left on the resin surface, which could continue to exchange nutrients. Finally, the cleaned resin probes were stored at 4 °C before being sent to Western Ag Innovations' laboratories (Saskatoon, SK, Canada) for complete analysis. At Western Ag Innovations' laboratories, inorganic N (ammonium and nitrate) in the eluant was determined colorimetrically using automated flow injection analysis (Skalar San++ Analyzer, Skalar Inc., Breda, The Netherlands). The remaining nutrients (P, K, S, Ca, Mg, Fe, Mn, Cu, and Zn) were measured using inductively coupled plasma (ICP) spectrometry (Optima ICP-OES 8300, PerkinElmer Inc., Waltham, MA, USA).

Figure 2. Experimental design with PRS® probes installed in the root zone and in root exclusion cylinders. The probes are shown here before they were completely inserted into the soil.

Table 2. Burial and retrieval dates for the PRS® probes and related periods.

Burial Date	Retrieval Date	Period
2019-07-10	2019-07-17	1
2019-07-17	2019-07-24	2
2019-07-24	2019-07-31	3
2019-07-31	2019-08-07	4
2019-08-07	2019-08-14	5
2019-08-14	2019-08-19	6

2.5. Lettuce Crop Management and Yield Evaluation

Small, five-day-old lettuce plants were transplanted on 1st July. All treatments were fertilized equally at planting with N-P-K (14.4-3.6-17.4) at a rate of 833 kg ha^{-1}. The plots were irrigated on 1st July (5 mm), 10th July (5 mm), and 6th August (15 mm). On 7th July, the first set of resin probes was installed in the field. Plot yields (fresh plant matter) were evaluated on 19th August, by weighing the lettuce harvested from one linear meter, repeated three times per treatment.

2.6. Soil Sampling and Analysis

Following the lettuce harvest, composite soil samples were taken at a depth of 0–20 cm at each of the locations previously occupied by the resin probes. The soil samples were then dried at 70 °C, sieved at 2 mm, and used for chemical analyses.

Nitrogen was extracted from the soil with water, using a 1:10 (*w/v*) soil:solution ratio. After an agitation period of 1 h at 200 rpm (19 mm circular orbit), the extract was centrifuged at 3000 rpm (2060× *g*) for 10 min and filtered through a No. 42 Whatman filter paper. Ammonium (NH_4^+) and nitrate (NO_3^-) concentrations were quantified with a Quikchem 8500 Series 2 system (Lachat Instruments, Loveland, CO, USA), using Quikchem methods 10-107-06-2-B and 12-107-04-1-F, respectively. Total N in the water extract was measured using the same system with prior persulfate oxidation, according to the procedure described by Qualls (1989) [32]. Soluble organic N (SON) was calculated by subtracting $N\text{-}NO_3^-$ and $N\text{-}NH_4^+$ from total N.

Mehlich III elements were extracted with a 1 g of soil to 30 mL of solution ratio, a 5-min agitation period at 200 rpm (19 mm circular orbit), and filtration through a No. 42 Whatman filter paper [33]. Elements were analyzed by inductively coupled plasma–mass spectrometry (Icap 6500 MK2 radial, ThermoFisher Scientific, USA) within 12 h of extraction to prevent organic compounds from precipitating with metals.

For the active carbon fraction, a 0.02 M $KMnO_4$/0.1 M $CaCl_2$ solution was used according to the procedure described in Weil et al. (2003) and Blair et al. (1995) [34,35]. Total organic carbon and total nitrogen concentrations were determined using a LCN-2000 dry combustion analyzer (Leco Corporation, St. Joseph, MI, USA).

2.7. Data Visualization and Statistics

Data visualization was performed using the RStudio ggplot2 and emmeans packages. Analyses of variance with the *post hoc* Tukey HSD test were computed using the mixed procedure (nlme library) (RStudio Team, 2020). A natural logarithmic transformation was used when needed to respect the normality and homogeneity of variance assumptions.

3. Results

3.1. Yield (Fresh Mass)

Lettuce fresh mass decreased by 35% in the miscanthus-amended treatment and by 14% in the willow-amended treatment, although these differences were not statistically significant (Figure 3, Table 3).

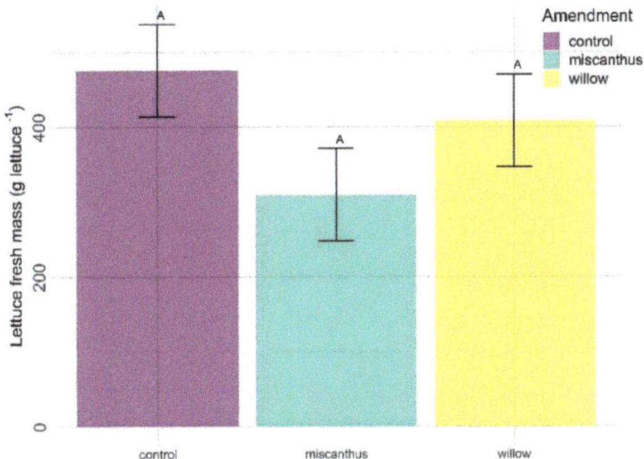

Figure 3. Lettuce fresh mass per plant as a function of soil amendment. Different letters indicate significance at $p < 0.05$ using the *post hoc* Tukey HSD test.

Table 3. Tukey's p-values for lettuce yield and all nutrients analyzed with the PRS®.

	Amendments	Sampling Zone	Time	Amendment X Sampling Zone	Amendment X Time	Sampling Zone X Time
Lettuce yield	0.175	NA	NA	NA	NA	NA
N	<0.0001	0.0012	<0.0001	0.0134	0.9938	0.5338
P	0.0027	0.4923	0.0001	0.2515	0.4876	0.4075
K	0.4556	0.8435	<0.0001	0.1581	0.9557	0.5026
Ca	0.6404	<0.0001	<0.0001	0.3570	0.0148	0.0489
Mg	0.0906	<0.0001	<0.0001	0.8260	0.0165	0.0058
S	0.0005	0.0412	<0.0001	0.3465	0.7285	0.0173
Fe	0.4555	0.0002	<0.0001	0.1651	0.8025	<0.0001
Mn	0.3166	0.0039	<0.0001	0.0541	0.8456	0.0008
Cu	0.3731	0.0110	<0.0001	0.1131	0.9251	0.1421
Zn	0.9325	0.0275	<0.0001	0.2657	1.0000	0.1650

3.2. PRS® Probes

Overall, the addition of plant-based amendments significantly decreased the N and P supply rates and significantly increased the S supply rate (Table 3). Significant differences were observed between the sampling zones, with N, Ca, Mg, and S showing significant decreases and Fe, Mn, Cu, and Zn showing significant increases in the root zone as compared to the bulk soil.

The following sections present more detailed results for each of the aforementioned nutrients.

3.2.1. Nitrogen

N supply rates differed significantly with amendments, sampling zone, time, and interaction between sampling zone and time (Table 3). The N supply rate decreased significantly in both the miscanthus- (−76%) and willow- (−75%) amended treatments as compared to the control (Figures 4 and 5). Overall, the N supply rate decreased by 90% from the beginning to the end of lettuce growth and was 34% higher in the bulk soil than in the root zone. The significant interaction between amendment and sampling zone showed that the N supply rates were significantly lower in the root zone as compared to the bulk soil in the control and miscanthus-amended soil but not in the willow-amended soil (Figure 5). Finally, the plant N uptake index, calculated as the difference between the N supply rate in the bulk soil and that in the root zone (Figure 6), revealed a clear difference between the control treatment and the amended soils. While the plant N uptake index for the control treatment increased and then decreased markedly during lettuce growth, the index showed a consistent decrease in the miscanthus treatment for the same period. In the willow treatment, where the difference between the root zone and the bulk soil was not significant, the plant N uptake index was negative at the beginning of lettuce growth and increased slowly over time during lettuce growth.

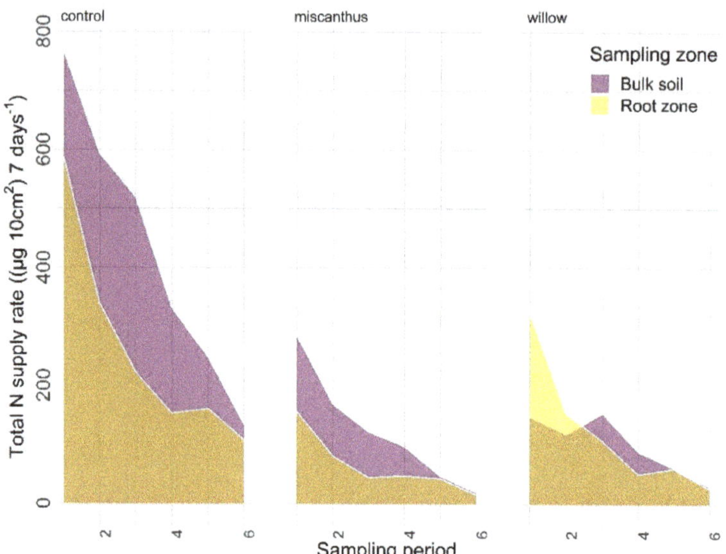

Figure 4. Total N supply rates measured in the bulk soil and the root zone as a function of time and amendment.

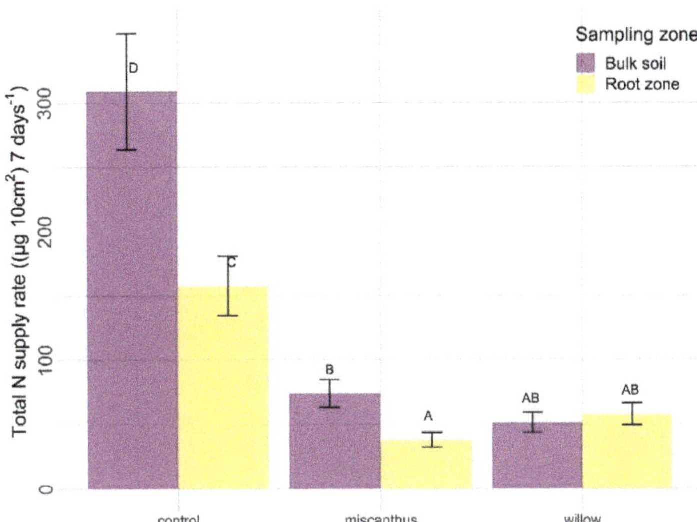

Figure 5. Statistical differences for total N supply rates between sampling zones and treatments. Different letters indicate significance at *p* < 0.05 using the *post hoc* Tukey HSD test.

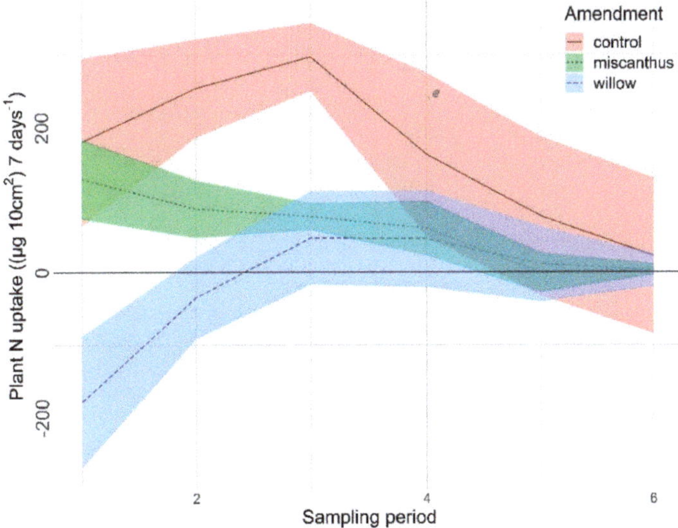

Figure 6. Plant N uptake index as a function of time and amendment. The colored area represents the error envelope as plotted with the geom_ribbon function in the ggplot2 R package.

3.2.2. Other Nutrients

Phosphorus. For P, amendments and time showed significant differences (Table 3). The P supply rate was significantly lower (−24%) in the willow treatment than in the miscanthus treatment and the control (Figure 7, left). In addition, the P supply rate varied over the lettuce growth period (Figure 7, right).

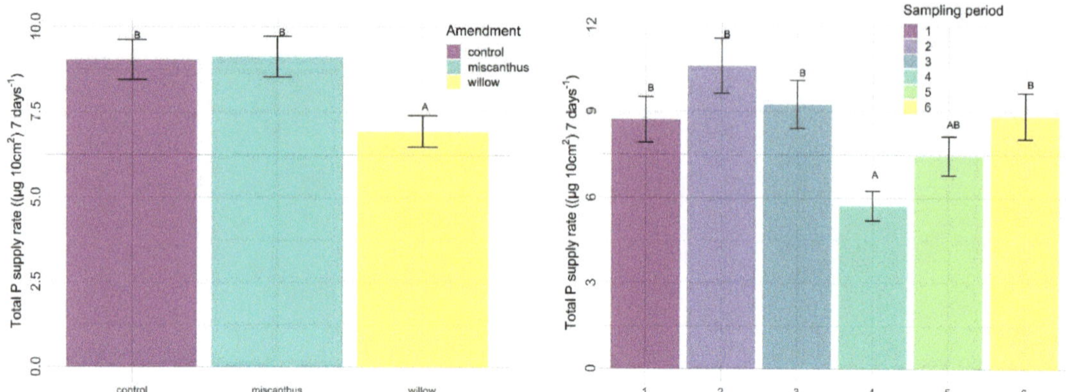

Figure 7. P supply rates as a function of amendment (**left**) and time (resin sampling period) (**right**). Different letters indicate significance at $p < 0.05$ using the *post hoc* Tukey HSD test.

Potassium. For K, only the PRS® probe sampling period showed a statistically significant difference, with a decrease over time (Table 3, Figure 8). From the beginning to the end of lettuce growth, K supply rates decreased by 30%.

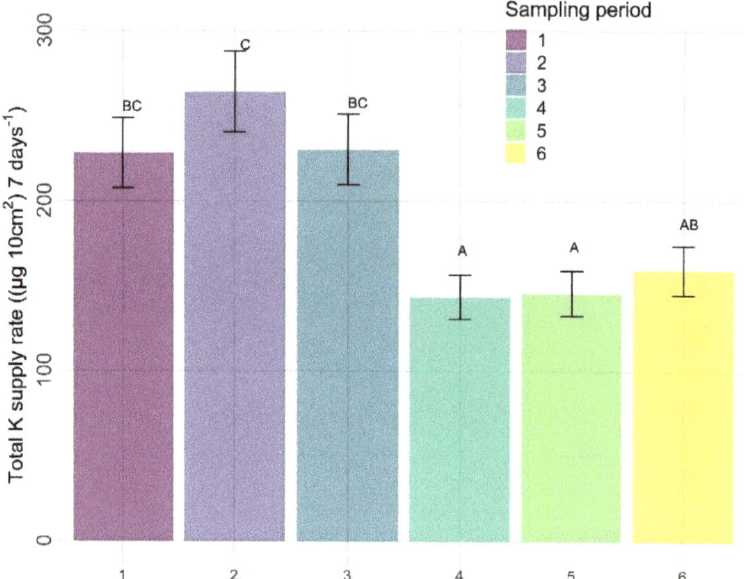

Figure 8. K supply rates as a function of time (resin sampling period). Different letters indicate significance at $p < 0.05$ using the *post hoc* Tukey HSD test.

Calcium. For Ca, sampling zone, time, amendment X time, and sampling zone X time all showed significant differences (Table 3). Although the amendment X time interaction was significant, it was not relevant from a practical point of view, as, for example, the control in the first sampling period differed from the miscanthus in the fifth sampling period. Ca supply rates decreased by an average of 30% across amendments and sampling zones from the beginning to the end of lettuce growth (Figure 9). In addition, the average difference between root zone and bulk soil sampling was 10%; the decrease in the root zone was significant for resin sampling periods 3 and 4, with differences of 18% and 17%, respectively, between the root zone and the bulk soil (Figure 9).

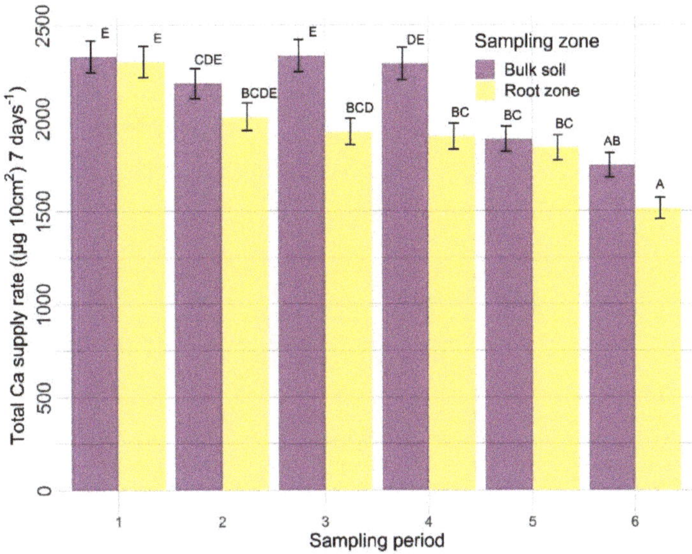

Figure 9. Ca supply rates as a function of sampling zone and time. Different letters indicate significance at $p < 0.05$ using the *post hoc* Tukey HSD test.

Magnesium. The behavior of Mg closely followed that of Ca. Sampling zone, time, amendment X time, and sampling zone X time all showed significant differences (Table 3). As mentioned above for Ca, the amendment X time interaction was significant but not relevant in practical terms. Mg supply rates decreased by an average of 18% across amendments and sampling zones by the end of lettuce growth (Figure 10). In addition, the average difference between root zone and bulk soil sampling was 12% and the decrease was significant for the third resin sampling period, with a 21% difference between the root zone and the bulk soil (Figure 10).

Sulfur. S supply rates differed significantly by amendment, sampling zone, time, and interaction between sampling zone and time (Table 3). The S supply rate increased significantly in both the miscanthus (+38%) and the willow (+59%) treatments (Figure 11, left). Overall, the S supply rate increased by 42% from the beginning to the end of lettuce growth and was 18% higher in the bulk soil than in the root zone. However, there was a significant interaction between the period and sampling zone, and the S supply rate was significantly higher in the bulk soil compared to the root zone in the fourth resin sampling period (Figure 11).

Iron. For Fe, sampling zone, time, and sampling zone X time all showed significant differences (Table 3). On average, Fe supply rates decreased by 43% across amendments and sampling zones from the beginning to the end of lettuce growth (Figure 12). In addition, the average difference between root and bulk soil sampling was 32%. The interaction between time and sampling zone showed that in the fourth period of resin sampling, the Fe supply rate was significantly higher (+244%) in the root zone as compared to the bulk soil (Figure 12).

Manganese. The behavior of Mn closely followed that of Fe. Sampling zone, time, and sampling zone X time all showed significant differences (Table 3). Mn supply rates had decreased by an average of 47% across amendments and sampling zones by the end of lettuce growth (Figure 13). In addition, the average difference between root zone and bulk soil sampling was 26%. The interaction between time and sampling zone showed that in the fourth and fifth resin sampling periods, Mn supply rates in the root zone were, respectively, 125% and 93% higher than in the bulk soil (Figure 13).

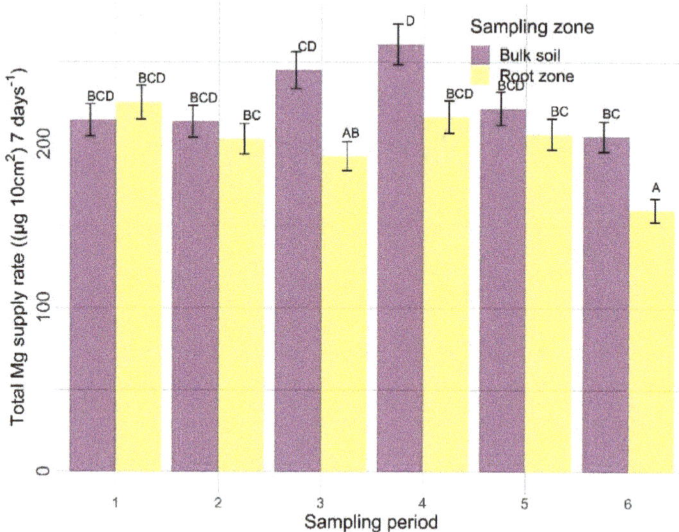

Figure 10. Mg supply rates as a function of sampling zone and time. Different letters indicate significance at *p* < 0.05 using the *post hoc* Tukey HSD test.

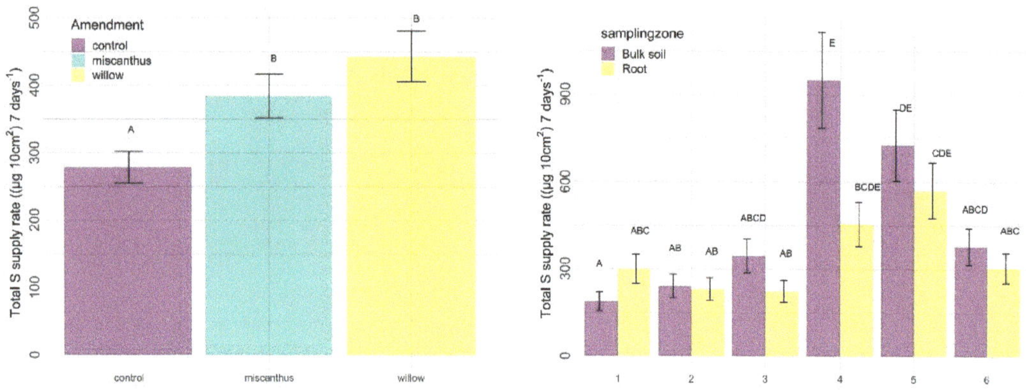

Figure 11. S supply rates as a function of amendment (**left**) and interaction between sampling zone and time (**right**). Different letters indicate significance at *p* < 0.05 using the *post hoc* Tukey HSD test.

Copper. Cu supply rates differed significantly by sampling zone and time (Table 3). Overall, Cu supply rates were 24% higher in the root zone than in the bulk soil (Figure 14, left). In addition, a significant decrease (50%) was observed between the beginning and the end of lettuce growth (Figure 14, right).

Zinc. Zn supply rates differed significantly by sampling zone and time (Table 3). Overall, Zn supply rates were 96% higher in the root zone than in the bulk soil (Figure 15, left). In addition, a significant decrease (61%) was observed between the beginning and the end of lettuce growth (Figure 15, right).

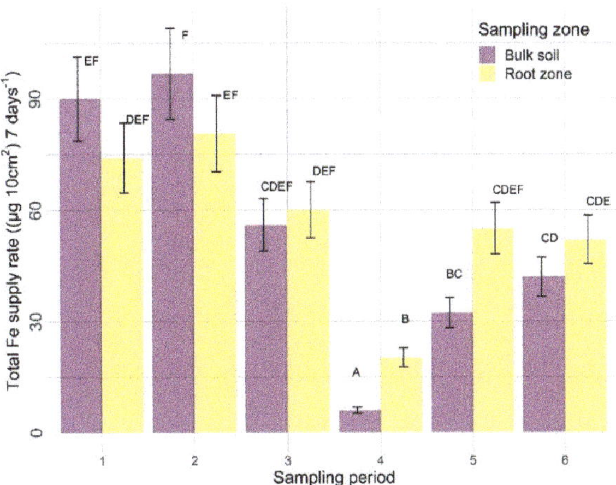

Figure 12. Fe supply rates as a function of sampling zone and time. Different letters indicate significance at $p < 0.05$ using the *post hoc* Tukey HSD test.

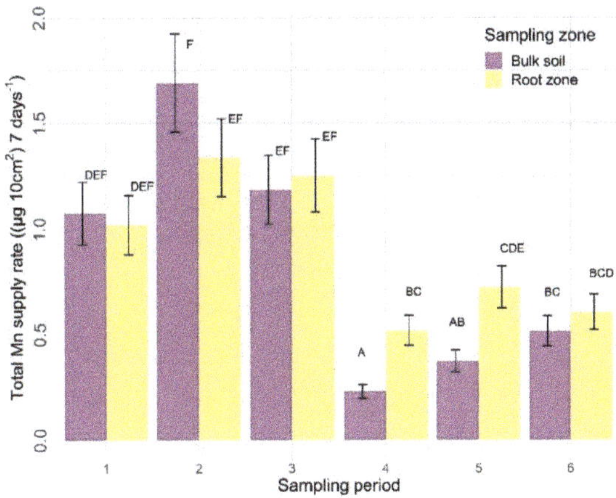

Figure 13. Mn supply rates as a function of sampling zone and time. Different letters indicate significance at $p < 0.05$ using the *post hoc* Tukey HSD test.

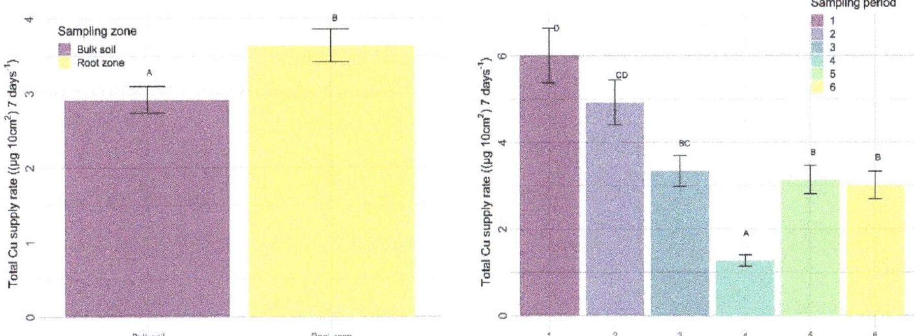

Figure 14. Cu supply rates as a function of sampling zone (**left**) and time (sampling period) (**right**). Different letters indicate significance at $p < 0.05$ using the *post hoc* Tukey HSD test.

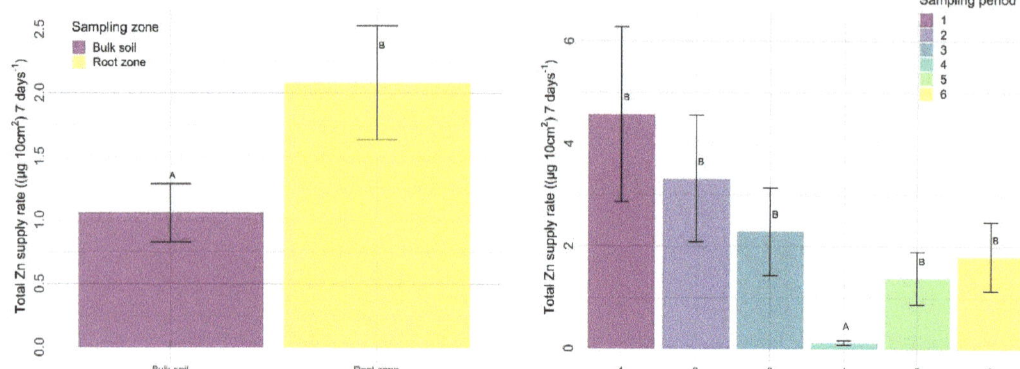

Figure 15. Zn supply rates as a function of sampling zone (**left**) and interaction between sampling zone and time (**right**). Different letters indicate significance at $p < 0.05$ using the *post hoc* Tukey HSD test.

3.3. Soil Samples following Lettuce Harvest

After harvesting the lettuce, soil samples (0–15 cm) were taken and analyzed (Table 4). The results showed that soluble organic N was the only parameter significantly influenced by the amendments, with significant decreases in both the miscanthus and the willow treatments. A tendency for Cu to be lower under willow treatment and for Fe to be higher under miscanthus treatment was also observed ($p = 0.055$).

Table 4. Mean $\pm$ (se) and p-values (Tukey's test) for different soil parameters measured in the soil at the end of lettuce growth as a function of amendment.

	Control	Miscanthus	Willow	p-Values
Mineral N	60.4 ± (7.5)	58.1 ± (10.0)	62.4 ± (6.2)	0.933
Soluble organic N	**70.1 ± (9.6)**	**21.0 ± (3.2)**	**30.1 ± (6.2)**	**0.005**
Active C	10 604.0 ± (172.0)	10 779.0 ± (189.4)	12 430.7 ± (1487.2)	0.331
C	45.9 ± (0.4)	46.5 ± (0.2)	46.6 ± (0.4)	0.331
N	1.9 ± (0.0)	2.0 ± (0.0)	2.0 ± (0.0)	0.617
C/N	23.7 ± (0.1)	23.9± (0.1)	23.5 ± (0.2)	0.164
P	314.7 ± (39.3)	219.7 ± (25.3)	201.7 ± (35.5)	0.110
K	1 91.2 ± (66.4)	324.3 ± (19.4)	328.0 ± (33.3)	0.120
Ca	14 033.3 ± (550.7)	13 891.0 ± (497.6)	13 886.7 ± (456.6)	0.970
Mg	1 107.3 ± (34.9)	1 135.0 ± (62.5)	1 138.3 ± (15.1)	0.970
Fe	900.0 ± (40.3)	754.7 ± (9.9)	831.7 ± (38.9)	0.055
Mn	16.3 ± (1.6)	17.4 ± (0.8)	21.5 ± (1.9)	0.106
Cu	26.8 ± (3.4)	26.6 ± (1.7)	37.8 ± (3.2)	0.055
Zn	22.2 ± (3.3)	27.3 ± (2.9)	34.6 ± (3.9)	0.105

4. Discussion

4.1. Nitrogen

The use of plant-based, high-C/N-ratio amendments has been proposed as a conservation strategy in cultivated peatland (Dessureault-Rompré et al., 2020). In the present study, PRS® probes were used to evaluate the impact of incorporating miscanthus straw and willow chip amendments in the field on nutrient availability in the rhizosphere of a lettuce crop. The field was fertilized as usual by the grower. Miscanthus and willow amendments decreased fresh lettuce yield by 35% and 14%, respectively. Although not statistically significant, such a yield decrease is necessarily relevant for growers as it could have substantial economic consequences.

Straw mulches from different plant sources have been studied and used in vegetable crop production for years, with both positive and negative effects on crop yield [36–38]. In the present study, however, the miscanthus straw and willow chips were not used as a mulch but were incorporated into the soil as an organic amendment. Few studies have investigated the effect of straw incorporation on vegetable crop yields. More numerous studies have focused on the effect of straw incorporation on wheat or wheat-rice rotation [39–41] and corn yields [42–44], with most reporting a neutral to positive crop response. In the present study, however, the quality of the straw was different, and straw was added at a much higher rate.

The present study highlighted the significant impact of miscanthus straw and willow chip incorporation on the field soil nutrient supply rate. Nitrogen was by far the most affected by the soil amendments, with a 75% reduction in the N supply rate. In a modeling study on the effect of 8 t ha^{-1} of wheat straw incorporation on nitrogen dynamics, Garnier et al. [45] observed a 13% decrease in the net amount of nitrogen mineralized 13 months after the straw was incorporated. In an incubation experiment, both wheat straw and spruce sawdust added at a rate of 4.5 t C ha^{-1} [46] were found to cause nitrogen immobilization, reaching a maximum level of 42 kg N ha^{-1} and then decreasing to 8–15 kg N ha^{-1} after a few weeks. In a microcosm study, rice straw incorporated at a rate of 10 t ha^{-1} was found to immobilize 16–39% of applied N over the course of a 160 days incubation period [47]. In the present study, in the bulk soil, the difference between amended treatment and the nonamended soil did not attenuate over time (42 days) for willow and increased in the miscanthus-amended treatment. In the rhizosphere, the difference between the amended and nonamended treatments increased over time for both amendments (Figure 4). In a 56 days incubation study, Bourdon et al. [11] observed that the addition of 11 t ha^{-1} of miscanthus straw and willow chips to a moderately decomposed histosol reduced mineral N from a KCl extract by 60% and 54%, respectively, overall, with the effect of these two amendments increasing over the course of the incubation period. In a 182 days incubation experiment, Marmier [20] observed that following miscanthus straw and willow chip amendments at a rate of 15 t ha^{-1}, N immobilization remained stable, increasing or decreasing over the course of the incubation period, depending on the type of histosol and amendment. Finally, the plant N uptake index in the control was found to increase until week 3 and then decrease until the end of lettuce growth. The plant N uptake index was much lower in the miscanthus treatment than in the control treatment and decreased from the beginning to the end of lettuce growth. The index for the willow treatment was initially negative and increased slightly until week 3. These results showed that miscanthus and willow amendments, by their negative impact on the supply rate of N, decrease the capacity of the lettuce crop to absorb N, which is consistent with the impact observed on lettuce yield. Interestingly, soil extract at the end of the growing season showed no difference in mineral N concentrations between the control and amended soil. However, a significant decrease in soluble organic N was observed. Dynamic changes in the fast-cycling mineral N pool might be better captured using PRS® probes than soil extract; however, future research should explore the time evolution between mineral and organic N pools under field-amended cultivated peatland.

4.2. Other Nutrients

4.2.1. Amendments

For nutrients other than N, the effect of the amendments was significant only on P and S, with the willow treatment decreasing the phosphorous supply rate and both the willow and miscanthus treatments increasing the sulfur supply rate (Figures 7 and 11). Several authors have reported P immobilization with the incorporation of straw under different pedo-climatic conditions [48–50], while others have noted increased available P or P losses [51–53]. Bourdon et al. (2021) pointed out that because the P content of Quebec peat soils can be excessively high [54], P immobilization could help reduce eutrophication. Although Bourdon et al. (2021) found S immobilization to occur during a 56 days incubation experiment with miscanthus and willow straw, the mineralization–immobilization of S has been shown to be related to the C:S ratio of the amendment [55]. In the present field study using PRS® probes, the S supply was found to increase substantially, by 38% (miscanthus) and 59% (willow). The S supply changed over time, however, increasing to a peak in the fourth sampling period and then decreasing, indicating that S was mineralized and then immobilized during lettuce growth.

4.2.2. Sampling Zone

Significant differences were found in the root zone for all nutrients analyzed, except P and K. The effect of sampling zone showed decreases in N, Ca, and Mg in the root zone as compared to the bulk soil zone. The opposite effect was noted for S, Fe, Mn, Cu, and Zn. Depletion and accumulation zones have often been observed in the rhizosphere [56]. Lettuce crops have shown important quantitative differences in root exudates depending on soil type, even under well-fertilized conditions [57]. Although we did not find studies on root exudates from lettuce grown in cultivated peatland, root exudation and activities are known to be intimately related to plant nutrition [58] and, therefore, might have contributed to the micro-nutrient mobilization observed in this study.

4.2.3. Time

Overall, the nutrient supply rate decreased over time, with few exceptions. Initial fertilization, lettuce uptake, and soil processes such as immobilization, mineralization, and adsorption to the soil matrix could all have been implicated to different degrees in the results obtained in the present study.

4.3. Retrospective on the Initial Hypotheses

The present study was developed based on three main hypotheses. The first one stated that both amendments similarly reduced the availability of nutrient, mainly nitrogen. This is actually true for nitrogen where the overall impact of miscanthus and willow on the reduction of the N supply rate was −76% for miscanthus and −75% for willow. However, more precise information came from the investigation of this effect at the root zone scale and allowed us to put forward that the supply of N under the willow amendment seemed slower for the first half of the lettuce growth period (see Figure 6) as compared to the N supply rate observed under the miscanthus treatment. In addition, few other nutrients were impacted by the amendment (only P and S).

The second hypothesis stated that the intensity of the reduction observed decreased over time. In the present study, we distinguished this effect for the bulk soil and root zone. In the bulk soil, the difference between the amended treatment and the nonamended soil did not attenuate over time for willow and increased in the miscanthus-amended treatment. In the rhizosphere, the difference between the amended and nonamended treatments increased over time for both amendments (see Figure 4).

The third and last hypothesis stated that rhizosphere strategies compensate for the effect of amendment. At the temporal (weekly) and spatial (root zone and bulk soil) scales used in this study, we did observe that in the rhizosphere zone under willow amendment, the N supply rate was higher as compared to the N supply rate observed in the bulk soil for the first two weeks and half of the lettuce growth, which could lead to a nutrient mobilization strategy (Figure 4). However, with the data in hand, we cannot distinguish between a mobilization strategy or a slower N uptake by the lettuce root. In addition, the difference between the root zone and the bulk soil under miscanthus treatment and for the control treatment was identical (−49% in the root zone as compared to the bulk soil).

5. Overall Perspective: Found and Missing Pieces of the Proposed Conservation Strategy Puzzle

The use of plant-based amendments with high C/N ratios as a conservation strategy in cultivated peatland is still in its infancy. This field study over one lettuce growth cycle examined the effects of such amendments in a real farming situation to evaluate the potential use of this conservation practice for sustainable vegetable crop production on cultivated peatlands. Although the pieces of the puzzle are starting to fit into place [11,12,19,20], this first field study revealed that some important information is still missing. Using PRS® probes under field conditions, this study confirmed that the main impact on crop growth resulting from the incorporation of miscanthus straw and willow chips is linked to N availability, with a consequent reduction in fresh lettuce yield. The plant N uptake index indicated how miscanthus and willow amendments can impact N uptake by the lettuce crop, pointing toward the need for fertilization adjustments under this conservation strategy.

As for the missing pieces, a number of unanswered questions remain with regard to optimizing the application of such amendments, such as the optimal time (fall versus spring) of application, the effects of combining this strategy with other conservation strategies such as cover crops (for example, a fall application of the amendment together with a cover crop), the effects of composting miscanthus straw and willow chips before incorporating them in the soil, and also the possibility of using these plant-based amendments as a mulch for the first year before incorporating them into the soil. Additional work is also needed to investigate the effects of consecutive years of soil amendment on different vegetable crops and in different types of cultivated peatlands to confirm and generalize the findings of the present study. Future field studies should also explore the long-term carbon dynamics in peatland amended with plant-based, high-C/N-ratio amendments under real field conditions to determine if this strategy could offset annual C losses and ensure muck soil conservation for generations to come.

Author Contributions: Conceptualization: J.D.-R.; methodology, J.D.-R. and A.G.; formal analysis, J.D.-R. and A.G.; investigation, A.G.; resources, J.D.-R. and J.C.; data curation, J.D.-R. and A.G.; writing—original draft preparation, J.D.-R. and A.G.; writing—review and editing, J.D.-R., A.G. and J.C.; visualization, J.D.-R.; supervision, J.D.-R. and J.C.; project administration, J.D.-R. and J.C.; funding acquisition, J.C. All authors have read and agreed to the published version of the manuscript.

Funding: We would like to acknowledge the financial support of the *Natural Sciences and Engineering Research Council of Canada (NSERC)* through a *Collaborative Research and Development Grant* in partnership with *Productions Horticoles Van Winden Inc., Les Fermes Hotte et Van Winden Inc., Maraîchers J.P.L. Guérin et fils Inc., Delfland inc., Vert Nature Inc., Isabelle Inc., La Production Barry Inc., Le Potager Montréalais ltée., Les Jardins A. Guérin et fils INC., Les Fermes du Soleil Inc., Les Fermes R.R. et fils inc., Productions Maraîchères Breizh Inc.,* and *R. Pinsonneault et fils ltée.*

Institutional Review Board Statement: Not applicable.

Acknowledgments: We would like to acknowledge Moranne Béliveau, Charles Gauthier-Marcil, Simon Corbeil and Nicolas Shooner, who helped with the field study, particularly the cleaning of the resin probes.

Conflicts of Interest: The authors declare no conflict of interest.

Sample Availability: Samples of the compounds are not available from the authors.

References

1. Bonn, A.; Allott, T.; Evans, M.; Joosten, H.; Stoneman, R. *Peatland Restoration and Ecosystem Services: Science, Policy and Practice*; Cambridge University Press: Cambridge, UK, 2016.
2. Verhoeven, J.T.; Setter, T.L. Agricultural use of wetlands: Opportunities and limitations. *Ann. Bot.* **2010**, *105*, 155–163. [CrossRef] [PubMed]
3. Ilnicki, P.; Zeitz, J. *Organic Soils and Peat Materials for Sustainable Agriculture*; CRC Press: Boca Raton, FL, USA, 2002; pp. 15–32.
4. Elder, J.W.; Lal, R. Tillage effects on gaseous emissions from an intensively farmed organic soil in North Central Ohio. *Soil Tillage Res.* **2008**, *98*, 45–55. [CrossRef]
5. Kroetsch, D.J.; Geng, X.; Chang, S.X.; Saurette, D.D. Organic soils of Canada: Part 1. Wetland organic soils. *Can. J. Soil Sci.* **2011**, *91*, 807–822. [CrossRef]
6. Kløve, B.; Berglund, K.; Berglund, Ö.; Weldon, S.; Maljanen, M. Future options for cultivated Nordic peat soils: Can land management and rewetting control greenhouse gas emissions? *Environ. Sci. Policy* **2017**, *69*, 85–93. [CrossRef]
7. Joosten, H.; Tapio-Biström, M.-L.; Tol, S. *Peatlands: Guidance for Climate Change Mitigation through Conservation, Rehabilitation and Sustainable Use*; Food and Agriculture Organization of the United Nations Rome: Rome, Italy, 2012.

8. Ferré, M.; Muller, A.; Leifeld, J.; Bader, C.; Müller, M.; Engel, S.; Wichmann, S. Sustainable management of cultivated peatlands in Switzerland: Insights, challenges, and opportunities. *Land Use Policy* **2019**, *87*, 104019. [CrossRef]
9. Grootjans, A.P. Paludiculture—Productive Use of Wet Peatlands. *Restor. Ecol.* **2017**, *25*, 661–663. [CrossRef]
10. Joosten, H.; Gaudig, G.; Krawczynski, R.; Tanneberger, F.; Wichmann, S.; Wichtmann, W. 25 Managing Soil Carbon in Europe: Paludicultures as a New Perspective for Peatlands. *Soil Carbon Sci. Manag. Policy Mult. Benefits* **2014**, *71*, 297.
11. Bourdon, K.; Fortin, J.; Dessureault-Rompré, J.; Caron, J. Agricultural peatlands conservation: How does the addition of plant biomass and copper affect soil fertility? *Soil Sci. Soc. Am. J.* **2021**, *85*, 1242–1255. [CrossRef]
12. Dessureault-Rompré, J.; Libbrecht, C.; Caron, J. Biomass crops as a soil amendment in cultivated histosols: Can we reach carbon equilibrium? *Soil Sci. Soc. Am. J.* **2020**, *84*, 597–608. [CrossRef]
13. Parent, L.; Millette, J.; Mehuys, G. Subsidence and erosion of a Histosol. *Soil Sci. Soc. Am. J.* **1982**, *46*, 404–408. [CrossRef]
14. Regina, K.; Sheehy, J.; Myllys, M. Mitigating greenhouse gas fluxes from cultivated organic soils with raised water table. *Mitig. Adapt. Strateg. Glob. Change* **2015**, *20*, 1529–1544. [CrossRef]
15. Lamontagne, L.; Martin, A.; Nolin, M.C. *Étude Pédologique du Comté de Napierville (Québec)*; Laboratoires de pédologie et d'agriculture de précision, Centre de recherche de St-Jean sur le Richelieu, Agriculture et Agroalimentaire Canada: Québec, QC, Canada, 2014.
16. MAPAQ. Portrait-Diagnostic Sectoriel des Légumes Frais au Québec. 2018. Available online: https://www.mapaq.gouv.qc.ca/fr/Publications/Portraitsectoriellegumesfrais.pdf (accessed on 20 September 2021).
17. Rodriguez, A.F.; Gerber, S.; Daroub, S. Modeling soil subsidence in a subtropical drained peatland. The case of the everglades agricultural Area. *Ecol. Model.* **2019**, *415*, 108859. [CrossRef]
18. Grégoire, V. L'apport de Biomasse Végétale et L'amélioration des Propriétés Hydrauliques des sols Organiques Cultivés. Thèse de Maîtrise, Université Laval, Québec, QC, Canada, 2020.
19. Bourdon, K.; Fortin, J.; Dessureault-Rompré, J.; Caron, J. Can We Slow Down Cultivated Peatland Respiration without Changing Land Use? In Proceedings of the ASA, CSSA and SSSA International Annual Meetings, Online, 9–13 November 2020.
20. Marmier, V. *Mineralization of Nitrogen and Release of Dissolved Phosphorus and Carbon in Plant-Amended Organic Soils*; Department of Environmental Sciences, ETH Zürich: Zürich, Switzerland, 2021.
21. Amichev, B.Y.; Hangs, R.D.; Konecsni, S.M.; Stadnyk, C.N.; Volk, T.A.; Bélanger, N.; Vujanovic, V.; Schoenau, J.J.; Moukoumi, J.; Van Rees, K.C. Willow Short-Rotation Production Systems in Canada and Northern United States: A Review. *Soil Sci. Soc. Am. J.* **2014**, *78*, S168–S182. [CrossRef]
22. Erickson, L.E.; Pidlisnyuk, V. (Eds.) *Phytotechnology with Biomass Production: Sustainable Management of Contaminated Sites*, 1st ed.; CRC Press: Boca Raton, FL, USA, 2021. [CrossRef]
23. Stolarski, M.J.; Szczukowski, S.; Tworkowski, J.; Klasa, A. Willow biomass production under conditions of low-input agriculture on marginal soils. *For. Ecol. Manag.* **2011**, *262*, 1558–1566. [CrossRef]
24. Emmerling, C.; Pude, R. Introducing Miscanthus to the greening measures of the EU Common Agricultural Policy. *Gcb Bioenergy* **2017**, *9*, 274–279. [CrossRef]
25. Fairchild, D. *Ground Water Quality and Agricultural Practices*; CRC Press: Boca Raton, FL, USA, 1987.
26. Erickson, M. Horticulture production, food safety and organic amendments: Effect on public health. *Acta Hortic.* **2015**, *1076*, 201–214. [CrossRef]
27. Strack, M. *Peatlands and Climate Change*; IPS, International Peat Society Jyväskylä: Jyväskylä, Finland, 2008; pp. 44–69.
28. Staff, S.S. Soil taxonomy: A basic system of soil classification for making and interpreting soil surveys. *Agric. Handb.* **1999**, *436*, 886.
29. Miller, J.J.; Bremer, E.; Curtis, T. Influence of Organic Amendment and Compaction on Nutrient Dynamics in a Saturated Saline-Sodic Soil from the Riparian Zone. *J. Environ. Qual.* **2016**, *45*, 1437–1444. [CrossRef] [PubMed]
30. Harrison, D.J.; Maynard, D.G. Nitrogen mineralization assessment using PRS™ probes (ion-exchange membranes) and soil extractions in fertilized and unfertilized pine and spruce soils. *Can. J. Soil Sci.* **2014**, *94*, 21–34. [CrossRef]
31. Drohan, P.J.; Merkler, D.J.; Buck, B.J. Suitability of the Plant Root Simulator Probe for Use in the Mojave Desert. *Soil Sci. Soc. Am. J.* **2005**, *69*, 1482–1491. [CrossRef]
32. Qualls, R. Determination of Total Nitrogen and Phosphorus in Water Using Persulfate Oxidation: A Modification for Small Sample Volumes Using the Method of Koroleff (1983). The Biogeochemical Properties of the Retention of Nitrogen, Phosphorus, and Carbon. Ph.D. Thesis, University of Georgia Institute of Ecology, Athens, GA, USA, 1989.
33. Centre d'Expertise en Analyse Environnementale du Québec. Détermination des métaux et du phosphore assimilables: Méthode par spectrométrie de masse à source ionisante au plasma d'argon. In *MA. 200–Mét-Pass. 1.0, Centre d'Expertise en Analyse Environnementale du Québec*; Ministère de L'environnement et de la Lutte aux Changements Climatique: Quebec, QC, Canada, 2014; 15p.
34. Weil, R.R.; Islam, K.R.; Stine, M.A.; Gruver, J.B.; Samson-Liebig, S.E. Estimating active carbon for soil quality assessment: A simplified method for laboratory and field use. *Am. J. Altern. Agric.* **2003**, *18*, 3–17.
35. Blair, G.; Lefroy, R.; Lisle, L. Soil carbon fractions based on their degree of oxidation, and the development of a carbon management index for agricultural systems. *Aust. J. Agric. Res.* **1995**, *46*, 1459–1466. [CrossRef]
36. Dessureault-Rompré, J.; Gloutney, A.; Caron, J. Nutrient Availability under Lettuce Grown in Rye Mulch in Histosols. *Nitrogen* **2020**, *1*, 12. [CrossRef]

37. Chopra, M.; Koul, B. Comparative assessment of different types of mulching in various crops: A review. *Plant Arch.* **2020**, *20*, 1620–1626.

38. Kumar, V.; Kumar, M.; Rai, A.P. Mulching in fruit and vegetable crop production under rainfed conditions: A review. *J. Soil Water Conserv.* **2016**, *15*, 313. [CrossRef]

39. Xia, L.; Wang, S.; Yan, X. Effects of long-term straw incorporation on the net global warming potential and the net economic benefit in a rice–wheat cropping system in China. *Agric. Ecosyst. Environ.* **2014**, *197*, 118–127. [CrossRef]

40. Song, K.; Yang, J.; Xue, Y.; Lv, W.; Zheng, X.; Pan, J. Influence of tillage practices and straw incorporation on soil aggregates, organic carbon, and crop yields in a rice-wheat rotation system. *Sci. Rep.* **2016**, *6*, 36602. [CrossRef] [PubMed]

41. Zhao, X.; Yuan, G.; Wang, H.; Lu, D.; Chen, X.; Zhou, J. Effects of Full Straw Incorporation on Soil Fertility and Crop Yield in Rice-Wheat Rotation for Silty Clay Loamy Cropland. *Agronomy* **2019**, *9*, 133. [CrossRef]

42. Jiang, C.; Yu, W.; Ma, Q.; Xu, Y.; Zou, H. Alleviating global warming potential by soil carbon sequestration: A multi-level straw incorporation experiment from a maize cropping system in Northeast China. *Soil Tillage Res.* **2017**, *170*, 77–84. [CrossRef]

43. Zheng, J.; Qu, Y.; Kilasara, M.M.; Mmari, W.N.; Funakawa, S. Nitrate leaching from the critical root zone of maize in two tropical highlands of Tanzania: Effects of fertilizer-nitrogen rate and straw incorporation. *Soil Tillage Res.* **2019**, *194*, 104295. [CrossRef]

44. Wang, X.; Jia, Z.; Liang, L.; Zhao, Y.; Yang, B.; Ding, R.; Wang, J.; Nie, J. Changes in soil characteristics and maize yield under straw returning system in dryland farming. *Field Crop. Res.* **2018**, *218*, 11–17. [CrossRef]

45. Garnier, P.; Néel, C.; Aita, C.; Recous, S.; Lafolie, F.; Mary, B. Modelling carbon and nitrogen dynamics in a bare soil with and without straw incorporation. *Eur. J. Soil Sci.* **2003**, *54*, 555–568. [CrossRef]

46. Reichel, R.; Wei, J.; Islam, M.S.; Schmid, C.; Wissel, H.; Schröder, P.; Brüggemann, N. Potential of wheat straw, spruce sawdust, and lignin as high organic carbon soil amendments to improve agricultural nitrogen retention capacity: An incubation study. *Front. Plant Sci.* **2018**, *9*, 900. [CrossRef] [PubMed]

47. Said-Pullicino, D.; Cucu, M.A.; Sodano, M.; Birk, J.J.; Glaser, B.; Celi, L. Nitrogen immobilization in paddy soils as affected by redox conditions and rice straw incorporation. *Geoderma* **2014**, *228–229*, 44–53. [CrossRef]

48. Fei, C.; Zhang, S.; Wei, W.; Liang, B.; Li, J.; Ding, X. Straw and optimized nitrogen fertilizer decreases phosphorus leaching risks in a long-term greenhouse soil. *J. Soils Sediments* **2019**, *20*, 1199–1207. [CrossRef]

49. Yan, C.; Yan, S.; Hou, Z.; Dong, S.; Gong, Z.; Zhang, Z. Phosphorus adsorption characteristics of soil with rice straw retention in northeast china. *Ekoloji* **2019**, *28*, 1671–1678.

50. Tian, G.; Kang, B.; Brussaard, L. Biological effects of plant residues with contrasting chemical compositions under humid tropical conditions—decomposition and nutrient release. *Soil Biol. Biochem.* **1992**, *24*, 1051–1060. [CrossRef]

51. Wang, J.; Wang, D.; Zhang, G.; Wang, Y.; Wang, C.; Teng, Y.; Christie, P. Nitrogen and phosphorus leaching losses from intensively managed paddy fields with straw retention. *Agric. Water Manag.* **2014**, *141*, 66–73. [CrossRef]

52. Gupta, R.; Singh, Y.; Ladha, J.; Singh, B.; Singh, J.; Singh, G.; Pathak, H. Yield and Phosphorus Transformations in a Rice-Wheat System with Crop Residue and Phosphorus Management. *Soil Sci. Soc. Am. J.* **2007**, *71*, 1500–1507. [CrossRef]

53. Li, F.; Liang, X.; Zhang, H.; Tian, G. The influence of no-till coupled with straw return on soil phosphorus speciation in a two-year rice-fallow practice. *Soil Tillage Res.* **2019**, *195*, 104389. [CrossRef]

54. Guérin, J.; Parent, L.; Abdelhafid, R. Agri-environmental Thresholds using Mehlich III Soil Phosphorus Saturation Index for Vegetables in Histosols. *J. Environ. Qual.* **2007**, *36*, 975–982. [CrossRef]

55. Eriksen, J. Gross sulphur mineralisation–immobilisation turnover in soil amended with plant residues. *Soil Biol. Biochem.* **2005**, *37*, 2216–2224. [CrossRef]

56. Kuzyakov, Y.; Razavi, B.S. Rhizosphere size and shape: Temporal dynamics and spatial stationarity. *Soil Biol. Biochem.* **2019**, *135*, 343–360. [CrossRef]

57. Neumann, G.; Bott, S.; Ohler, M.; Mock, H.-P.; Lippmann, R.; Grosch, R.; Smalla, K. Root exudation and root development of lettuce (Lactuca sativa L. cv. Tizian) as affected by different soils. *Front. Microbiol.* **2014**, *5*, 2. [CrossRef] [PubMed]

58. Mimmo, T.; Pii, Y.; Valentinuzzi, F.; Astolfi, S.; Lehto, N.; Robinson, B.; Brunetto, G.; Terzano, R.; Cesco, S. Nutrient availability in the rhizosphere: A review. *Acta Hortic.* **2018**, *1217*, 13–28. [CrossRef]

 *Nitrogen*

Article

Combination of Inorganic Nitrogen and Organic Soil Amendment Improves Nitrogen Use Efficiency While Reducing Nitrogen Runoff

Ian Phillips [1], Chanyarat Paungfoo-Lonhienne [2,*], Iman Tahmasbian [1], Benjamin Hunter [1], Brianna Smith [1], David Mayer [1] and Matthew Redding [1]

[1] Department of Agriculture and Fisheries, Queensland Government, P.O. Box 102, Toowoomba, QLD 4350, Australia; Ian.Phillips@daf.qld.gov.au (I.P.); Iman.Tahmasbian@daf.qld.gov.au (I.T.); benchunter1@gmail.com (B.H.); Brianna.Smith@daf.qld.gov.au (B.S.); David.Mayer@daf.qld.gov.au (D.M.); Matthew.Redding@daf.qld.gov.au (M.R.)

[2] School of Agriculture and Food Science, University of Queensland, St. Lucia, QLD 4072, Australia

* Correspondence: chanyarat@uq.edu.au; Tel.: +61-04-4878-5658

Abstract: Improved nitrogen fertiliser management and increased nitrogen use efficiency (NUE) can be achieved by synchronising nitrogen (N) availability with plant uptake requirements. Organic materials in conjunction with inorganic fertilisers provide a strategy for supplying plant-available N over the growing season and reducing N loss. This study investigated whether a combined application of inorganic N with an organic soil amendment could improve nitrogen use efficiency by reducing N loss in runoff. Nitrogen runoff from a ryegrass (*Lolium multiflorum*) cover was investigated using a rainfall simulator. Nitrogen was applied at low, medium and high (50, 75 and 100 kg/ha) rates as either $(NH_4)_2SO_4$ or in combination with a poultry manure-based organic material. We showed that the NUE in the combination (58–75%) was two-fold greater than in $(NH_4)_2SO_4$ (24–42%). Furthermore, this combination also resulted in a two-fold lower N runoff compared with the inorganic fertiliser alone. This effect was attributed to the slower rate of N release from the organic amendment relative to the inorganic fertiliser. Here, we demonstrated that the combined use of inorganic and organic N substrates can reduce nutrient losses in surface runoff due to a better synchronisation of N availability with plant uptake requirements.

Keywords: rainfall simulator; nutrient runoff; ammonium; nitrate; nitrogen use efficiency

Citation: Phillips, I.; Paungfoo-Lonhienne, C.; Tahmasbian, I.; Hunter, B.; Smith, B.; Mayer, D.; Redding, M. Combination of Inorganic Nitrogen and Organic Soil Amendment Improves Nitrogen Use Efficiency While Reducing Nitrogen Runoff. *Nitrogen* **2022**, *3*, 58–73. https://doi.org/10.3390/nitrogen3010004

Academic Editor: Jacynthe Dessureault-Rompré

Received: 19 November 2021
Accepted: 18 January 2022
Published: 20 January 2022

Publisher's Note: MDPI stays neutral with regard to jurisdictional claims in published maps and institutional affiliations.

1. Introduction

The efficient use of fertilisers relies on optimising the time of application to meet crop nutrient requirements, and minimising nutrient losses through processes such as leaching and surface water runoff. With the intensification of Australian agriculture in response to increased global food requirements, the use of nitrogen-based fertilisers has concomitantly increased dramatically [1–4]. However, <50% of applied nitrogen (N) is effectively utilised by the growing plant in many cropping situations, and this unutilised N represents a significant economic loss and can pose a high risk of environmental pollution [4–8].

At the same time, the need for developing more agricultural land to meet global food requirements has seen an increased research focus on the role of sandy soils in cropping systems, and on identifying management strategies to overcome cropping constraints [9]. Despite sands being perceived as highly leachable, this characteristic is of less importance in shallow sands underlain by low-permeability layers due to increasing clay content and/or mechanical impedance/compaction [10]. Therefore, any decline in infiltrability or deep drainage can encourage other water loss mechanisms, such as runoff [11].

There has been renewed interest in the use of organic fertilisers for supplying plant nutrients [12,13], and as a feedback mechanism for improving soil health [14]. The term "organic fertiliser" covers a wide range of substrates including manures, composts, and plant stubble and root residues [13,15,16], and these substrates vary considerably in nutrient content in both form and concentration [17]. However, there are issues relating to the slow rate of N release from organic substrates. The subsequent slow kinetics of mineralisation and nitrification rates have brought into question the sole reliance on organic fertilisers for satisfying the plant nutrient requirements [18,19]. Paungfoo-Lonhienne et al. (2019) [20] reported less biomass for kikuyu grass (*Pennisetum clandestinum*) grown in a poultry manure-based organic fertiliser (CropUp^{TM}) compared with an inorganic fertiliser (urea). These researchers suggested that the rate of N supply (including resident NH_4 and NO_3, and mineralised N) from the organic substrate alone was insufficient to meet the N demand of the growing plant. To address this N limitation, plants were grown in soil receiving a combination of 50% urea and 50% CropUp^{TM}. This combination significantly decreased mineral N in leachates compared to urea alone. However, in contrast to leachate N, the soil mineral N levels for the combined (organic + urea N sources) and urea-only treatments were similar, indicating that little N mineralisation from the organic substrate had occurred. Although organic fertilisers are capable of supplying N for plant use, factors such as the organic substrate and soil chemical and physical characteristics, as well as microbial diversity and functionality, play key roles in the mineralisation and nutrient release rates from these organic materials [13,15,16,19].

The retention of inorganic and organic forms of N closer to the soil surface due to reduced leaching and mineralisation rates can expose this nutrient to loss mechanisms such as wind and water erosion, depending on climatic and soil physical conditions. Soluble forms of both organic and inorganic N are the most prone to loss, particularly in surface runoff. As global demand for agricultural land increases, sandy soils with low nutrient and water retention properties are expected to play an increasing role in food production. The combined use of inorganic and organic sources of N may provide one strategy to minimise nutrient loss and maximise N use efficiency. Although the leaching of N supplied as combined inorganic/organic substrates has been investigated [20], losses from runoff, particularly for sandy soil, are scarce [21–23].

The aim of this study was to investigate the suitability of a combined inorganic and organic fertiliser to improve nitrogen use efficiency (NUE) in a sand. Specifically, this study targeted (1) N losses in surface runoff from well-controlled rainfall simulation experiments for a combined (organic + inorganic sources) and an inorganic N-based fertiliser, and (2) laboratory-scale N release from inorganic and organic N sources.

2. Materials and Methods

2.1. Rainfall Simulation Trial

Rainfall runoff trials were undertaken using a nutritionally deficient (low total and available N) coarse-textured sand (Table 1). This material was selected as the growing medium to reflect the potential N runoff loss from a sandy-textured soil, to minimise any potential extraneous N sources for plant uptake (e.g., mineralisation of resident organic N), and to minimise suspended colloid concentrations that could preferentially transport fertiliser N in runoff via ion adsorption. This coarse sandy material therefore maximised the potential N losses in runoff as a means of evaluating the benefits of organic versus inorganic sources of N in maximising plant uptake.

Table 1. Selected properties of the sand. EC = electrical conductivity, ECEC = effective cation exchange capacity.

Parameter	Sand
$pH_{1:5}$	6.15
$EC_{1:5}$ (dS/m)	0.02
Total Carbon (%)	0.22
Total Nitrogen (%)	0.022
Organic Carbon (%)	0.22
NH_4-N (mg/kg)	3
NO_3-N (mg/kg)	<2
Exchangeable Cations ($cmol_+$/kg)	
Ca	0.634
Mg	0.373
K	0.114
Na	<0.080
ECEC	1.12
Particle Size Distribution (%)	
Coarse Sand	62.5
Fine Sand	30.2
Silt	6.3
Clay	3.6
Texture	coarse sand

The sand (total sand $\approx$ 93%) was slightly acidic (pH 6.15), non-saline, and contained extremely low organic carbon (organic C = 0.22%) and nitrogen (total N = 0.022%) concentrations. The effective cation exchange capacity (ECEC = 1.12 $cmol_+$/kg) was very low, with calcium being the dominant exchangeable cation, along with lesser concentrations of magnesium and potassium.

Nitrogen was added as either a poultry-based organic fertiliser (CropUp^{TM}) or as inorganic ammonium sulphate ((NH_4)$_2SO_4$). CropUp^{TM} is a mixture of composted manure, molasses, humates and natural minerals (14.5% zeolite), and is slightly soluble in water (Sustainable Organic Solutions Pty Ltd., Brisbane, QLD, Australia—Safety Data Sheet). CropUp^{TM} contains 3.07% N (as total N; LECO CN Dumas analyser, St Joseph, MI, USA), 23.25% C (as total C; LECO CN Dumas analyser), a C:N ratio of 7.56, 4544.49 mg/kg of available NH_4^+ (2M KCl-extractable; [24]; Method 7C2), 40.67 mg/kg of available NO_3^- (2M KCl-extractable; [24]; Method 7C2), a pH of 8.41 (1:5 CropUp^{TM}:18.2 MΩ deionised water) and an EC of 9.18 dS/m (1:5 CropUp^{TM}:18.2 MΩ deionised water).

Ammonium sulphate, rather than urea [20], was used as the inorganic N source to minimise the potential for NH_3 volatilisation. Inorganic N was applied at rates of 0 (Control), 50 (ASLow), 75 (ASMedium) and 100 (ASHigh) kg N/ha as (NH_4)$_2SO_4$ (Table 2), which reflects the application rates used by previous workers [20]. The poultry manure-based organic material CropUp^{TM} was applied to achieve N rates of 25 (CULow), 37.5 (CUMedium) and 50 (CUHigh) kg/ha, and was supplemented with (NH_4)$_2SO_4$ (25, 37.5 and 50 kg N/ha, respectively) to match the amount of N added in inorganic N treatments, and to ensure that N availability was not limited during the plant establishment and growth phases.

Table 2. Treatments used in the rainfall simulation experiment. "AS" indicates inorganic N source $(NH_4)_2SO_4$ and "CU" indicates combined $(NH_4)_2SO_4 + CropUp^{TM}$.

Treatment	$(NH_4)_2SO_4$ (kg/ha)	CropUp (kg/ha)	$(NH_4)_2SO_4$ (g/kg Soil)	CropUp (g/kg Soil)
Control	0	0	0.000	0.000
ASLow	50	0	0.546	0.000
CULow	25	25	0.273	2.593
ASMedium	75	0	0.819	0.000
CUMedium	37.5	37.5	0.409	3.890
ASHigh	100	0	1.092	0.000
CUHigh	50	50	0.546	5.187

The sand was packed into stainless steel trays ($1045 \times 457 \times 40$ mm; $n = 3$) to achieve an approximate bulk density (ρ_b) of 1110 kg/m^3. The sand was initially packed to a height of 30 mm, and the various treatments (Table 2) were uniformly surface applied. The treatments were covered with an additional 7 mm of sand, and lightly compacted to produce a relatively uniform surface. Ryegrass (*Lolium multiflorum*) seed was spread across the sand surface at a rate, equivalent on a surface area basis, of 200 kg/ha. The grass seed was covered with an additional 3 mm of sand, and the soil tray was slowly moistened with water. Water was applied to achieve an approximate gravimetric water content (θ_g) corresponding to 60% of field capacity (θ_{fc}). The soil trays were maintained at this moisture content for a period of 42 days prior to undertaking rainfall simulation trials.

To minimise the likelihood of nutrient deficiencies limiting the ryegrass growth, each tray received a basal nutrient application equivalent, on a surface area basis, to 20 kg P/ha, 100 kg K/ha, 28 kg Mg/ha, 70 kg S/ha, 0.43 kg Cu/ha, 0.84 kg Zn/ha, 7.7 kg Mn/ha, 0.97 kg B/ha, 0.33 kg Mo/ha and 30 kg Ca/ha.

Nutrient runoff from the treated soil trays was generated using a rainfall simulator built in accordance with published specifications [25], using a similar procedure described by [26]. The rainfall simulator was positioned centrally over two flumes, and the simulation was conducted at a nozzle pressure of 28 kPa (the design pressure required to deliver a rainfall intensity of 70 mm/h) over a runoff period of 20 min.

Prior to commencing each simulation run, the soil surface was photographed as a measure of grass cover, as this parameter can influence surface runoff rates, and hence the nutrient loading of the runoff water. The photographs were taken orthogonally to the soil surface under uniform light conditions. The software was written in Python using the OpenCV library (Python Software Foundation, 2019) to enable a uniform set of pixel colours (defined by hue, saturation, and value, the standard HSV digital colour space) to be selected as either soil or plant matter across all the images after gamma-balancing each image. Plant coverage was calculated from the ratio of background to plant pixel counts and validated automatically via the total percentage of area covered by soil and the percentage of area covered by plant pixels. The set of pixel characteristics selected was modified to minimise discrepancies across these three methods iteratively, and then applied uniformly across all images.

After being placed in the flume, each treatment tray was manually wet up to saturation prior to commencing rainfall. A sample of rainfall water was collected for analysis as outlined below for the runoff samples. A composite sample of the water exiting the flume of each treatment was automatically collected (WS750 water sampler, Global Water Instrumentation Inc., College Station, TX, USA) at the commencement of flow and subsequently every 5 min (100 mL aliquot composited for each 5 min collection event). The height of the water at the flume discharge point was measured at each 5 min sampling period, and the volume flow rate (V) was calculated using [27]:

$$V = 341 \times H^{2.31} \tag{1}$$

where V = volume flow rate (L/s) and H = head of water (m).

At the end of each simulation period, eight cores (internal diameter = 110 mm) were removed from the treatment tray. The cumulative area of the eight cores was, on a surface area basis, approximately 8% of the treatment area. The composited sand core materials (soil + grass) were thoroughly mixed and then stored at <4 °C prior to analysis.

The runoff water, soil and plant materials were analysed as follows. An aliquot of each sample of runoff water was initially filtered (<0.45 μm), and samples of unfiltered and filtered water were analysed for total N (APHA 5310B). The filtered samples were also analysed colorimetrically for NH_4^+ [28] and $NO_3^- + NO_2^-$ using a modified Griess method [29] with a microtiter plate reader (BioTek EPOCH^2 Microplate Reader) at a wavelength of 625 and 540 nm, respectively. Runoff nutrient and particulate concentrations were converted to mass loss to account for small variations in flow rate between the two flumes using Equation (1).

Sub-samples of soil ($n = 3$) were analysed for mineral N (2M KCl extractable; [24]; Method 7C2). Plant material (above- and below-ground material) was separated from the soil by washing with Milli-Q deionised water. The retained plant material was oven-dried at 60 °C, weighed to estimate the dry matter (DM for whole plant biomass), and ground prior to analysis for total N and total C by high-temperature combustion (LECO CN Analyser), and for aluminium, boron, calcium, copper, iron, potassium, magnesium, manganese, sodium, phosphorus, sulphur and zinc by nitric acid digestion and ICPOES.

2.2. Nutrient Release Trial

This experiment was a laboratory-scale leaching study to investigate N mineralisation kinetics using coarse-textured sand (Table 1). Nitrogen was added at a rate of 291 kg/ha as either CropUp^{TM} (equivalent to 8.675 mg/g soil) or as ammonium chloride (NH_4Cl) (equivalent to 0.9274 mg/g soil). This rate is higher than that used in the runoff trial (50–100 kg N/ha) to ensure that effects of dissolution and transformation were measurable. Sand with no N applied served as a control. Leaching was undertaken in 50 mL polypropylene centrifuge tubes with a small hole (5 mm diameter) drilled through the base to allow for drainage of the leachate. A disc of Whatman glass fibre paper was placed inside the tube to cover the hole so as to avoid soil loss during leaching.

Air-dried (<2 mm) sand (equivalent to 40 g on an oven-dry weight basis) was placed into the respective centrifuge tube. Initially, approximately 80–85% of the total soil mass was added. The respective treatments were then added to the sand surface, with CropUp^{TM} as solid material and NH_4Cl in 1 mL of solution. The remaining 15–20% of the sand was then placed above the soil/treatment interface. The sand was lightly compacted by dropping the tube 10–15 times vertically from a 2 cm height. This packing procedure resulted in the treatments residing ≈10 mm below the sand surface.

Each sand column was leached with the equivalent of 1.5 pore volumes of 0.005 M $CaCl_2$ (≈12.7 mL) on days 1, 3, 10, 17, 24, 33 and 42 of the study. Dilute $CaCl_2$ was chosen, rather than water, as the leaching solution to simulate the soil solution's ionic composition and ionic strength, and to minimise the likelihood of, albeit a low amount of, mobilised clay clogging the filter disc.

Leaching occurred over a 2–3 h period (under gravity/free drainage), and the leachates were collected in 70 mL polypropylene containers. After drainage ceased, each leaching tube was placed under a slight vacuum for 3 to 5 s to remove excessive solution from the base of the tube and to avoid generating anaerobic conditions. The drainage collected under vacuum was added to the gravity drainage, and the volume of the leachate was estimated by weighing.

All the leachates were filtered (<0.45 μm cellulose acetate filter membranes) and stored frozen before being analysed colormetrically for NO_x and NH_4 using a microtiter plate reader (BioTek EPOCH^2 Microplate Reader) at a wavelength of 625 and 540 nm, respectively. Following each leaching event, the tubes were re-capped (lids remained loose to maintain aerobic conditions) and incubated at 25 °C.

After the final leaching (day 42), the remaining available mineral N was extracted from the sand using 2 M KCl (2 h end-over-end shaking at a 1:5 soil-to-solution ratio). After extraction, the supernatant was removed, centrifuged (10,000 rpm for 20 min), filtered (<0.45 µm) and stored frozen prior to NO_x and NH_4 analysis. The concentrations of NO_x and NH_4 were corrected by subtracting the amount of each ion retained in the entrained solution.

2.3. Statistical Analysis

General linear mixed models (GLMs) were used to analyse the rainfall simulator data, using restricted maximum likelihood (REML) in GenStat (2018). Residual plots were used to confirm the assumptions of homogeneous variances and low skewness, with flow rate as the standardising covariate. The treatments were initially analysed as 13 discrete levels, and subsequently as the factorial structure of N-rates by N-sources. A post hoc comparison between the adjusted means was performed using protected least significant difference testing at a significance level of 5% ($p < 0.05$). Nitrogen release data were analysed using analysis of variance (general linear model; *Statistix* version 10). The significant differences among the main treatments were separated by LSD ($p < 0.05$). The relationships between cumulative leachate NH_4 and NO_x with time were adequately described (r^2 for NH_4 = 0.85–0.95 and r^2 for NO_x = 0.75–0.89) by an equation of the form:

$$N_i = k \times \ln t + N_0 \tag{2}$$

where N_i = leachate N (mg NH_4 or NO_x) at time i, t = time (days), N_0 = leachate N (mg NH_4 or NO_x) at time 0 and k = first-order rate coefficient (d^{-1}).

3. Results

A rainfall simulation trial was undertaken to evaluate the impacts of combined inorganic and organic fertiliser on nitrogen runoff losses compared with conventional inorganic fertiliser ((NH_4)$_2$$SO_4$) alone. A nitrogen release experiment was concurrently undertaken to investigate N mineralisation kinetics.

3.1. Nitrogen Release from Organic and Inorganic Sources

A leaching experiment was undertaken to investigate the N mineralisation kinetics in the coarse-textured sand used in the rainfall simulation trial. The results show that negligible NH_4 and NO_x were leached over the 42-day period in the control treatment (no N applied), confirming the very low mineral N status of the sand (Figure 1). The leachate from the CropUp^{TM} treatment contained very low amounts of NH_4 (<0.7 mg), which were often not significantly ($p < 0.05$) different from that found in the control. For example, leachate NH_4 initially increased from 0.09 mg on day 1 to 0.67 mg on day 3, after which leachate NH_4 decreased steadily to background (control) levels. Leachate NO_x from CropUp^{TM} remained very low over the initial 3 days, but increased to 0.17 mg by day 10, and 0.28 mg by day 17. By day 24, the amount of NO_x in the leachate was not significantly different from that found in the control.

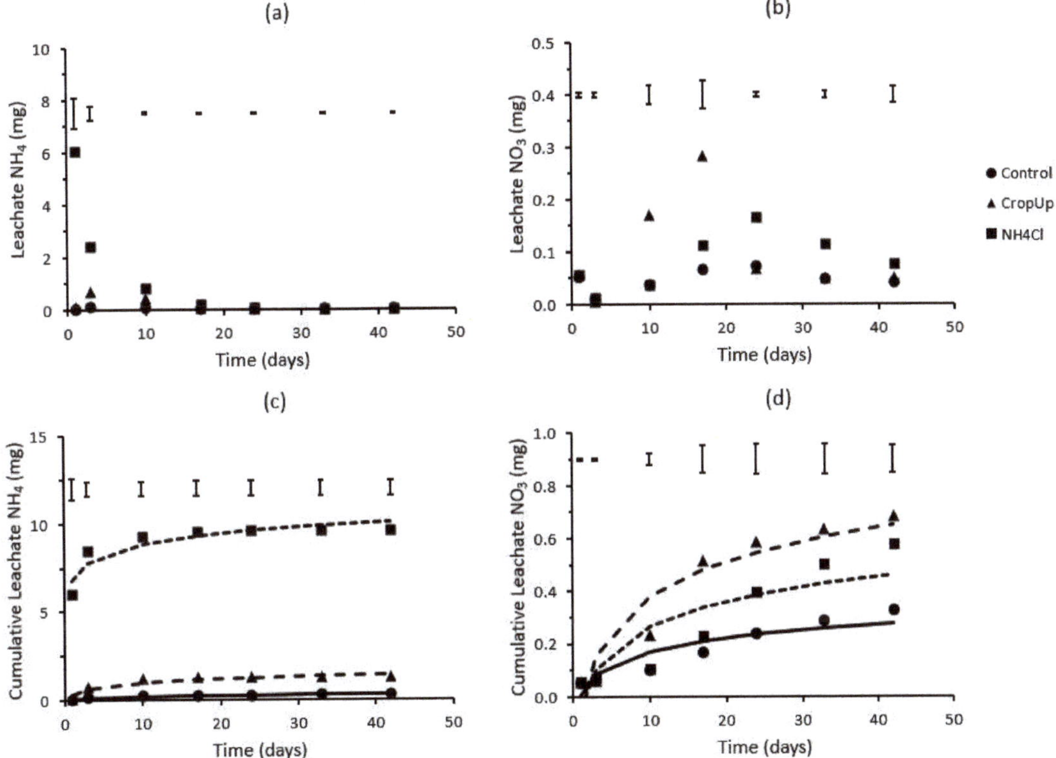

Figure 1. Leachate (**a**) NH_4 and (**b**) NO_x, and cumulative leachate (**c**) NH_4 and (**d**) NO_x as a function of time. Bars represent LSD values ($p < 0.05$). The solid and dashed lines in (**c**) and (**d**) were predicted using Equation (2).

The majority (>95%) of the mineral N leached from the NH_4Cl treatment was in the NH_4 form with little conversion to NO_x (Figure 1). Of this NH_4, $\approx$96% leached from the sand column within the initial 10 days of the study (corresponding to 2–3 pore volumes of leachate). Levels of NO_x ranging from 0.11 to 0.16 mg were measured between day 17 and day 24 (corresponding to 3–4 pore volumes), but declined to background levels with increased leaching. The distributions of NH_4 and NO_x in the leachate showed that the majority of NH_4 was readily leached from the sand, but a small amount of residual NH_4 was nitrified during the later stage of the study.

Over the 42-day incubation period, 0.26, 1.26 and 9.64 mg of NH_4 and 0.33, 0.68 and 0.58 of NO_x leached from the control, CropUp^{TM} and NH_4Cl treatments, respectively (Figure 1c,d). The amount of cumulative mineral N (NH_4 + NO_x) that leached from the control, CropUp^{TM} and NH_4Cl treatments was 0.59, 1.94 and 10.22 mg, respectively (Table 3). The ammonium and NO_3 release rate coefficients (Equation (2)) for the control, CropUp^{TM} and NH_4Cl treatments were 0.062, 0.302 and 0.884 d^{-1}, and 0.073, 0.187 and 0.137 d^{-1}, respectively.

After 42 days, much of the (2M KCl-extractable) mineral N was present as NH_4, with negligible NO_x, irrespective of the treatment (Figure 2). Mean NH_4 concentrations ranged from 12.49 mg/kg in the control to 13.60 mg/kg in the NH_4Cl treatment, but these were not significantly different from each other. Mean NO_x concentrations ranged from $\approx$0 mg/kg in the control and NH_4Cl treatments to 1.05 mg/kg in the CropUp^{TM} treatment, with the majority of extracted NO_x present in the pore water.

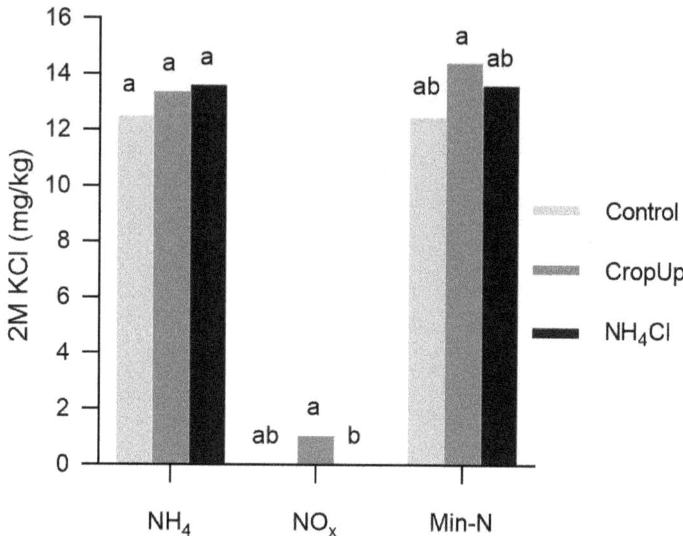

Figure 2. Mean 2M KCl-extractable NH$_4$ and NO$_x$, and mineral N (NH$_4$ + NO$_x$) concentrations for the control, CropUp^{TM} and NH$_4$Cl treatments. Parameters with the same letter are not significantly different at $p < 0.05$ from each other.

The amounts of mineral N (i.e., NH$_4$ + NO$_x$) recovered for the control, CropUp^{TM} and NH$_4$Cl treatments were 1.14, 2.58 and 10.84 mg, respectively (Table 3). Of these amounts, 2M KCl-extractable NH$_4$ was $\approx$0.5 mg for all treatments, suggesting that little to no NH$_4$ from either CropUp^{TM} or NH$_4$Cl was retained by the sand. This small quantity of NH$_4$ may represent a background concentration that cannot be displaced from the cation exchange sites by K-NH$_4$ or Ca-NH$_4$ exchange.

3.2. Rainfall Runoff Trial

The plant dry matter displayed a strong positive, linear relationship with the plant cover (Figure 3; *plant cover* = 0.626*DM* + 1.95; r^2 = 0.94). The treatments receiving inorganic N fertiliser in the form of (NH$_4$)$_2$SO$_4$ (ASLow, ASMedium and ASHigh) contained significantly lower plant dry matter and percentage cover than the treatments receiving the combined ((NH$_4$)$_2$SO$_4$ + CropUp^{TM}) fertiliser (CULow, CUMedium and CUHigh). The ammonium sulphate treatments produced a plant cover of approximately 40–45%, while the combined (NH$_4$)$_2$SO$_4$ + CropUp^{TM} produced plant covers ranging from 70 to 90% (Figure 3). The lower plant cover was consistent with the relatively poor ryegrass (*Lolium multiflorum*) seed germination observed during the 42-day growth stage.

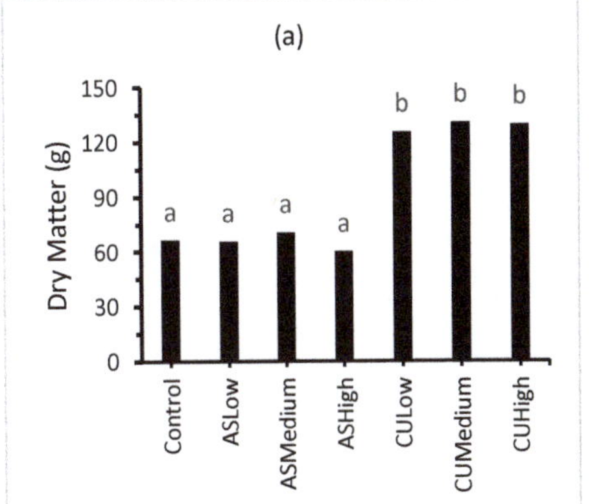

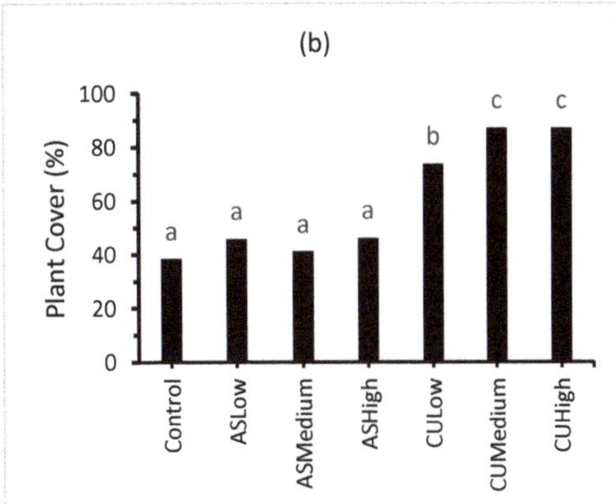

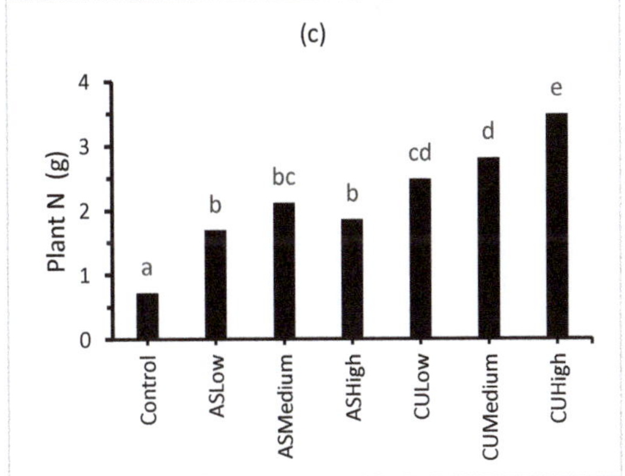

Figure 3. Effect of treatment on: (**a**) percentage plant cover; (**b**) plant biomass; and (**c**) plant total N. Bars represent the mean of three replicates ($n = 3$) and bars designated by the same letter are not significantly different at ($p < 0.05$).

The ryegrass treatments (CULow, CUMedium and CUHigh) contained significantly ($p < 0.05$) more plant C and plant N than ASLow, ASMedium and ASHigh (Table 4 and Figure 3). The plant macronutrient (P, K, Ca, Mg and S) and micronutrient (Al, B, Fe, Mn, Na and Zn) content was significantly higher for the ryegrass grown in the presence of CropUp^{TM} compared with the control and $(NH_4)_2SO_4$ treatments (Table 4).

Table 3. Distribution of NH_4, NO_x and mineral N in the soil and leachate, and percentage recovery of added N.

Treatment	2M KCl (mg)		Leachate (mg)		Sum (mg)		Min-N (mg)	Increase in (mg)			% Recovery		
	NH_4	NO_x	NH_4	NO_x	NH_4	NO_x		NH_4	NO_x	Min-N	NH_4	NO_x	Min-N
Control	0.511	0.037	0.258	0.329	0.770	0.366	1.136						
CropUp[TM]	0.547	0.088	1.260	0.683	1.808	0.771	2.579	1.038 ± 0.106	0.405 ± 0.145	1.443 ± 0.155	10.684	4.171	14.855
NH_4Cl	0.574	0.049	9.641	0.577	10.215	0.626	10.841	9.446 ± 0.937	0.259 ± 0.012	9.705 ± 0.939	97.220	2.672	99.892

Table 4. Nutrient content in the grass cover (Lolium multiflorum) after the 42-day growing period for each treatment. Parameters designated by the same letter are not significantly different at $p < 0.05$.

Treatment	TC	TN	Al	B	Ca	Cu	Fe	K	Mg	Mn	Na	P	S	Zn	NUE *
							(g)								(%)
Control	23565 b	691 e	203.5 b	0.40 c	189.69 c	6.04 a	213.5 b	208.8 b	84.1 b	7.08 b	35.97 b	58.61 d	70.0 f	78.9 d	-
ASLow	26643 b	1690 d	236.5 b	0.39 c	143.86 cd	4.23 b	200.2 b	313.9 b	70.7 bc	6.60 b	29.63 bc	74.69 cd	157.6 e	206.7 c	42
CULow	46982 a	2480 bc	392.7 a	0.84 b	432.60 b	5.99 a	374.8 a	721.2 a	150.8 a	13.59 a	58.50 a	164.79 b	225.4 bc	296.9 b	75
ASMedium	28696 b	2112 cd	236.0 b	0.40 c	137.39 cd	4.31 b	203.7 b	418.8 ab	72.9 bc	7.04 b	31.90 bc	97.35 c	197.6 cd	317.6 ab	40
CUMedium	48472 a	2805 bc	438.0 a	0.85 b	418.99 b	6.12 a	400.6 a	462.5 ab	153.6 a	13.70 a	60.23 a	181.34 b	254.4 b	325.3 ab	59
ASHigh	24442 b	1856 d	201.6 b	0.34 c	119.09 d	3.42 b	189.2 b	426.6 ab	62.5 c	6.33 b	27.04 c	86.12 cd	172.1 de	284.8 b	24
CUHigh	50864 a	3477 a	382.9 a	1.00 a	496.22 a	6.24 a	376.1 a	628.1 b	164.0 a	15.22 a	58.72 a	220.23 a	317.4 a	380.4 a	58

* NUE = nitrogen use efficiency calculated as: ((Plant N per treatment − Plant N in control) ÷ N added in Fertiliser) × 100 [30–32].

The nitrogen use efficiency (NUE) was estimated as the proportion of applied N taken up by the ryegrass (Table 4). The NUE for the $(NH_4)_2SO_4$-only treatments ranged from 58% to 75% in the $(NH_4)_2SO_4$ + CropUp^{TM} treatments, and was two-fold greater than for the $(NH_4)_2SO_4$) treatment (24–42%).

The masses of total N, NH_4, NO_x and mineral N (NH_4 + NO_x) in the runoff from the Control, ASLow, ASMedium, ASHigh, CULow, CUMedium and CUHigh treatments are presented in Figure 4. The masses of runoff total N, NH_4 and mineral N for the inorganic N fertiliser ($(NH_4)_2SO_4$) were higher ($p < 0.05$) than for the $(NH_4)_2SO_4$ + CropUp^{TM} treatments. For example, $\geq$200 mg of total N was collected for the inorganic N treatments compared with $\approx$100 mg for the $(NH_4)_2SO_4$ + CropUp^{TM} treatments, which represented a two-fold reduction in N runoff. The runoff NO_3 content was very low for all treatments.

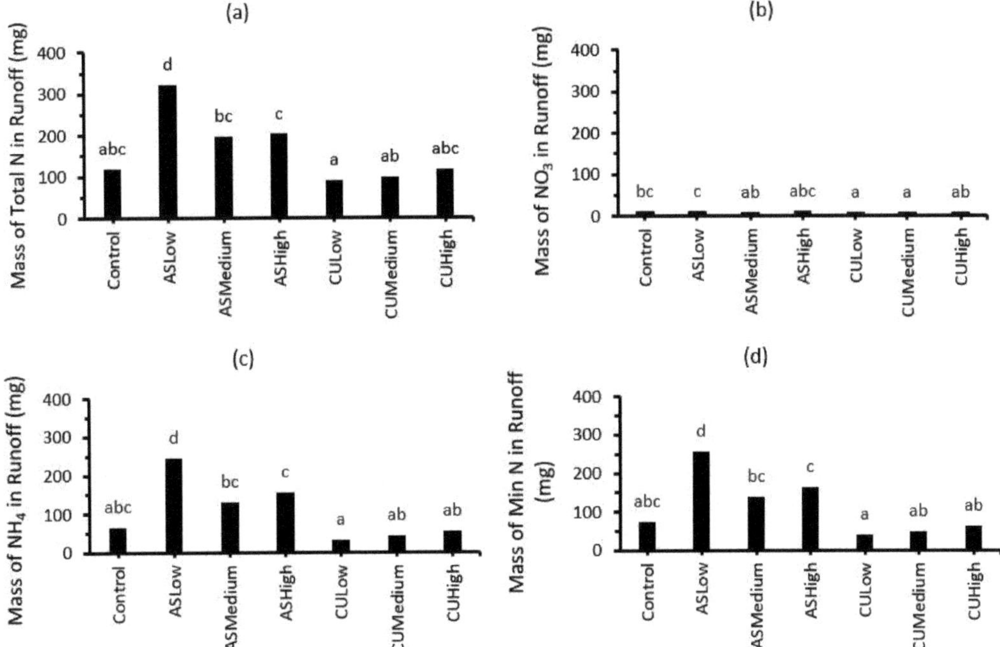

Figure 4. Masses of (**a**) total N, (**b**) NO_x, (**c**) NH_4 and (**d**) mineral N (NH_4 + NO_x) in runoff from the Control, ASLow, ASMed, ASHigh, CULow, CUMed and CUHigh treatments. Bars represent the mean of three replicates (n = 3). Same lowercase alphabets indicate statistically non-significant differences between treatments at $p < 0.05$.

4. Discussion

4.1. Nitrogen Release from Organic and Inorganic Sources

Leaching with dilute 0.005 M $CaCl_2$ displaced >99% of the N added as NH_4Cl but only 15% of that added as CropUp^{TM}, with NH_4 being the dominant form of mineral N in the leachate. This low recovery of mineral N from CropUp^{TM} results from the slow rate of nutrient release from this organic material (i.e., 0.3 d^{-1}) relative to readily available inorganic substrates (i.e., 0.9 d^{-1}). Importantly, this slow rate of release of N from CropUp^{TM} would be highly beneficial for the nutrient dynamics in terms of optimising N availability relative to plant uptake requirements and reducing the N loading in the soil solution where it is susceptible to leaching and runoff loss, such as in sandy soils. Furthermore, although the leachate NO_3 concentrations were very low for all N substrates, CropUp^{TM} tended to encourage nitrification relative to the control and NH_4Cl treatments, possibly through the introduction of nitrifying bacteria within its organic substrate, or by stimulating the resident, albeit limited, nitrifying bacteria population within the sand. Stimulating the microbial community in terms of functionality and diversity is highly beneficial in sandy soil, particularly in terms of nutrient cycling and immobilisation, which can act as a mechanism for nutrient storage and subsequent release, and for reducing leaching losses in low-cation exchange capacity (CEC) materials [33–35].

The loss of mineral N primarily as NH_4 demonstrates the sand's poor ability to retain this cation, thereby encouraging potential losses through leaching and runoff. The cation exchange capacity (CEC) of the sand is 1.1 $cmol_+$/kg (equivalent to 6.16 mg of negative charge per 40 g of sand expressed as N (Table 1)). The 0.005M $CaCl_2$ leaching solution provided a ready supply of Ca capable of competing with NH_4 for the limited cation exchange sites. Over the leaching study, approximately 0.35 mmoles of Ca was added to the sand, being equivalent to ≈10 mg of N. This mass of Ca exceeded the amount of negative charge (6.16 mg) available to retain the 9 mg of N added in the NH_4Cl and CropUp^{TM} treatments. Furthermore, the rapid loss of NH_4 from the sand during the early leaching events suggests that the sand's cation exchange sites may have a stronger preference for Ca relative to NH_4, thereby favouring NH_4 displacement and leaching with the 0.005 M $CaCl_2$ solution. Although this effect was more pronounced for the NH_4Cl treatment, similar behaviour may be expected for the CropUp^{TM} treatment. Therefore, the extremely low CEC of the sand coupled with a higher preference for Ca may explain the similar 2M KCl-NH_4 concentrations measured between treatments (i.e., 0.5 $cmol_+$/kg). The poor NH_4 retention by sandy soils would not only encourage N leaching, but would also result in a loss of this nutrient in runoff and lateral soil water flow (see below Rainfall runoff trial).

The mass of mineral N recovered from the CropUp^{TM} and NH_4Cl treatments in excess of the control was 1.04 and 9.45 mg, respectively (Table 3). Given each treatment received 9.72 mg of N, the percentage recovery of N added to the CropUp^{TM} and NH_4Cl treatments was approximately 15% and 100%, respectively. This suggests that nearly all of the N added as NH_4Cl but only 15% of that added in CropUp^{TM} was readily available over the 42-day study (Table 3). The mineralisation rates for pelleted poultry manure have been reported to be approximately 10%, 23% and 36% of initial N (total N = 2–4%) after 1, 4 and 8 weeks' incubation (25 °C), respectively [36]. These mineralisation (or N release) rates are not dissimilar to those reported in this study and show CropUp^{TM} as a supplementary N source to inorganic fertiliser, which would have been most beneficial for ryegrass establishment and sustainability during the latter stage of the 42-day growth period when inorganic N sources had been depleted and the grass growth rates were high.

CropUp^{TM} contains approximately 4500 and 40 mg/kg of 2M KCl extractable NH_4 and NO_x, respectively. Of the 9.72 mg total N added in 0.347 g of CropUp^{TM}, 1.56 mg was as NH_4, and 0.01 mg as NO_x. Therefore, approximately 8.1 mg of N was present in non-mineral forms, probably as organic N. Over the 42-day leaching study, approximately 1.44 mg of mineral N ($NH_4 + NO_x$) was recovered (>92%; Table 3), which was very similar to that added in CropUp^{TM}. This shows that leaching with 0.005 M $CaCl_2$ effectively removed mineral N from CropUp^{TM} with negligible contributions from other fractions, such as organic matter. CropUp^{TM} has a C:N ratio of 7.6 and a total N content of 3%, which may be expected to encourage the mineralisation of organic N to NH_4 [37]. The sand may therefore lack a significantly large functional microbial community capable of organic matter mineralisation and subsequent nitrification. Limited nitrification in the sand is supported by the absence of NO_3 in the NH_4Cl treatment (Figure 1). However, organic inputs to sands have been shown to steadily increase organic matter and to alleviate poor microbial activity [12,19,33].

4.2. Rainfall Runoff Trial

The addition of CropUp^{TM} (in conjunction with $(NH_4)_2SO_4$) improved plant growth, plant cover and, consequently, N uptake relative to the treatments based on $(NH_4)_2SO_4$ alone. Importantly, a higher uptake effectively removed N from the soil pore water where it was more susceptible to loss through runoff. The reason for the lower N uptake and lower plant cover for the inorganic N treatments is unclear, but may be a consequence of: (a) the high solubility and mobility of mineral N following the solubilisation of $(NH_4)_2SO_4$ (i.e., 76.4 g/100 g water; [38] relative to CropUp^{TM} (Figure 1)); (b) the possibility that, in the CULow, CUMedium and CUHigh treatments, some of the NH_4 released from $(NH_4)_2SO_4$ may have subsequently been retained by CropUp^{TM}; and/or (c) the plant was able to extract nutrients from CropUp^{TM} over the growing period.

First, following dissolution, soluble NH_4 is highly mobile due to the very low CEC of the sand (Table 1 and Figure 1), and this cation (along with other essential nutrients added to the basal solution) may have been transported with surface-applied water deeper into the soil tray and away from the ryegrass seeds and emerging plant roots. If the availability of N and other nutrients that are critical for seedling survival was inadequate, then poor grass establishment and survival would be expected. This may explain the poor establishment growth and percent coverage of the ryegrass cover for the treatments receiving $(NH_4)_2SO_4$ alone.

Second, the inclusion of ($\approx$14.5% by weight) zeolite in CropUp^{TM}, coupled with slower dissolution rates, may have enabled the retention of some of the solubilised NH_4 closer to the point of soil incorporation, and hence closer to the developing roots. The effect of the added zeolite on NH_4 behaviour was not investigated specifically in this study but based on the strong preference of zeolite for NH_4 [39,40]. A proportion of the added N may have been preferentially adsorbed and removed from the pore water over the 42-day growth period. If the zeolite retained a proportion of the added NH_4 close to the source, and a proportion of this cation was subsequently available for plant uptake, this may have aided higher plant growth in the treatments that received CropUp^{TM}.

Third, plants have strategies to enhance nutrient uptake from the rhizosphere, such as by exuding organic acidic anions (citrate, malate and carboxylates) from plant roots [41–44], by the release of H ions from the roots to maintain cation/anion balance during nutrient uptake [42,45], or by stimulating nutrient mineralisation of the organic substrate [46–48]. The increased availability of one nutrient (e.g., P) by plant root exudates can cause a concomitant increase in the availability of co-precipitated or complexed nutrients (e.g., Mn) [43]. Although the reason for the enhanced nutrient uptake in the presence of CropUp^{TM} was not investigated in this study, other researchers have reported that ryegrass (*Lolium rigidum* and *Lolium perenne* L.) roots can encourage the dissolution of phosphate and other nutrients from sparingly soluble inorganic and organic soil fractions roots due to proton excretion [44,45].

Given the slow release of N from CropUp^{TM} (Figure 1), much of the mineral N measured in the runoff from the $(NH_4)_2SO_4$ + CropUp^{TM} treatments most likely originated from the inorganic $(NH_4)_2SO_4$ component, which had not been taken up by the ryegrass during the growth period. Furthermore, the loss from the treatments receiving CropUp^{TM} was not significantly different from that from the control, suggesting that N had been released from CropUp^{TM} at a rate that matched the plant N uptake requirements. This match between N availability and plant uptake is supported by the significantly ($p < 0.05$) higher N content of the ryegrass grown in the presence of CropUp^{TM} (Figure 3). Therefore, the slower release rate of N in the presence of an established plant cover clearly demonstrates the beneficial effect of CropUp^{TM} on increasing plant N uptake and reducing N runoff losses compared with inorganic N fertiliser alone. The higher plant uptake of macronutrients and micronutrients demonstrates that CropUp^{TM} may also act as a slow-release fertiliser.

The nitrogen use efficiency (NUE) was higher in the treatments incorporating CropUp^{TM}, demonstrating the benefits of organic materials in N fertiliser management. The NUE values for the ryegrass that received inorganic N fertiliser only are similar to the NUE reported for the urea N applied to the ryegrass, being 33–47% for N rates of 17–50 kg/ha [49]. The recovery of fertiliser N as determined from the NUE of major agricultural systems in Australia ranges from 28% in sugarcane to 45–62% for irrigated cotton and 78% for dairy [30], compared with an estimated global NUE of 33% [32]. Clearly, the inclusion of organic materials as an N source can be highly beneficial for increasing NUE in pasture and cropping systems, and represents a key strategy to better manage N in the environment with the intensification of agriculture [31,32,50,51]. Importantly, the inclusion of organic materials capable of supplying plant nutrients may represent one step in the development of cropping systems on marginal soils such as sands.

Our findings extend previous research on the mechanisms involved in the beneficial effects of combined organic and inorganic fertilisers. Several studies have shown that the addition of organic fertiliser to inorganic fertiliser has a positive effect on soil chemical and biological properties [52,53]. Wen et al. (2016) [54] showed that this fertilisation regime increases nutrient uptake via stimulating root growth. It also reduces N leaching and enhances denitrifier activity [55]. Here, we provide further evidence that this combination benefits plant N use efficiency via reduced N runoff losses. This has important implications particularly for increased NUE of farming systems in high rainfall geographical locations.

5. Conclusions

The combination of inorganic and organic N substrates reduced the total and mineral N runoff losses compared with inorganic N alone in coarse-textured sand. The higher N uptake by the plant cover provides strong evidence that the combined use of inorganic and organic sources has the potential for increasing NUE by reducing fertiliser losses in surface runoff and synchronising better the N availability with plant requirements. Inorganic N fertiliser can be applied at low rates to provide sufficient N at the early growth stage, with organic substrates, such as composted manures, supplying N at later stages, particularly when the plant growth rate is high.

Furthermore, the potentially slower release of N from organic materials such as CropUp^{TM} may reduce the impacts of other loss pathways (e.g., leaching and sub-surface lateral flow) relative to highly soluble inorganic N fertilisers, particularly when the release rate better matches plant N uptake requirements. These organic materials have widespread applications to sandy soils not only in agriculture, but also in other activities that involve a high N application to sand, such as golf courses and sporting ovals.

Author Contributions: Conceptualisation, I.P., C.P.-L. and M.R.; methodology, I.P., M.R. and D.M.; formal analysis, I.P. and D.M.; investigation, I.P.; resources, C.P.-L., I.T., B.S. and B.H.; data curation, I.P.; writing—original draft preparation, I.P.; writing—review and editing, I.P., C.P.-L., M.R., I.T., B.S., D.M. and B.H.; supervision, M.R.; funding acquisition, C.P.-L. All authors have read and agreed to the published version of the manuscript.

Funding: This research was funded by the Australian Government's Cooperative Research Centres Projects, grant number CRCPFIVE000015.

Institutional Review Board Statement: Not applicable.

Informed Consent Statement: Not applicable.

Data Availability Statement: Not applicable.

Acknowledgments: We would like to thank the Cooperative Research Centres Projects (grant number CRCPFIVE000015) for funding this project.

Conflicts of Interest: The authors declare no conflict of interest.

References

1. Cassman, K.G.; Dobermann, A. Nitrogen and the Future of Agriculture: 20 Years On. *Ambio* **2022**, 17–24. [CrossRef]
2. Chen, D.; Suter, H.C.; Islam, A.; Edis, R.; Freney, J.; Walker, C. Prospects of Improving Efficiency of Fertiliser Nitrogen in Australian Agriculture: A Review of Enhanced Efficiency Fertilisers. *Aust. J. Soil Res.* **2008**, *46*, 289–301. [CrossRef]
3. Conant, R.T.; Berdanier, A.B.; Grace, P.R. Patterns and Trends in Nitrogen Use and Nitrogen Recovery Efficiency in World Agriculture. *Glob. Biogeochem. Cycles* **2013**, *27*, 558–566. [CrossRef]
4. Lassaletta, L.; Billen, G.; Grizzetti, B.; Anglade, J.; Garnier, J. 50 Year Trends in Nitrogen Use Efficiency of World Cropping Systems: The Relationship between Yield and Nitrogen Input to Cropland. *Environ. Res. Lett.* **2014**, *9*, 105011. [CrossRef]

5. Thorburn, P.; Biggs, J.; Attard, S.; Kemei, J. Environmental Impacts of Irrigated Sugarcane Production: Nitrogen Lost through Runoff and Leaching. *Agric. Ecosyst. Environ.* **2011**, *144*, 1–12. [CrossRef]

6. Timilsena, Y.P.; Adhikari, R.; Casey, P.; Muster, T.; Gill, H.; Adhikari, B. Enhanced Efficiency Fertilisers: A Review of Formulation and Nutrient Release Patterns. *J. Sci. Food Agric.* **2015**, *95*, 1131–1142. [CrossRef] [PubMed]

7. Webster, A.; Bartley, R.; Armour, J.; Brodie, J.; Thorburn, P. Reducing Dissolved Inorganic Nitrogen in Surface Runoff Water from Sugarcane Production Systems. *Mar. Pollut. Bull.* **2012**, *65*, 128–135. [CrossRef]

8. Zerulla, W.; Barth, T.; Dressel, J.; Erhardt, K.; von Locquenghien, K.H.; Pasda, G.; Rädle, M.; Wissemeier, A. 3, 4-Dimethylpyrazole Phosphate (DMPP)–a New Nitrification Inhibitor for Agriculture and Horticulture. *Biol. Fertil. Soils* **2001**, *34*, 79–84. [CrossRef]

9. Leonard, E. *Spread, Delve, Spade, Invert: A Best Practice Guide to the Addition of Clay to Sandy Soils*; Grains Research and Development Corporation: Barton, ACT, Australia, 2011; ISBN 1-921779-27-6.

10. Hamblin, A.; Richards, Q.; Blake, J. Crop Growth across a Toposequence Controlled by Depth of Sand over Clay. *Soil Res.* **1988**, *26*, 623–635. [CrossRef]

11. Wu, Y.; Hasi, E.; Wu, X. Characteristics of Surface Runoff in a Sandy Area in Southern Mu Us Sandy Land. *Chin. Sci. Bull.* **2012**, *57*, 270–275. [CrossRef]

12. Bolan, N.S.; Szogi, A.A.; Chuasavathi, T.; Seshadri, B.; Rothrock, M.J., Jr.; Panneerselvam, P. Uses and Management of Poultry Litter. *Worlds Poult. Sci. J.* **2010**, *66*, 673–698. [CrossRef]

13. Paungfoo-Lonhienne, C.; Visser, J.; Lonhienne, T.G.A.; Schmidt, S. Past, Present and Future of Organic Nutrients. *Plant Soil* **2012**, *359*, 1–18. [CrossRef]

14. Haynes, R.; Naidu, R. Influence of Lime, Fertilizer and Manure Applications on Soil Organic Matter Content and Soil Physical Conditions: A Review. *Nutr. Cycl. Agroecosystems* **1998**, *51*, 123–137. [CrossRef]

15. Chang, E.-H.; Chung, R.-S.; Wang, F.-N. Effect of Different Types of Organic Fertilizers on the Chemical Properties and Enzymatic Activities of an Oxisol under Intensive Cultivation of Vegetables for 4 Years. *Soil Sci. Plant Nutr.* **2008**, *54*, 587–599. [CrossRef]

16. van Zwieten, L. The Long-Term Role of Organic Amendments in Addressing Soil Constraints to Production. *Nutr. Cycl. Agroecosystems* **2018**, *111*, 99–102. [CrossRef]

17. Quilty, J.; Cattle, S. Use and Understanding of Organic Amendments in Australian Agriculture: A Review. *Soil Res.* **2011**, *49*, 1–26. [CrossRef]

18. Angle, J.; Singh, U.; Dimkpa, C.; Heliums, D.; Bindraban, P. *Role of Fertilisers for Climate-Resilient Agriculture*; International Fertiliser Society: Colchester, UK, 2017; pp. 1–44.

19. Chen, Y.; Camps-Arbestain, M.; Shen, Q.; Singh, B.; Cayuela, M.L. The Long-Term Role of Organic Amendments in Building Soil Nutrient Fertility: A Meta-Analysis and Review. *Nutr. Cycl. Agroecosystems* **2018**, *111*, 103–125. [CrossRef]

20. Paungfoo-Lonhienne, C.; Redding, M.; Pratt, C.; Wang, W. Plant Growth Promoting Rhizobacteria Increase the Efficiency of Fertilisers While Reducing Nitrogen Loss. *J. Environ. Manage.* **2019**, *233*, 337–341. [CrossRef] [PubMed]

21. Brown, K.W.; Thomas, J.C.; Duble, R.L. Nitrogen Source Effect on Nitrate and Ammonium Leaching and Runoff Losses from Greens1. *Agron. J.* **1982**, *74*, 947–950. [CrossRef]

22. Salo, T.; Turtola, E. Nitrogen Balance as an Indicator of Nitrogen Leaching in Finland. *Agric. Ecosyst. Environ.* **2006**, *113*, 98–107. [CrossRef]

23. Zhang, M.; Wang, L.; He, Z. Spatial and Temporal Variation of Nitrogen Exported by Runoff from Sandy Agricultural Soils. *J. Environ. Sci.* **2007**, *19*, 1086–1092. [CrossRef]

24. Rayment, G.E.; Lyons, D.J. *Soil Chemical Methods—Australasia*; CSIRO Publishing: Melbourne, Australia, 2010.

25. Humphry, J.; Daniel, T.; Edwards, D.; Sharpley, A. A Portable Rainfall Simulator for Plot-Scale Runoff Studies. *Appl. Eng. Agric.* **2002**, *18*, 199. [CrossRef]

26. Redding, M.R. Bentonites and Layered Double Hydroxides Can Decrease Nutrient Losses from Spent Poultry Litter. *Appl. Clay Sci.* **2011**, *52*, 20–26. [CrossRef]

27. Walkowiak, D.K. *ISCO Open Channel Flow Measurement Handbook*; ISCO: Lincoln, NE, USA, 2006; ISBN 0-9622757-3-5.

28. Kandeler, E.; Gerber, H. Short-Term Assay of Soil Urease Activity Using Colorimetric Determination of Ammonium. *Biol. Fertil. Soils* **1988**, *6*, 68–72. [CrossRef]

29. Schnetger, B.; Lehners, C. Determination of Nitrate plus Nitrite in Small Volume Marine Water Samples Using Vanadium(III)Chloride as a Reduction Agent. *Mar. Chem.* **2014**, *160*, 91–98. [CrossRef]

30. Antille, D.L.; Moody, P.W. Nitrogen Use Efficiency Indicators for the Australian Cotton, Grains, Sugar, Dairy and Horticulture Industries. *Environ. Sustain. Indic.* **2021**, 100099. [CrossRef]

31. Congreves, K.A.; Otchere, O.; Ferland, D.; Farzadfar, S.; Williams, S.; Arcand, M.M. Nitrogen Use Efficiency Definitions of Today and Tomorrow. *Front. Plant Sci.* **2021**, *12*, 912. [CrossRef]

32. Raun, W.R.; Johnson, G.V. Improving Nitrogen Use Efficiency for Cereal Production. *Agron. J.* **1999**, *91*, 357–363. [CrossRef]

33. Banning, N.C.; Sawada, Y.; Phillips, I.R.; Murphy, D.V. Amendment of Bauxite Residue Sand Can Alleviate Constraints to Plant Establishment and Nutrient Cycling Capacity in a Water-Limited Environment. *Ecol. Eng.* **2014**, *62*, 179–187. [CrossRef]

34. Phillips, I.R.; Courtney, R. Long Term Field Trials Demonstrate Sustainable Nutrient Supply and Uptake in Rehabilitated Bauxite Residue. *Sci. Total Environ.* **2022**, *804*, 150134. [CrossRef]

35. Teutscherova, N.; Houška, J.; Navas, M.; Masaguer, A.; Benito, M.; Vazquez, E. Leaching of Ammonium and Nitrate from Acrisol and Calcisol Amended with Holm Oak Biochar: A Column Study. *Geoderma* **2018**, *323*, 136–145. [CrossRef]

36. Gaskell, M.; Smith, R. Nitrogen Sources for Organic Vegetable Crops. *HortTechnology Hortte* **2007**, *17*, 431–441. [CrossRef]
37. Sumner, M.E. *Handbook of Soil Science*; CRC Prss: Boca Raton, FL, USA, 1999; ISBN 0-8493-3136-6.
38. Lide, D.R. *CRC Handbook of Chemistry and Physics: A Ready-Reference Book of Chemical and Physical Data*; CRC Press: Boca Raton, FL, USA, 2017.
39. Phillips, I.; Sheehan, K. Importance of Surface Charge Characteristics When Selecting Soils for Wastewater Re-Use. *Soil Res.* **2005**, *43*, 915–927. [CrossRef]
40. Watanabe, Y.; Yamada, H.; Tanaka, J.; Komatsu, Y.; Moriyoshi, Y. Ammonium Ion Exchange of Synthetic Zeolites: The Effect of Their Open-window Sizes, Pore Structures, and Cation Exchange Capacities. *Sep. Sci. Technol.* **2005**, *39*, 2091–2104. [CrossRef]
41. Hinsinger, P. Bioavailability of Soil Inorganic P in the Rhizosphere as Affected by Root-Induced Chemical Changes: A Review. *Plant Soil* **2001**, *237*, 173–195. [CrossRef]
42. Rengel, Z. Availability of Mn, Zn and Fe in the Rhizosphere. *J. Soil Sci. Plant Nutr.* **2015**, *15*, 397–409. [CrossRef]
43. Teste, F.P.; Veneklaas, E.J.; Dixon, K.W.; Lambers, H. Complementary Plant Nutrient-acquisition Strategies Promote Growth of Neighbour Species. *Funct. Ecol.* **2014**, *28*, 819–828. [CrossRef]
44. Touhami, D.; McDowell, R.W.; Condron, L.M. Role of Organic Anions and Phosphatase Enzymes in Phosphorus Acquisition in the Rhizospheres of Legumes and Grasses Grown in a Low Phosphorus Pasture Soil. *Plants* **2020**, *9*, 1185. [CrossRef]
45. Hinsinger, P.; Gilkes, R. Dissolution of Phosphate Rock in the Rhizosphere of Five Plant Species Grown in an Acid, P-Fixing Mineral Substrate. *Geoderma* **1997**, *75*, 231–249. [CrossRef]
46. Foereid, B. Nutrients Recovered from Organic Residues as Fertilizers: Challenges to Management and Research Methods. *World J. Agri. Soil Sci.* **2019**, *1*, 1–7. [CrossRef]
47. Liljeroth, E.; Bååth, E.; Mathiasson, I.; Lundborg, T. Root Exudation and Rhizoplane Bacterial Abundance of Barley (Hordeum Vulgare, L.) in Relation to Nitrogen Fertilization and Root Growth. *Plant Soil* **1990**, *127*, 81–89. [CrossRef]
48. Robinson, D.; Griffiths, B.; Ritz, K.; Wheatley, R. Root-Induced Nitrogen Mineralisation: A Theoretical Analysis. *Plant Soil* **1989**, *117*, 185–193. [CrossRef]
49. Suter, H.; Lam, S.; Walker, C.; Chen, D. Nitrogen Use Efficiency for Pasture Production–Impact of Enhanced Efficiency Fertilisers and N Rate. In *Building Productive, Diverse and Sustainable Landscapes, Proceedings of the 17th Australian Agronomy Conference, Hobart, Tasmania, Australia, 21–24 September 2015*; Acuña, T., Moeller, C., Parsons, D., Harrison, M., Eds.; Australian Society of Agronomy: Australia, 2015.
50. Chmelíková, L.; Schmid, H.; Anke, S.; Hülsbergen, K.-J. Nitrogen-Use Efficiency of Organic and Conventional Arable and Dairy Farming Systems in Germany. *Nutr. Cycl. Agroecosystems* **2021**, *119*, 337–354. [CrossRef]
51. Hirel, B.; Tétu, T.; Lea, P.J.; Dubois, F. Improving Nitrogen Use Efficiency in Crops for Sustainable Agriculture. *Sustainability* **2011**, *3*, 1452–1485. [CrossRef]
52. Goyal, S.; Chander, K.; Mundra, M.C.; Kapoor, K.K. Influence of Inorganic Fertilizers and Organic Amendments on Soil Organic Matter and Soil Microbial Properties under Tropical Conditions. *Biol. Fertil. Soils* **1999**, *29*, 196–200. [CrossRef]
53. Kaur, K.; Kapoor, K.K.; Gupta, A.P. Impact of Organic Manures with and without Mineral Fertilizers on Soil Chemical and Biological Properties under Tropical Conditions. *J. Plant Nutr. Soil Sci.* **2005**, *168*, 117–122. [CrossRef]
54. Wen, Z.; Shen, J.; Blackwell, M.; Li, H.; Zhao, B.; Yuan, H. Combined Applications of Nitrogen and Phosphorus Fertilizers with Manure Increase Maize Yield and Nutrient Uptake via Stimulating Root Growth in a Long-Term Experiment. *Pedosphere* **2016**, *26*, 62–73. [CrossRef]
55. Kramer, S.B.; Reganold, J.P.; Glover, J.D.; Bohannan, B.J.M.; Mooney, H.A. Reduced Nitrate Leaching and Enhanced Denitrifier Activity and Efficiency in Organically Fertilized Soils. *Proc. Natl. Acad. Sci. USA* **2006**, *103*, 4522–4527. [CrossRef]

 *Nitrogen*

Article

Degradation of *N*-(*n*-butyl) Thiophosphoric Triamide (NBPT) with and without Nitrification Inhibitor in Soils

Ahmed A. Lasisi * and Olalekan O. Akinremi

Department of Soil Science, University of Manitoba, Winnipeg, MB R3T 5H3, Canada; wole.akinremi@umanitoba.ca

* Correspondence: greatahmlas@gmail.com

Abstract: Recent studies have shown that nitrification inhibitor (NI) impairs the efficacy of urease inhibitor, *N*-(*n*-butyl) thiophosphoric triamide (NBPT), in reducing ammonia volatilization and urea hydrolysis rate. A laboratory study was conducted to evaluate the influence of NI (specifically 3,4-dimethyl pyrazole phosphate) on the degradation of NBPT in six soils. Soils were amended with either NBPT (10 mg NBPT kg^{-1} soil) or NBPT plus NI (DI; 10 mg NBPT + 2.5 mg NI kg^{-1} soil), incubated at 21 °C, and destructively sampled eight times during a 14-day incubation period. The degradation of NBPT in soil was quantified by measuring NBPT concentration with high-performance liquid chromatography-mass spectrometry, and the degradation rate constant was modeled with an exponential decay function. The study showed that the persistence of NBPT in soil was not influenced by the presence of NI, as the NBPT degradation rate constant across soils was 0.5 d^{-1} with either NBPT or DI. In contrast, the degradation rate constant was significantly dependent on soils, with values ranging from 0.4 to 1.7 d^{-1}. Soil pH was the most important variable affecting the persistence of NBPT in soils. The half-life of NBPT was 0.4 d in acidic soil and 1.3 to 2.1 d in neutral to alkaline soils. The faster degradation of NBPT in acidic soils may explain its reduced efficacy in such soils.

Keywords: NBPT; nitrification inhibitor; half-life; degradation rate constant

Citation: Lasisi, A.A.; Akinremi, O.O. Degradation of *N*-(*n*-butyl) Thiophosphoric Triamide (NBPT) with and without Nitrification Inhibitor in Soils. *Nitrogen* **2022**, 3, 161–169. https://doi.org/10.3390/nitrogen3020012

Academic Editor: Jacynthe Dessureault-Rompré

Received: 19 February 2022
Accepted: 31 March 2022
Published: 2 April 2022

Publisher's Note: MDPI stays neutral with regard to jurisdictional claims in published maps and institutional affiliations.

1. Introduction

Globally, urea is the predominant form of granular nitrogen (N) fertilizer used to supplement soil N availability to crops. It is relatively safe to handle and contains high N content (46%). When urea is applied to soils, it becomes hydrolyzed in the presence of the urease enzyme into one bicarbonate ion and two molecules of ammonia [1]. The process of urea hydrolysis increases the soil pH around urea, which drives the equilibrium between ammonia and ammonium toward ammonia, thereby resulting in the volatilization of ammonia. Ammonia volatilization from urea when it is surface-applied without incorporation could be greater than 20% of applied N and is one of the reasons for low urea-N use efficiency [2]. Apart from being an economic loss to the farmers, ammonia volatilization has a deleterious effect on the environment and human health [3].

N-(*n*-butyl) thiophosphoric triamide (NBPT) is a urease inhibitor that has been widely reported to decrease ammonia volatilization from urea-based fertilizers under different soil and environmental conditions [2,4,5]. The NBPT is usually used to coat urea granules or mixed with liquid urea-based fertilizers, such as urea ammonium nitrate. The NBPT reduces ammonia volatilization by suppressing the activity of urease enzymes responsible for the hydrolysis of urea [6]. To suppress urease activity, NBPT becomes converted to either *N*-(*n*-butyl) thiophosphoric diamide (NBPD) or *N*-(*n*-butyl) phosphoric triamide (NBPTO) in soils [1,7]. The NBPD and NBPTO become hydrolyzed into monoamido thiophosphoric acid and diamido phosphoric acid, respectively. These then block the two nickel ions' active sites of urease enzymes via two oxygen atoms and one amide group [1]. The global efficiency of NBPT in reducing ammonia volatilization from urea relative to untreated urea has been estimated to be 52% [2].

Several studies have reported a reduction in the efficiency of NBPT when combined with a nitrification inhibitor (NI) in reducing ammonia volatilization from urea [8–13]. The decrease in NBPT efficiency when used as NBPT plus NI (double inhibitor, DI) was attributed to the persistence of ammonium by the NI. Recent studies to elucidate the mechanism of the reduced NBPT efficacy in decreasing ammonia volatilization with DI showed that NI, specifically 3,4-dimethyl pyrazole phosphate, significantly impaired the inhibitory effect of NBPT on urea hydrolysis rates across several soils and temperatures [14,15]. For example, NI was found to reduce the inhibitory effect of NBPT on the urea hydrolysis rate by 21% across five soils at 21 °C [14]. The effectiveness of NBPT in reducing the urea hydrolysis rate decreases as temperature increases [15]. The action of NBPT on urea has been shown to be more effective in reducing ammonia volatilization during fall than spring seasons on the Canadian prairies [13]. Moreover, studies have also shown that the rate of NBPT degradation in soils was greater in acidic than alkaline soils [16,17]. However, there is a lack of information on the influence of NI on the degradation of NBPT with or without urea in soils. This study was conducted to evaluate the influence of NI, particularly 3,4-dimethyl pyrazole phosphate, on the degradation rate of NBPT without urea in six soils. We hypothesized that NI would interfere with the persistence of NBPT in soils.

2. Materials and Methods

2.1. Soil Characteristics

An incubation study was conducted on soils (0–15 cm depth) that were collected from six locations in Manitoba, Canada. The locations were Carman (Soil 1; 49°29′6″N, 98°02′2″W), Carberry (Soil 2; 49°53′7″N, 99°22′29″W), Deerwood (Soil 3; 49°22′1″N, 98°23′34″W), High Bluff (Soil 4; 50°01′ 2″N, 98°08′9″W), Beausejour (Soil 5; 50°05′13″N, 96°29′58″W), and Portage la prairie (Soil 6; 49°57′9″N, 98°16′0″W). These were the same six soils used in two previous studies [14,15]. In the Canadian soil classification system, all soils are classified as Chernozems (an equivalent of Chernozem in the FAO classification system) except Soil 4, which is classified as a Regosol (an equivalent of Regosol in the FAO classification system) [18]. The soils were air-dried and ground to pass through a 2-mm sieve. A subsample of each soil was analyzed (Table 1) for organic matter by the wet oxidation method [19], cation exchange capacity by ammonium acetate method [20], urease activity [21], soil texture by pipette method [22], field capacity [23], and pH (soil/water, 1:2) and electrical conductivity with a combined conductivity and pH meter (Orion versaStar, ThermoFisher Scientific Inc., Waltham, MA, USA).

2.2. Experimental Design and Treatment Applications

The experimental setup was a completely randomized design containing two inhibitor treatments of six soils, a factorial layout for eight sampling periods, and was replicated three times for a total of 288 experimental units. The inhibitor treatments were NBPT (10 mg NBPT kg^{-1} soil) and NBPT plus NI (DI; 10 mg NBPT + 2.5 mg NI kg^{-1} soil). We used analytical grades of NBPT (CAS: 94317-64-3) and NI (3,4-dimethyl pyrazole phosphate; CAS: 202842-98-6) in this study.

Ten grams of each soil was weighed in 50 mL centrifuge tubes. The soil was wetted to 75% field capacity based on soil mass, capped, and left to equilibrate for 24 h at room temperature. Twenty-four hours after wetting, the soils in the centrifuge tubes were spiked with 0.5 mL of a solution containing either 200 mg NBPT L^{-1} (NBPT inhibitor treatment) or 200 mg NBPT + 50 mg NI L^{-1} (DI inhibitor treatment). The ratio of NBPT to NI in the DI inhibitor treatment was the same as the ratio of NBPT to NI in the double inhibitor formulation used in our previous studies [14,15]. However, the current study did not include urea with the inhibitors, as we discovered that the presence of urea interfered with the analytical procedure for NBPT. The tubes were recapped and placed in an incubator (Isotope Incubator, Model 304, Fisher Scientific, Hampton, NH, USA) set at 21 °C. On days

0, 0.5, 1, 2, 4, 7, 10, and 14 after treatment application, three replicates or samples of each soil by inhibitor treatment (i.e., six soils × two inhibitor treatments × three replicates for a total of 36 samples) were removed (destructive sampling) from the incubator for NBPT extraction and analysis. Day 0 was immediately after the soil was spiked with the inhibitor treatments.

Table 1. Selected soil (0–15 cm) properties.

Soil Property	Soil 1	Soil 2	Soil 3	Soil 4	Soil 5	Soil 6
Soil classification [a]	Orthic Black Chernozem	Orthic Black Chernozem	Orthic Dark Gray Chernozem	Gleyed Cumulic Regosol	Gleyed Rego Black Chernozem	Gleyed Rego Black Chernozem
Soil series	Hibsin	Fairland	Dezwood	High Bluff	Dencross	Neurhorst
Soil pH$_{water}$	5.51	6.65	6.62	7.46	7.76	7.96
Electrical conductivity (μS cm^{-1})	394	228	1853	899	1377	596
Organic matter (g kg^{-1})	27	33	34	45	88	71
Available N (mg kg^{-1})	31	15	186	58	22	82
Field capacity (m m^{-3})	0.35	0.34	0.36	0.41	0.61	0.44
Urease activity (mg NH$_4^+$-N kg^{-1} soil hr^{-1})	11	17	24	57	63	88
Cation exchange capacity (cmol kg^{-1})	16	14	23	28	47	36
Soil texture	Sandy loam	Sandy loam	Loam	Loam	Clay	Clay loam
Sand (g kg^{-1})	711	764	465	427	108	269
Silt (g kg^{-1})	123	128	318	325	322	343
Clay (g kg^{-1})	166	108	217	248	570	388

[a] Canadian soil classification system.

2.3. Extraction and Analysis of NBPT

On each sampling day, 25 mL of deionized water was dispensed on the sampled centrifuge tubes and shaken on a reciprocating shaker for 30 min at 120 excursions per minute. After 30 min of shaking, the samples were centrifuged for 5 min at 10,000× g to allow soil residues to settle to the bottom. Immediately after centrifugation, about 4 mL aliquot was transferred using a 0.2 µm syringe filter (Basix™ Syringe Filters, ThermoFisher Scientific, Waltham, MA, USA) into a 20 mL vial. This was followed by transferring 1 mL of the filtered aliquot into a 2 mL high-performance liquid chromatography (HPLC) vial (9 mm surestop screw vial, ThermoFisher Scientific, Waltham, MA, USA) containing 0.1 mL dimethyl sulfoxide for NBPT analysis with HPLC-mass spectrometry (HPLC-MS), as described by Engel et al. [17].

The HPLC-MS (Bruker Compact QqTOF, Billerica, MA, USA) used was equipped with an electrospray source that operated in the positive ionization mode. The nebulizer pressure of the source was 0.3 bar with 5 L min^{-1} of N$_2$ drying gas at 200 °C. The capillary voltage was 3500 V, and the capillary exit voltage was 70 V. Reverse-phase chromatography was used to separate NBPT using an Intensity Solo C18 (100 × 2.1 mm, 2 µm) HPLC column (Bruker Daltonik, Billerica, MA, USA). The column was maintained at 35 °C with a flow rate of 300 µL min^{-1}. The mobile phase consisted of formic acid 0.1% in Milli-Q water for Channel "A" and acetonitrile for Channel "B". A 2 µL aliquot of the sample was injected into the column and kept at 80% B from 0 to 3 min. From 3 to 4 min, the gradient was linearly ramped to 20% B, where it was kept for 1.5 min. Then, the gradient was linearly ramped to 80%, and it was held for 2.5 min at 80% for re-equilibration. The NBPT was eluted at approximately 3.3 min.

Data quantitation was performed using Bruker Daltonic QuantAnalysis (ver. 4.4) software (Billerica, MA, USA). The ion chromatograms for NBPT were defined as [M + H]$^+$ (168.0719 m/z). The concentrations of NBPT recovered in soil were determined from a calibration curve of known standard solutions of NBPT and their corresponding peak areas. The quantity of NBPT recovered was expressed as a percentage of NBPT applied to the soils.

2.4. Kinetics and Statistical Analysis

Model fitting and statistical analysis were performed with SAS software (SAS Institute 2014, ver. 9.4 [24]). PROC NLIN was used to fit an exponential decay function (Equation (1)) to determine the degradation rate constant (k) of NBPT in the soils as follows:

$$Y = b_0[\exp(-kt)] \tag{1}$$

where Y is the % of NBPT recovered in soils at time t, t is the time in days, k is the NBPT degradation rate constant, and b_0 is an empirical constant.

For ease of interpretation, the generated k was used to calculate the half-life ($t_{1/2}$) of NBPT in the inhibitor treatments using Equation (2):

$$t_{1/2} = \ln(2)/k \tag{2}$$

We used PROC GLIMMIX (beta distribution) for repeated measure analysis to determine the significant effect of time, inhibitor treatments, and their interaction on the % of NBPT recovered in each soil. Furthermore, analysis of variance with PROC GLIMMIX (gamma distribution) was performed on the degradation rate constant and half-life of the NBPT across soils and inhibitor treatments. The fixed effects in the model were soil and inhibitor treatment. Mean comparisons were deemed significant at a probability level of 0.05 Fishers' protected least-significant difference. The goodness of fit for the exponential decay model was tested using the Nash–Sutcliffe model efficiency (ME) and root means square error [25]. Stepwise regression with PROC REG was used to analyze the influence of soil properties on the half-life of NBPT in soils.

$$ME = 1 - \frac{\sum_{i=1}^{n} \left(Y_i^m - Y_i^p\right)^2}{\sum_{i=1}^{n} \left(Y_i^m - \overline{Y}\right)^2} \tag{3}$$

where Y_i^m is the measured NBPT recovered in soil, Y_i^p is the predicted NBPT recovered in soil, and $\overline{Y}$ is the mean of measured NBPT recovered in soil. When $ME = 1$, there is a perfect relationship measured and predicted NBPT recovery in soil; and when $ME = 0$, the model has the same precision as the mean of measured NBPT recovered.

3. Results

3.1. NBPT Recovery

The interaction of treatment and time did not significantly affect % NBPT recovered, except in Soil 2 (Table 2). The significant interaction in Soil 2 was because of greater NBPT recovered in NBPT only with DI on 0.5 and 2 d (Figure 1). The % of NBPT recovered immediately after treatment application (time = 0 d) was less than 70% in all the soils except in Soil 1 (Figure 1). The low recovery of NBPT on 0 d might be due to other NBPT species (e.g., [M + Na]$^+$) of the ion chromatograms that were not accounted for. As expected, the % of NBPT recovered significantly decreased with time in an exponential decay order (Figure 1). The persistence of NBPT was shortest in Soil 1, with the NBPT recovery reaching the lowest point (3%) by 2 d in both inhibitor treatments. In contrast, the % of NBPT recovered from neutral to alkaline soils on 2 d ranged from 22 to 44% in both inhibitor treatments (Figure 1). By 7 d, NBPT was below the detection limit in all soils. The NBPT recovery was well predicted by the exponential decay function as indicated by the Nash–Sutcliffe model efficiency, which ranged from 0.93 to 0.99 across the soils.

Table 2. Effect of inhibitor treatment and time on % of NBPT recovered in soils.

Model Effect	df	Soil 1	Soil 2	Soil 3	Soil 4	Soil 5	Soil 6
		Probability values					
Inhibitor treatment (I)	1	0.8588	0.8945	0.8266	0.813	0.9467	0.9731
time (t)	7	<0.0001	<0.0001	<0.0001	<0.0001	<0.0001	<0.0001
I × t	7	1.0000	0.0199	0.3853	0.8640	0.4941	0.7716

Probability values are significant at <0.05. df, degree of freedom.

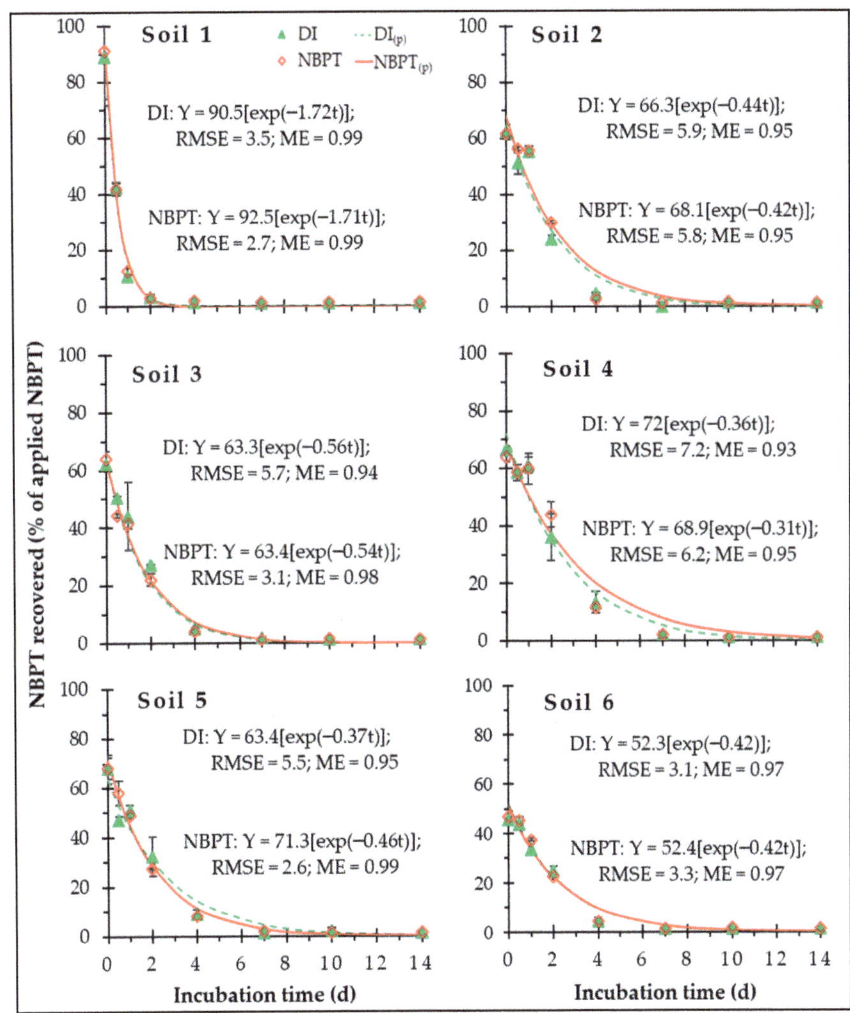

Figure 1. The percentage recovery of NBPT in soils. Error bars are standard errors of the mean (*n* = 3). NBPT and NBPT(p) are measured and predicted % of NBPT recovered in NBPT inhibitor treatment; DI and DI(p) are measured and predicted % of NBPT recovered in DI inhibitor treatment. *N*-(*n*-butyl) thiophosphoric triamide; DI, double inhibitor [NBPT + 3,4-dimethyl pyrazole phosphate]; Y, % of applied NBPT recovered; RMSE, Root mean square error, ME, Nash–Sutcliffe model efficiency.

3.2. Kinetics of NBPT Degradation

The NBPT degradation rate constant was not significantly affected by inhibitor treatment or the interaction of soil and inhibitor treatment (Table 3). As such, the half-life of NBPT in each soil was not affected by the type of inhibitor treatment (NBPT only versus DI; Table 3). Averaged across soils, the half-life of NBPT in either inhibitor treatment was 1.3 d. The lack of a significant difference in the half-life of NBPT between the two inhibitor treatments did not agree with our hypothesis. Our previous study had found that NI reduced the half-life of NBPT-treated urea in soils by 1 d at 21 °C [15].

Table 3. Effect of inhibitor treatment and soil on degradation rate constant (k) and half-life ($t_{1/2}$) of NBPT.

Group Means	k	$t_{1/2}$
Inhibitor treatment (I)	d	d^{-1}
NBPT	0.54 a	1.32 a
DI	0.54 a	1.33 a
Soil (S)		
Soil 1	1.72 a	0.44 d
Soil 2	0.43 c	1.61 b
Soil 3	0.55 b	1.30 c
Soil 4	0.33 d	2.09 a
Soil 5	0.41 c	1.71 b
Soil 6	0.42 c	1.66 b
Model effects	Probability values	
I	0.8784	0.8876
S	<0.0001	<0.0001
I × S	0.5452	0.5021

Note. Means with different letters within a column are significantly different at a probability value of <0.05 using Fisher protected LSD.

Unlike the inhibitor treatment, there was a significant effect of soil on the half-life of NBPT in soils. Soil 1, which was the acidic soil, had the shortest half-life (0.4 d), while Soil 4, which was slightly alkaline, had the longest half-life (2.1 d) when averaged across inhibitor treatments (Table 3). NBPT is *N*-(*n*-butyl) thiophosphoric triamide, DI is double inhibitor [*N*-(*n*-butyl) thiophosphoric triamide + 3,4-dimethyl pyrazole phosphate]. The shortest half-life of NBPT in Soil 1 was consistent with previous studies that found that the NBPT degradation rate was faster in acidic than alkaline soils [16,17]. Similarly, the shortest half-life of the inhibitor treatments in Soil 1 corroborated our previous studies that used the same soils and found the half-life of urea treated with either NBPT or DI to be shorter in Soil 1 than in other soils [14,15]. Additionally, other studies had also reported a lower NBPT inhibition of urea hydrolysis in acidic than alkaline soils [26,27]. The lack of NI on NBPT degradation in this study was probably because of the absence of urea. This is because the soil pH around applied urea changes during the hydrolysis of urea and the nitrification process. This implies that NI did not affect the persistence of NBPT in soil but rather impaired the inhibitory effect of NBPT on urea hydrolysis. With no effect of NI on NBPT degradation, the observed inhibition of NBPT to reduce urea hydrolysis by NI, as noted in the studies [14,15] might have been because of the soil acidification during nitrification [28]. While hydrolysis of NBPT in soils to form NBPTO and NBPD is required to inhibit the process of urea hydrolysis [1,7], rapid hydrolysis of NBPT, as shown in the case of acidic soil, may be counter-effective. For example, NBPTO and phenyl phosphorodiamidate are potent urease inhibitors with greater inhibition of urease than NBPT under a buffered solution, but their reduced persistence in soils makes them less effective in reducing ammonia volatilization when compared to NBPT [16,29–31].

Stepwise regression analysis showed that soil pH, organic matter, and urease activities accounted for 91% of the variation in the half-life of NBPT in soil. Of these soil properties, soil pH was the most predictive factor of NBPT half-life, as indicated in Equation (4).

The persistence of NBPT in soils increased as the soil pH increased from strongly acidic to neutral soil pH and then decreased from neutral soil pH to slightly to moderately alkaline soil pH (Figure 2; pH classification based on USDA). An earlier study had shown that the half-life of NBPT in acidic soil (pH = 4.9) could be extended by 2.5 d when the soil pH was increased to neutral pH (6.9) using calcium hydroxide [16]. Despite the reported reduced persistence of NBPT in acidic than alkaline soils, the reduction of ammonia volatilization by NBPT relative to untreated urea is not always lower in acidic than alkaline soils [32].

$$\text{Half-life} = -5.874 + 1.221(\text{pH}) - 0.0141(\text{urease activity}) - 0.1163(\text{organic matter}) \quad (4)$$

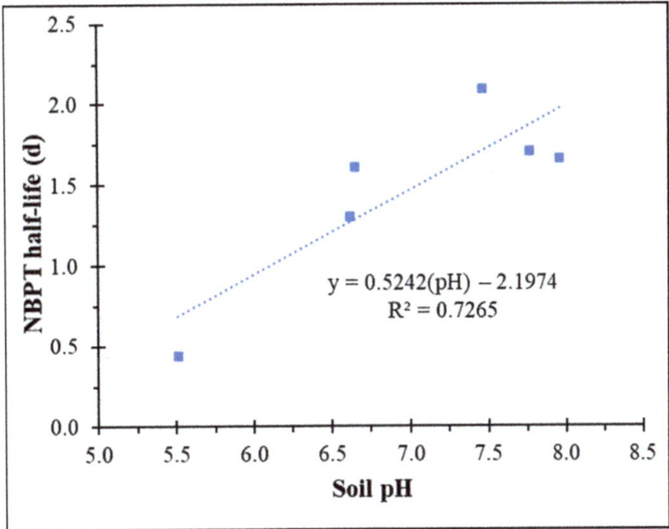

Figure 2. Relationship between soil pH and NBPT half-life. Y is NBPT half-life in soil.

4. Conclusions

The urease inhibitor, NBPT, plays an important role in conserving applied urea-N in the soil. While NI is known to impair the inhibitory effect of NBPT on urea hydrolysis, our study showed that NI did not interfere with the persistence of NBPT in soil. Instead, the persistence of NBPT in soil was mainly influenced by soil pH. We found that the degradation of NBPT was two to four times greater in acidic than neutral to alkaline soils. The half-life of NBPT was 0.4 d in acidic soil and 1.3 to 2.1 d in neutral to alkaline soils. As such, N management with NBPT may be more suitable for alkaline than acidic soils, and alkaline soils thereby provide more flexibility in precipitation or irrigation scheduling to incorporate urea into the soil while reducing N losses. Future studies will need to evaluate how the interaction between urea, NI, and NBPT affect the persistence of NBPT over a wide range of soils and environmental conditions.

Author Contributions: Conceptualization, A.A.L.; Data curation, A.A.L.; Formal analysis, A.A.L.; Funding acquisition, O.O.A.; Investigation, A.A.L.; Methodology, A.A.L.; Project administration, O.O.A.; Supervision, O.O.A.; Validation, A.A.L.; Visualization, A.A.L.; Writing—original draft, A.A.L.; Writing—review and editing, A.A.L. and O.O.A. All authors have read and agreed to the published version of the manuscript.

Funding: This research received no external funding.

Data Availability Statement: Data available upon request.

Acknowledgments: The authors appreciate the technical support of Emily Komatsu in method development for measuring NBPT with the HPLC-MS.

Conflicts of Interest: The authors declare no conflict of interest.

References

1. Mazzei, L.; Cianci, M.; Contaldo, U.; Ciurli, S. Insights into urease inhibition by *N*-(*n*-Butyl) phosphoric triamide through an integrated structural and kinetic approach. *J. Agric. Food Chem.* **2019**, *67*, 2127–2138. [CrossRef] [PubMed]
2. Cantarella, H.; Otto, R.; Soares, J.R.; De Brito Silva, A.G. Agronomic efficiency of NBPT as a urease inhibitor: A review. *J. Adv. Res.* **2018**, *13*, 19–27. [CrossRef] [PubMed]
3. Sheppard, S.C.; Bittman, S.; Bruulsema, T.W. Monthly ammonia emissions from fertilizers in 12 Canadian Ecoregions. *Can. J. Soil Sci.* **2010**, *90*, 113–127. [CrossRef]
4. Modolo, L.V.; Da-Silva, C.J.; Brandão, D.S.; Chaves, I.S. A minireview on what we have learned about urease inhibitors of agricultural interest since mid-2000s. *J. Adv. Res.* **2018**, *13*, 29–37. [CrossRef]
5. Lasisi, A.A.; Akinremi, O.O.; Kumaragamage, D. Efficacy of a new *N*-(*n*-butyl) thiophosphoric triamide formulation in reducing ammonia volatilization from urea-based fertilizers. *Can. J. Soil Sci.* **2019**, *99*, 395–405. [CrossRef]
6. Sigurdarson, J.J.; Svane, S.; Karring, H. The molecular processes of urea hydrolysis in relation to ammonia emissions from agriculture. *Rev. Environ. Sci. Biotechnol.* **2018**, *17*, 241–258. [CrossRef]
7. Creason, G.L.; Schmitt, M.R.; Douglass, E.A.; Hendrscrson, L.L. Urease inhibitory activity associated with *N*-butyl thiophosphoric triamide is due to formation of its oxon analog. *Soil Biol. Biochem.* **1990**, *22*, 209–211. [CrossRef]
8. Gioacchini, P.; Nastri, A.; Marzadori, C.; Giovannini, C.; Vittori Antisari, L.; Gessa, C. Influence of urease and nitrification inhibitors on N losses from soils fertilized with urea. *Biol. Fertil. Soils* **2002**, *36*, 129–135. [CrossRef]
9. Zaman, M.; Nguyen, M.L.; Blennerhassett, J.D.; Quin, B.F. Reducing NH_3, N_2O and NO_3^--N losses from a pasture soil with urease or nitrification inhibitors and elemental S-amended nitrogenous fertilizers. *Biol. Fertil. Soils* **2008**, *44*, 693–705. [CrossRef]
10. Soares, J.R.; Cantarella, H.; Menegale, M.L.d.C. Ammonia volatilization losses from surface-applied urea with urease and nitrification inhibitors. *Soil Biol. Biochem.* **2012**, *52*, 82–89. [CrossRef]
11. Frame, W. Ammonia volatilization from urea treated with NBPT and two nitrification inhibitors. *Agron. J.* **2017**, *109*, 378–387. [CrossRef]
12. Mariano, E.; De Sant Ana Filho, C.R.; Bortoletto-Santos, R.; Bendassolli, J.A.; Trivelin, P.C.O. Ammonia losses following surface application of enhanced-efficiency nitrogen fertilizers and urea. *Atmos. Environ.* **2019**, *203*, 242–251. [CrossRef]
13. Lasisi, A.A.; Akinremi, O.O.; Zhang, Q.; Kumaragamage, D. Efficiency of fall versus spring applied urea-based fertilizers treated with urease and nitrification inhibitors I. Ammonia volatilization and mitigation by NBPT. *Soil Sci. Soc. Am. J.* **2020**, *84*, 949–962. [CrossRef]
14. Lasisi, A.A.; Akinremi, O.O.; Kumaragamage, D. Nitrification inhibitor reduces the inhibitory effect of *N*-(*n*-butyl) thiophosphoric triamide (NBPT) on the hydrolysis of urea. *Soil Sci. Soc. Am. J.* **2020**, *84*, 1782–1794. [CrossRef]
15. Lasisi, A.A.; Akinremi, O.O. Kinetics and thermodynamics of urea hydrolysis in the presence of urease and nitrification inhibitors. *Can. J. Soil Sci.* **2021**, *101*, 192–202. [CrossRef]
16. Hendrickson, L.; Douglass, E. Metabolism of the urease inhibitor *N*-(*n*-butyl) thiophosphoric (NBPT) in soils. *Soil Biol. Biochem.* **1993**, *25*, 1613–1618. [CrossRef]
17. Engel, R.E.; Towey, B.D.; Gravens, E. Degradation of the urease inhibitor NBPT as affected by soil pH. *Soil Sci. Soc. Am. J.* **2015**, *79*, 1674–1683. [CrossRef]
18. MAFRI Manitoba Agriculture, Food and Rural Initiative. *Soil Series Descriptions*; MAFRI: Carman, MB, Canada, 2010; 208p.
19. Walkley, A.; Black, I.A. An examination of the Degtjareff method for determining soil organic matter, and a proposed modification of the chromic acid titration method. *Soil Sci.* **1934**, *37*, 29–38. [CrossRef]
20. Hendershot, W.H.; Lalande, H.; Duquette, M. Ion exchange and exchangeable cations. In *Soil Sampling and Methods of Analysis*; Carter, M.R., Gregorich, E.G., Eds.; Canadian Society of Soil Science: Boca Raton, FL, USA, 2008; pp. 173–178.
21. Tabatabai, M.; Bremner, J.M. Assay of urease activity in soils. *Soil Biol. Biochem.* **1972**, *4*, 479–487. [CrossRef]
22. Gee, G.W.; Bauder, J.W. Particle-size analysis. In *Methods of Soil Analysis, Agronomy Monograph 9*, 2nd ed.; Klute, A., Ed.; ASA: Winscosin, WI, USA, 1986; pp. 383–411.
23. Cassel, D.K.; Nielsen, D.R. Field capacity and available water capacity. In *Methods of Soil Analysis, Agronomy Monograph 9*, 2nd ed.; Klute, A., Ed.; ASA: Winscosin, WI, USA, 1986; pp. 901–926.
24. SAS Institute, Inc. *SAS/STAT 13.2 User's Guide, Version 9.4*; SAS Institute, Inc.: Cary, NC, USA, 2014.
25. Archontoulis, S.V.; Miguez, F.E. Nonlinear regression models and applications in agricultural research. *Agron. J.* **2015**, *107*, 786–798. [CrossRef]
26. Suter, H.C.; Pengthamkeerati, P.; Walker, C.; Chen, D. Influence of temperature and soil type on inhibition of urea hydrolysis by *N*-(*n*-butyl) thiophosphoric triamide in wheat and pasture soils in south-eastern Australia. *Soil Res.* **2011**, *49*, 315–319. [CrossRef]
27. Fan, X.; Yin, C.; Yan, G.; Cui, P.; Shen, Q.; Wang, Q.; Chen, H.; Zhang, N.; Ye, M.; Zhao, Y.; et al. The contrasting effects of *N*-(*n*-butyl) thiophosphoric triamide (NBPT) on N_2O emissions in arable soils differing in pH are underlain by complex microbial mechanisms. *Sci. Total Environ.* **2018**, *642*, 155–167. [CrossRef] [PubMed]

28. Sahrawat, K.L. Factors affecting nitrification in soils. *Commun. Soil Sci. Plant Anal.* **2008**, *39*, 1436–1446. [CrossRef]
29. Christianson, C.B.; Byrnes, B.H.; Carmona, G. A comparison of the sulfur and oxygen analogs of phosphoric triamide urease inhibitors in reducing urea hydrolysis and ammonia volatilization. *Fertil. Res.* **1990**, *26*, 21–27. [CrossRef]
30. Bremner, J.M.; McCarty, G.W.; Higuchi, T. Persistence of the inhibitory effects of phosphoroamides on urea hydrolysis in soils. *Commun. Soil Sci. Plant Anal.* **1991**, *22*, 1519–1526. [CrossRef]
31. Kot, M.; Zaborska, W.; Orlinska, K. Inhibition of Jack Bean urease by *N*-(*n*-butyl) thiophosphorictriamide and *N*-(*n*-butyl) phosphorictriamide: Determination of the inhibition mechanism. *J. Enzyme Inhib.* **2001**, *16*, 507–516. [CrossRef]
32. Engel, R.; Jones, C.; Wallander, R. Ammonia volatilization from urea and mitigation by NBPT following surface application to cold soils. *Soil Sci. Soc. Am. J.* **2011**, *75*, 2348–2357. [CrossRef]

Nitrogen

Article

Simulated Cropping Season Effects on N Mineralization from Accumulated No-Till Crop Residues

Rashad S. Alghamdi [1,*], Larry Cihacek [1] and Qian Wen [2]

[1] Department of Soil Science, North Dakota State University, Fargo, ND 58108, USA; larry.cihacek@ndsu.edu
[2] Department of Statistics, North Dakota State University, Fargo, ND 58108, USA; qian.wen@ndus.edu
*   Correspondence: rashad.alghamdi@ndsu.edu

Citation: Alghamdi, R.S.; Cihacek, L.; Wen, Q. Simulated Cropping Season Effects on N Mineralization from Accumulated No-Till Crop Residues. *Nitrogen* **2022**, *3*, 149–160. https://doi.org/10.3390/nitrogen3020011

Academic Editor: Jacynthe Dessureault-Rompré

Received: 1 March 2022
Accepted: 28 March 2022
Published: 31 March 2022

Publisher's Note: MDPI stays neutral with regard to jurisdictional claims in published maps and institutional affiliations.

Abstract: The adoption of no-till management practices has increased in the United States over the last decade. In the state of North Dakota, approximately 5.7 million hectares of cropland is managed under no-till or conservation tillage management practices. Even though conservation tillage is known for building soil health, increasing soil organic matter, capturing soil moisture, and reducing wind and water erosion, it also presents a unique best management practice since an increased mass of crop residue remains on the soil surface. Producers are concerned about whether plant needs are being met by nitrogen fertilizer that is currently being applied based on current North Dakota recommendations for long-term no-till systems. A Forman clay loam soil (fine-loamy, mixed, superactive, frigid Calcic Argiudolls) was used in this study, as it represented glacial till soils of the region. We examined whether N mineralization from surface-applied crop residue would result in similar or different results when compared to crop residue mixed into the soil. Soil freeze-thaw contribution to soil N mineralization was also evaluated. Six residue treatments with different C/N ratios including corn (*Zea mays* L.), soybean (*Glycine max* L.), forage radish (*Raphanus sativus* L.), winter pea (*Pisum sativum* L.), spring wheat (*Triticum aestivum* L.), and winter wheat (*Triticum aestivum* L.) were used. Five 10–14-week cycles with a three-week freeze period between each cycle at 0 °C were evaluated for NO_3-N production. Crop residues with a narrow C/N ratio contributed to greater instances of N mineralization during each incubation cycle, and the accumulation of crop residues with a wide C/N ratio over each incubation cycle following the first incubation did not offset the immobilization trends observed in the first incubation. A change in N mineralized in the untreated control soil during the last two incubation cycles may have been caused by freeze-thaw effects or a shift in microbial population due to a lack of fresh C inputs.

Keywords: N mineralization; C/N ratio; crop residue; N availability

1. Introduction

In a no-till agricultural system, crop residue remains on the soil surface, whereby in a drier environment with lower soil temperatures may slow and/or reduce nitrogen mineralization, with the potential to affect crop yields. In areas where there is a short growing season, no-till practices have been perceived by producers to delay planting and slow crop emergence as a result of delayed soil warming and drying [1–4]. In a frigid environment, such as that in the Northern Great Plains, these concerns are more pronounced, as crop residue tends to accumulate when high residue crops (corn, small grains) are left on the soil surface as a result of no-till practices. Alghamdi et al. [5] examined soil warming and drying in a frigid environment for corn-soybean systems and provided evidence to suggest that these perceived delays are not related to moisture and temperature of the soil. Daigh et al. [6] have also reported research on full-production scale farms in the Red River Valley of Minnesota and North Dakota (i.e., frigid environment), finding that producer perceptions of delayed warming and drying of the soil do not translate into yield losses. This begs the question of where these perceived losses are coming from.

Alghamdi et al. [7] concluded that any differences observed in yield were attributed to fertilizer application methods and rates. Many earlier studies concluded that an addition of nitrogen fertilizer may be necessary under a no-till system to increase nitrogen (N) in the deeper soil profile [8–10], yet modern recommendations often overlook previous decades of research [11]. The rate of soil N mineralization is dependent upon crop residue factors that include, but are not limited to, the type and quantity of residue and other soil factors, such as temperature and moisture in the residue environment and organic matter (OM) in the soil [12]. As residue continues to accumulate on the soil surface in a no-till system, it is necessary to determine if enough N is being provided to subsequent crops to offset the potential immobilization in the system. Schoenau and Campbell [13] reported that conservation tillage such as reduced and no-till systems results in greater surface accumulation of crop residue on top of the soil surface, which slows residue decomposition of wide C/N ratio crops. No-till systems provide added benefits of increased soil moisture and organic matter, although reduced availability of oxygen from aeration needed by microbes may slow the mineralization of nitrogen from the crop residue. In addition to increasing soil moisture, no-till systems can be susceptible to water saturation as a result of increased crop residue on the soil surface. Thus, the activity of microbes may be reduced along with soil shading and slower soil drying due to cooler temperatures [14]. A laboratory study on North Dakotan soils concluded that the type of crop residue and organic matter present may increase N immobilization when crop residue has a wide C/N ratio (>25:1) as compared to crop residue with a narrow C/N ratio (<25:1) [7]. Green and Blackmer [15] examined the rate of N fertilization and its effect on residue decomposition on corn–corn and corn–soybean fields in Iowa. They determined that N immobilization tended to decrease with increased N fertilization as a result of the fertilizer contributing N to the residue biomass. Increasing rates of N fertilization decreased the time required for N mineralization to occur. This process was expedited when soybean residue was the proceeding crop due to the nature of the soybean residue. Li et al. [16] found that N mineralization occurred more quickly with crop residue placed on the soil surface versus crop residues incorporated into the soil. Satchell [17] explained the biological process of decomposition, where plant tissues are broken down by microorganisms. These microorganisms can break down residue more quickly when plant residue is incorporated with the soil [18] as compared to remaining on the soil surface [19]. Coppens et al. [20] found that incorporated crop residues decomposed faster than leaving residue on the soil surface and that higher N fluxes were more pronounced with crop residue incorporation. For the residue that did remain on the soil surface, the absence of moisture was a greater limitation to decomposition than N itself. Often, to aid in the breakdown of high lignin content crop residue, such as corn, many producers apply liquid N to their fields after harvest. Al-Kaisi et al. [21] examined corn residue decomposition and the application rate of liquid urea ammonium nitrate (UAN) following harvest and concluded that increased rates of decomposition were not attributed directly to the application rates of N, but to increased air and soil temperature. They recommended the limiting of liquid N application under less-than-ideal air temperature conditions and/or limited availability of soil moisture during critical periods of residue decomposition. Additional studies have determined similar finding where decomposition effects were attributed to temperature and moisture availability [22–24]. In cooler climates of the Northern Great Plains, such as in North Dakota, highly variable precipitation and temperature conditions during crucial periods of crop N needs can lead to the rate of residue decomposition being less predictable, resulting in less certain N mineralization for crop needs. Vigil and Kissel [25] examined the impact of temperature on nitrogen mineralization and residue decomposition in Kansas soils. Their findings indicated that higher incubation temperatures resulted in increased mineralization. At temperatures lower than 35 °C, microbial activity decreased. Nitrogen from sorghum and soybean residues mineralized faster at 35 °C than 25 °C, 15 °C, and 5 °C, respectively. Aher et al. [26] examined C/N ratios and the mass of residue on the soil surface near Forman, ND, and indicated potential N deficits to succeeding crops ranging from 56 to 105 kg N ha^{-1} following winter weathering of the crop residue. North Dakota has a frigid climate where residue decomposition and nutrient mineralization from crop residues can be limited by the short frost-free period of 100–135 days [27,28]. In the Canadian prairie, an incubation study was conducted to examine the factors influencing the stability of microbes in the soil [29]. Results indicated that freezing and thawing periods had a strong effect on releasing N and other nutrients back into the soil when compared with wetting and drying cycles. The reason for this was that the freeze and thaw periods enabled microorganisms to feed on decomposed residue. During freezing and thawing, soil aggregates are disturbed, exposing sheltered residue to microbes that help release nutrients back to the soil [30,31] or may break up residue particles to provide a greater residue surface area for greater microbial access. During the winter, native soil organic matter mineralizes to release N [32] due to microorganism abate and releases amino acids and simple sugars back into the soil [33]. During thaw periods, surviving microorganisms are active and feed on the nutrients released from dead microorganisms [34]. As a result, many studies have noted the increase in microbial respiration following freezing and thawing periods [33–36]. Current recommendations for N management in North Dakota do not regard temporal residue decomposition and N mineralization and immobilization effects that vary with different types of crop residue in their recommendations. Instead, standardized recommendations are cited for long-term no-till systems (i.e., managed for six or more years) where producers are to take a 34 kg N ha^{-1} credit [37]. Over the last ten years, recommendations for North Dakota have varied. Standardized N recommendations are useful in warmer and moister climates, which allow for optimum N mineralization from crop residue. Management recommendations for nitrogen fertilization of soil should aim to address concerns of the producer and advance best management practices. The objectives of this study are: (1) evaluate if the type and quantity of crop residue are offsetting the N immobilization, (2) evaluate the influence of freezing and thawing on crop residue decomposition, and (3) evaluate changes in N mineralization due to repeated residue addition over several simulated growing seasons.

2. Materials and Methods

2.1. Experimental Design

A laboratory study using a randomized complete block design with three replicates was set up using one soil and six residue treatments, plus an untreated soil control (n = 21). The soil was a Forman soil (fine-loamy, mixed, superactive, frigid Calcic Argiudolls) [38], similar to many soils of glaciated parts of the region. The six residue treatments were corn (CR), soybean (S), forage radish (R), pea (P), spring wheat (SW), and winter wheat (WW) (Table 1). Fresh individual crop residues were collected immediately in the fall following harvest and analyzed for carbon (C) and nitrogen using an Elementar Vario Max® CN analyzer (Ronkonkoma, New York, NY, USA) during a previous study conducted by Aher et al. [26]. Upon collection, the residue was oven-dried at 60 °C and ground in a Wiley mill to pass a <2 mm screen. Bulk Forman soil was collected at the Conservation Cropping System Project site [26] near Forman, North Dakota (97°38′38″ N 46°05′05″ W). The bulk soil sample was air-dried, crushed, and sieved through a 2 mm screen. Residue treatments (n = 18) were prepared containing 15 g of soil mixed with 15 g of quartz sand (20 mesh) with 0.50 g of residue placed on the soil surface following the procedure of Stanford and Smith [39]. The quantity of residue represents the equivalent of 6.25 Mg/ha in a field environment [7]. Soil only (n = 3) contained 15 g of soil mixed with 15 g of sand. Samples were then transferred to labeled glass leaching tubes where the soil was placed into the tube, and crop residue was then placed on top of the soil surface to simulate no-till residue accumulation. Leaching tubes were kept in a constant temperature room at 22 °C simulating average soil temperatures in the region during the growing season and incubated as described by Stanford and Smith [39]. Before the soil samples were transferred to the leaching tubes, a small amount of glass wool was placed at the bottom of the leaching tubes to prevent soil sediment loss during leaching and was placed at the top to prevent soil and residue disturbance during leaching solution application.

Table 1. Six crop residue treatments used for the five incubation cycles, their scientific name, abbreviation, and C/N ratios.

Residue Treatment	Scientific Name	Abbreviation	C/N Ratio [1]
Corn	*Zea mays* L.	CR	73
Soybean	*Glycine max* L.	S	53
Forage radish	*Raphanus sativus* L.	R	8
Winter pea	*Pisum sativum* L.	P	18
Spring wheat	*Triticum aestivum* L.	SW	76
Winter wheat	*Triticum aestivum* L.	WW	101

[1] C/N ratios for corn, soybean, forage radish, winter pea, spring wheat, and winter wheat were determined using near-infrared reflectance (NIR).

Five incubation cycles and four freeze–thaw periods were conducted for the experiment. Each cycle represented an annual growing season. Between each cycle, the soil treatments were frozen for three weeks to represent an annual freeze–thaw cycle. Following each freeze ($n = 4$), an additional 0.5 g of individual crop residue was placed on the soil surface (total = 2.5 g per leaching tube over the course of the study). Biweekly leachings were conducted at 2, 4, 6, 8, and 10 weeks for the first cycle. For the second cycle, additional leaching was conducted at 12 weeks. For cycles three, four, and five, additional leachings were conducted at 12 and 14 weeks to determine the potential of cumulative effects of the simulated growing seasons. For the first leaching of the samples and due to the air-dry nature of the soil/sand/residue column, 50 mL of 0.01 M $CaCl_2$ was added to the glass tubes in 10 mL increments, followed by 10 mL of nutrient solution as described by Stanford and Smith [39] to remove ambient levels of NH_4^+-N and NO_3-N. Subsequent leachings used 30 mL of 0.01 M $CaCl_2$ and 10 mL nutrient solution. Parafilm® (Menasha, WI, USA) was used to cover the incubation tubes between leachings to prevent contamination and preserve soil water content during this study. Air vents were created in the film to allow for soil respiration. At the end of an incubation period, the leaching tubes were then transferred into the freezer for three weeks at 0 °C. Leachate was collected, covered, and refrigerated (8–24 h) if necessary until mineral N (NO_3-N and NH_4-N) analysis was conducted using a Timberline TL 2900 NH_4/NO_3 analyzer (Timberline Instruments Inc., Boulder, CO, USA). Results are adjusted for soil mass and reported as mg N kg^{-1} soil in order to relate our results to quantities that might be experienced in field environments.

2.2. Statistical Analysis

A repeated-measures analysis of variance (ANOVA) was used to determine the effects of crop residue treatment, cycle, incubation period, and their interactions on NO_3-N mineralization. As the measurements of NO_3-N mineralization collected from the same experimental tube unit are related over time, the model imposed a covariance structure on the error term of the model. Akaike's information criteria (AIC) was used to determine the appropriate covariance structure, and the smaller the AIC value, the better. Throughout this paper, AR(1) covariance structure was used due to it always producing the smallest AIC value. The least square mean (LS mean) of each level of the freeze and thaw periods was estimated, and the significance of the difference between all possible pairs of these LS means was identified with Tukey's honest significant difference (HSD) at a significance level of 0.05. In addition, LS mean of each individual crop residue treatment within each freeze and thaw cycle was estimated, and an HSD test with a significance level of 0.05 was performed to test the significant difference among all possible pairs of the LS means. Moreover, LS mean of each individual crop residue treatment for each incubation period within each freeze and thaw cycle was estimated, and an HSD test with a significance level of 0.05 was used to find the LS means that are significantly different from each other. In order to detect the change of NO_3-N mineralization for the bare, unamended soil over incubation cycles for each freeze and thaw period, a simple one-way ANOVA with a repeated measurements model was fitted. LS mean of each incubation period within each freeze and thaw cycle was estimated, and an HSD test was used to find the LS means that are significantly different from each other. All analyses were conducted using SAS version 9.4 [40].

3. Results and Discussion

Mean soil NO_3-N mineralization capacity for individual crop residue treatments varies among freeze and thaw periods from 1.88 mg NO_3-N kg^{-1} to 58.97 mg NO_3-N kg^{-1} (Table S1, Supplementary Materials). The mean soil NO_3-N mineralization capacity (range in means) peaks during cycle 3 while displaying the narrowest range in mean soil NO_3-N mineralization capacity in cycle 5. Freeze and thaw period means varied from 4.95 mg NO_3-N kg^{-1} to 17.84 mg NO_3-N kg^{-1}, with cycles 4 and 5 significantly higher than cycles 1, 2, and 3. For all freeze and thaw periods, the control (bare, unamended soil) means show net N mineralization varying from 1.74 mg NO_3-N kg^{-1} to 22.72 mg NO_3-N kg^{-1}. Both cycle 4 and cycle 5 display higher mean soil NO_3-N mineralization values for the bare, unamended soil (22.72 mg NO_3-N kg^{-1} and 18.45 mg NO_3-N kg^{-1}, respectively), and it is important to note that the higher control values contribute to the narrowing of the range mineralization capacity in the later freeze and thaw periods. This increase observed in the bare, unamended soil may be due to a natural microbial shift as an adaptation to the fact that no new carbon source was added to the soil only controls [41]. In other words, microbes in the bare, unamended soil recognize no carbon source addition and start to consume SOM in those controls as an energy source whereby NO_3-N is being released.

When examining the five incubation cycles, the forage radish and pea (narrow C/N ratio) are the only crop residue treatments that show a significant increase in the soil NO_3-N mean from the control (Figure 1).

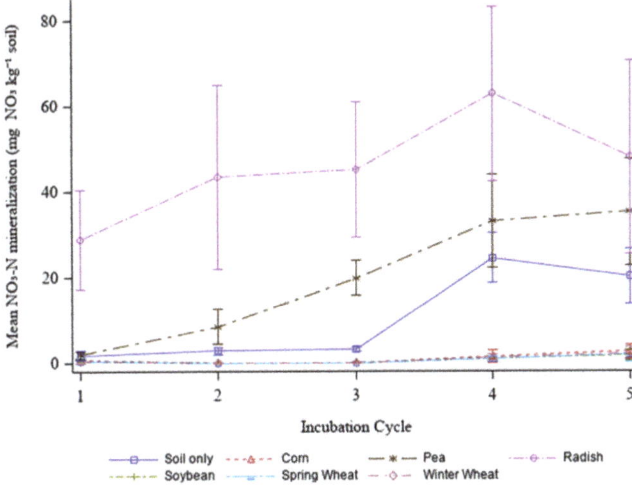

Figure 1. NO_3-N mineralization means and ranges in value for soil control and corn, pea, radish, soybean, spring wheat, and winter wheat crop residue treatment over five incubation cycles.

During cycle 1, the forage radish is the only crop residue treatment significantly different from the other crop residues treatment (Figure 2). In cycle 2, the soil NO_3-N mean increases in both the bare, unamended soil (control) and the pea crop residue treatment, although they are still significantly different from the forage radish crop residue treatment (Figure 2). For cycle 3, the forage radish crop residue treatment still remains significantly different from the control, although both are similar to the soil NO_3-N mean for the pea crop residue treatment (Figure 2). In cycle 4 and cycle 5, the bare, unamended soil (control) and the pea crop residue treatment are not significantly different from the forage radish, nor were they significantly different from the corn, soybean, spring wheat, and winter wheat crop residue treatments (Figure 2). By cycle 5, it is evident that the corn, soybean, spring wheat, and winter wheat crop residue treatments were not mineralizing soil NO_3-N and were always significantly different from the forage radish crop residue treatment. A reason for the observations in these mineralization patterns is the C/N ratio for each crop residue treatment influences the N mineralization characteristic of the residue [26].

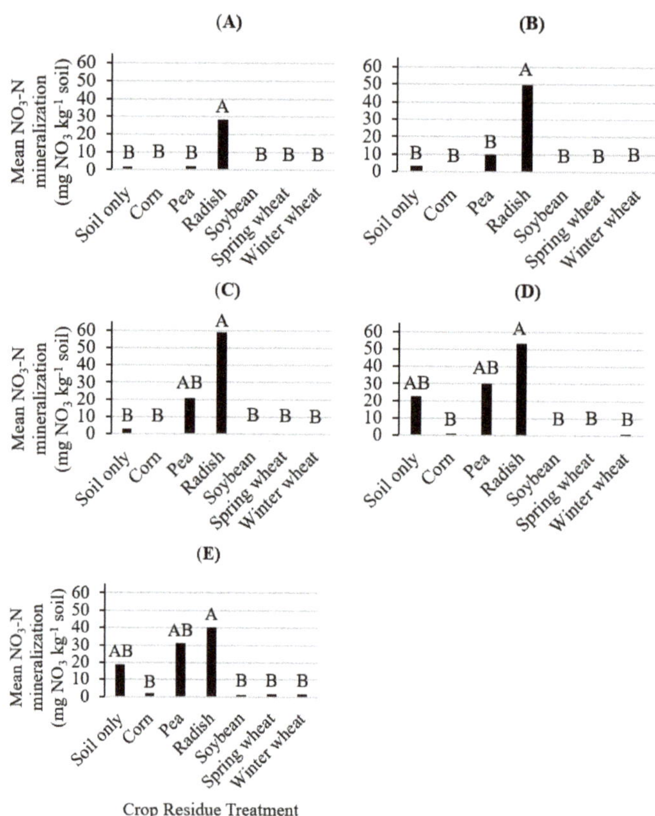

Figure 2. First incubation (**A**), second incubation (**B**), third incubation (**C**), fourth incubation (**D**), and fifth incubation (**E**) cycle mean soil NO_3-N mineralization patterns for control, corn, pea, radish, soybean, spring wheat, and winter wheat crop residue treatments. Different letters within each graph indicate significant difference at the 0.05 level using Tukey's comparison test.

For example, the forage radish (C/N = 8) NO_3-N mineralization mean is significantly higher when compared to each crop residue treatment and the bare, unamended soil during all five freeze and thaw periods. Pea (C/N = 18) shows increased N mineralization over the bare, unamended soil in all five freeze and thaw periods, with a steady increase in mineralization beginning in cycle 3 and beyond. All other crop residue treatments with higher C/N ratios (corn (C/N = 73), soybean (C/N = 53), spring wheat (C/N = 76), and winter wheat (C/N = 101)) exhibit patterns of soil NO_3-N immobilization.

A closer examination of the mean soil NO_3-N mineralization patterns over freeze and thaw periods (Figure 1) indicates a wide range in soil NO_3-N variation from the mean that is consistent for the forage radish crop residue treatment. This is because the forage radish crop residue possesses available N and other nutrients required by microorganisms, in addition to what the soil harbors itself; therefore, the microbes have plentiful food options based upon their wants and needs and N contained in the plant material. Conversely, the wider C/N ratio crops (corn, soybean, spring wheat, and winter wheat) exhibit narrower soil NO_3-N variation from the mean N mineralization from the soil and possess limited nutrient availability in the soil, creating an environment where it is much harder for microbes to extract N, illustrating the inverse relationship with the C/N ratio (i.e., as the C/N ratio decreases, range increases). With this information alone, it appears that freeze and thaw effects on mean soil NO_3-N mineralization vary based on crop residue treatment and the C/N ratio of the material.

The bare, unamended soil, absent of any crop residue treatment, shows a cumulative soil NO_3-N mineralization pattern that increases with freeze and thaw periods. Cumulative values of soil NO_3-N mineralization of the five cycles are 11, 19, 19, 147, and 122 mg NO_3-N kg^{-1}, respectively (Table S1). From this, it is evident that soil NO_3-N mineralization in the control itself is increasing in the absence of any crop residue on the soil surface. This cumulative build-up may be evidence of the microbial shift mentioned previously that is preferentially attacking the native SOM. Because there is no carbon source added to the soil, microbes compete for a relatively stable (C/N = 10–12) nutrient source, whereby the N is then released from stable SOM. When examining soil NO_3-N mineralization further over the incubation cycles for each crop residue treatment, mineralization shifts from cycles representing the early growing season towards the mid-growing season cycles with increased freeze and thaw periods. For example, during the first incubation cycle, on day 14, soil NO_3-N mineralization values are significantly higher than the other incubation cycles (days) for each crop residue treatment (Figure 3). During the second incubation cycle, on day 28, soil NO_3-N mineralization is significantly higher for all crop residue treatments and day 14 shows the lowest soil NO_3-N mineralization values overall (Figure 3). In cycle 3, no significant differences occur among incubation cycles for all crop residue treatments (Figure 3). The shift becomes evident in the fourth incubation cycle, where soil NO_3-N mineralization is significantly different on days 42, 56, and 70 from the other incubation cycles (Figure 3). In the fifth incubation cycle, soil NO_3-N mineralization is significantly higher in the mid-growing season (day 42) from all other incubation cycles, while day 84 indicates significantly lower soil NO_3-N mineralization values (Figure 3). This pattern of NO_3-N mineralization release with the increasing number of freeze and thaw periods is of importance to pinpoint when nutrients might be available to plants. From this analysis, increased freeze and thaw periods can contribute to nutrient availability in the mid-growing season period, when plants often need it the most.

When evaluating the individual crop residue treatments and their contributions to soil NO_3-N mineralization averaged over all incubation cycles these can be ranked as forage radish > pea > bare, unamended soil $\geq$ corn = winter wheat = spring wheat = soybean (47.80 > 19.86 > 10.57 > 0.94 = 0.76 = 0.70 = 0.66 mg NO_3-N kg^{-1}, respectively). Forage radish is significantly different from all crop residues treatments and the bare, unamended soil. The pea crop residue treatment is significantly different from the bare, unamended soil. The corn, soybean, spring wheat, and winter wheat crop residue treatments are always less than mg NO_3-N kg^{-1}, are similar, and significantly different from the forage radish, pea, and unamended soil control.

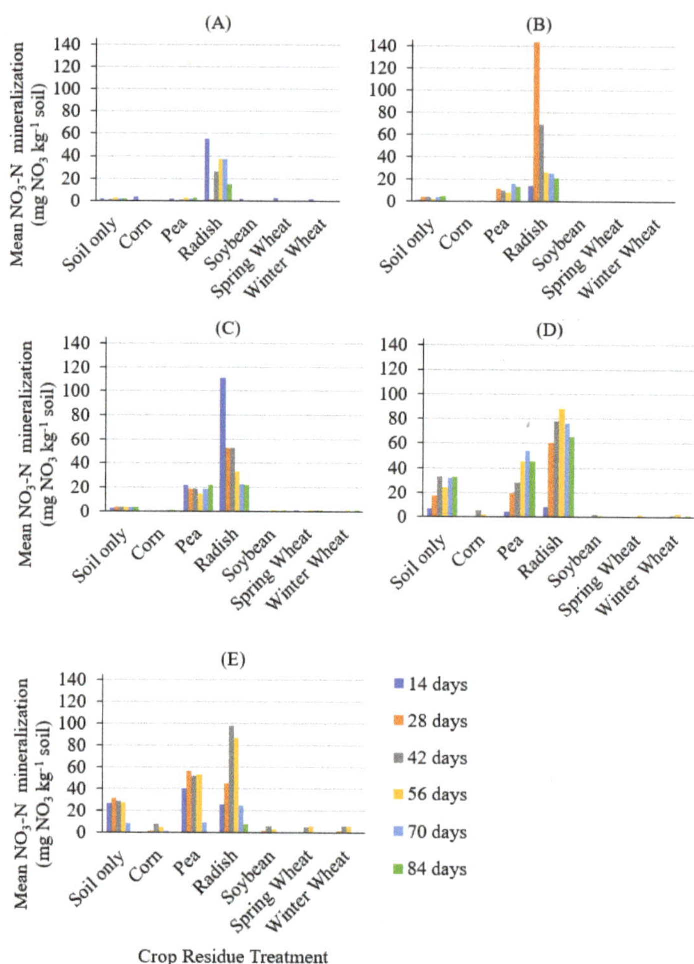

Figure 3. First incubation (**A**), second incubation (**B**), third incubation (**C**), fourth incubation (**D**), and fifth incubation (**E**) cycle mean soil NO_3-N mineralization patterns for control, corn, pea, radish, soybean, spring wheat, and winter wheat crop residue treatments by incubation period.

Figure 4 shows the average daily trends of soil NO_3-N mineralization and immobilization from the crop residue treatments during each incubation cycle ($n = 5$) to determine whether NO_3-N in the crop residue contributes to N availability in the soil. The bare, unamended soil is represented by the zero line. Above the zero line indicates N mineralization, while below the zero line suggests N immobilization. Mineralization/immobilization quantities differ based on the number of incubation cycles. Daily mineralization and immobilization NO_3-N values ranged from -0.20 to 2.53 mg NO_3-N kg^{-1} for the first cycle, from -0.34 to 18.61 mg NO_3-N kg^{-1} for the second cycle, from -0.30 to 7.74 mg NO_3-N kg^{-1} for the third cycle, from -2.32 to 4.60 mg NO_3-N kg^{-1} for the fourth cycle, and from -2.15 to 4.96 mg NO_3-N kg^{-1} for the fifth cycle. From these figures, it is evident that forage radish and pea are the only crops mineralizing at or above the bare, unamended soil control levels. All other crops (corn, soybean, winter wheat, and spring wheat) show N immobilization for each leaching of every incubation cycle.

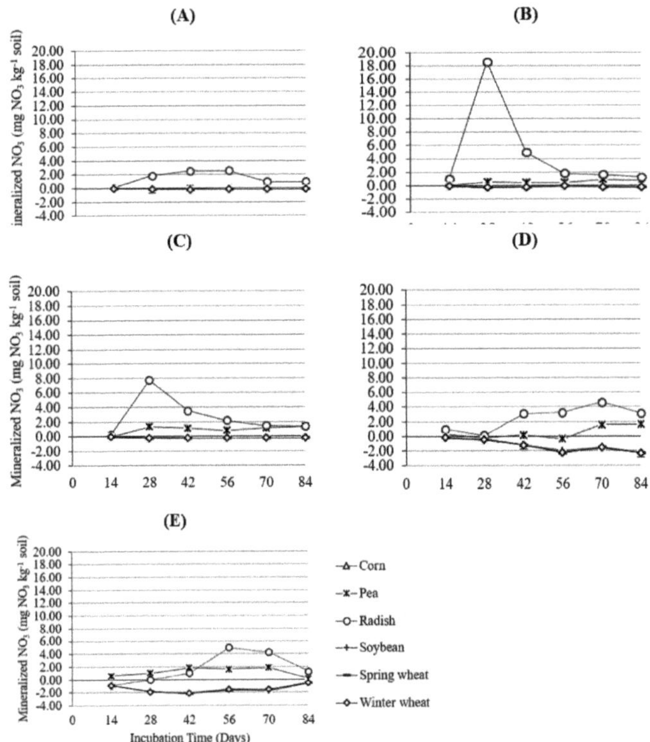

Figure 4. First incubation (**A**), second incubation (**B**), third incubation (**C**), fourth incubation (**D**), and fifth incubation (**E**) cycle daily mean mineralization/immobilization of soil NO_3-N.

These N mineralization/immobilization trends are consistent with the results provided earlier from crops with a narrower C/N ratio being the only ones to show mineralization above the bare, unamended soil. Patterns of the average daily NO_3-N mineralization/immobilization trends are also similar to the trends observed when examining the overall mean for each freeze and the cycle, where increases are observed in the earlier growing season for cycles 1 and 2 and where increases are observed in the mid to late growing season in cycles 4 and 5. These patterns are also generally consistent regardless if above or below the control, although they are heavily influenced by the forage radish results.

4. Conclusions

Accumulation of crop residue on the soil surface did not increase soil NO_3-N mineralization. Narrow C/N ratio crop residue treatments influenced the rate of soil NO_3-N mineralization, as was evident with the forage radish and pea crop residue treatments being the only ones to mineralize. This can help to inform freeze and thaw effects on NO_3-N mineralization in a frigid environment such as North Dakota, where producers are limited to short growing seasons. The number of incubation cycles may have had a direct effect on the mineralization process for all the crop residue treatments, where cycle 4 and cycle 5 were noted to release the most NO_3-N, particularly in the mid to late growing season as cycles increase. Wide C/N ratio crop residue such as corn, soybean, spring wheat, and winter wheat showed a net immobilization effect for every incubation cycle. The bare, unamended soil showed a cumulative increase in NO_3-N mineralization cycle-over-cycle. It would have been expected that the addition of a carbon source to the soil surface, in the form of crop residue, would have been at or above the rate of the control as carbon would act as a new nutrient source for microorganisms to decompose and release nutrients back to the soil (e.g., nitrate). This was not found to be true and indicates a state where degradation of the native SOM occurs, which may eventually contribute to soil degradation. In other words, the absence of a carbon source addition may lead microorganisms to seek a carbon source from the soil itself. When this occurs, it is due to the lack of nutrient substrate from organic matter. Thus, the microorganisms themselves release nutrients, including nitrogen, from the organic matter in the bare, unamended soil. Future studies may consider examining the addition of nitrogen fertilizer to determine whether it may aid in facilitating soil NO_3-N mineralization in fields under a long-term no-till system with wide C/N ratio crop residues, offsetting the mineralization of N by decomposing crop residues or releasing N from the death of microorganisms.

There are several directions in future research that can be informed by our results. Further studies should examine the effect of the freeze and thaw cycle as well as the microbial count and community to better understand soil NO_3-N mineralization source and sink. Microbes require nitrogen as a nutrient source; thus, limited nutrients in a cropping system may be observed where the volume of crop residue and the recommended fertilizer additions or credits are not equitable. In addition, further studies incorporating increased fertilizer may provide insights on the impacts of wide C/N ratio crop residues on NO_3-N mineralization source and sink. A limitation of this study was the homogenous crop residue on the soil surface, whereas in a field setting, a mixed residue application (i.e., a sequence of different crops in a rotation) would more closely align with field conditions in a production environment.

Supplementary Materials: The following supporting information can be downloaded at: https://www.mdpi.com/article/10.3390/nitrogen3020011/s1, Table S1: Mean and cumulative NO_3-N mineralization for nine incubation cycles and overall mean for corn, flax, pea, radish, soybean, spring wheat, and winter wheat crop residues for Heimdal-Emrick, Fargo, and Forman soil series in North Dakota.

Author Contributions: Conceptualization, L.C. and R.S.A.; methodology, L.C. and R.S.A.; validation, L.C.; formal analysis, Q.W.; investigation, R.S.A.; resources, L.C.; data curation, R.S.A.; writing—original draft preparation, R.S.A.; writing—review and editing, R.S.A., L.C., and Q.W.; supervision, L.C.; funding acquisition, L.C. All authors have read and agreed to the published version of the manuscript.

Funding: This research was funded by the North Dakota State Board of Research and Extension Corn Committee (grant # 18-5-0090 and 19-11-0090), Soybean Committee (grant #19-6-0139), and Wheat Committee (grant # 19-38-0162) with matching funding from the North Dakota Corn Council (grant # 18-5-0090 and 19-11-0090), North Dakota Soybean Council (grant #19-06-0139), and North Dakota Wheat Commission (grant #19-38-0162). This work was conducted in conjunction with NDSU Hatch Project No. FARG008572 contributing to Multistate Research Project NC-1178 (Land Use and Management Practices Impacts on Soil Carbon and Relate Agroecosystems Services) of the National Institute of Food and Agriculture, U.S. Department of Agriculture.

Institutional Review Board Statement: Not applicable.

Informed Consent Statement: Not applicable.

Data Availability Statement: Data supporting reported results can be accessed by contacting rashadalgamdi@hotmail.com.

Acknowledgments: This research is part of Alghamdi's doctoral thesis under the supervision of Cihacek, and it was published at North Dakota State University database.

Conflicts of Interest: The authors declare no conflict of interest. The funders had no role in the design of the study; in the collection, analyses, or interpretation of data; in the writing of the manuscript, or in the decision to publish the results.

References

1. Larney, F.J.; Lindwall, C.W.; Izaurralde, R.C.; Moulin, A.P. Tillage systems for soil and water conservation on the Canadian prairies. In *Conservation Tillage in Temperate Agroecosystems*; Carter, M.R., Ed.; Lewis Publishers: Boca Raton, FL, USA, 1994; pp. 305–328. ISBN 978-0873715713.
2. Kolberg, R.L.; Kitchen, N.R.; Westfall, D.G.; Peterson, G.A. Cropping Intensity and Nitrogen Management Impact of Dryland No-Till Rotations in the Semi-Arid Western Great Plains. *J. Prod. Agric.* **1996**, *9*, 517–521. [CrossRef]
3. Pedersen, P.; Lauer, J.G. Corn and Soybean Response to Rotation Sequence, Row Spacing, and Tillage System. *Agron. J.* **2003**, *95*, 965–971. [CrossRef]
4. Licht, M.A.; Al-Kaisi, M. Strip-tillage effect on seedbed soil temperature and other soil physical properties. *Soil Tillage Res.* **2005**, *80*, 233–249. [CrossRef]
5. Alghamdi, R.S.; Daigh, A.L.M.; DeJong-Hughes, J.; Wick, A.F. Soil temperature and water contents among vertical tillage, strip tillage, and chisel plowing in the Upper Great Plains. *Can. J. Soil Sci.* **2021**, *101*, 596–616. [CrossRef]
6. Daigh, A.L.M.; DeJong-Hughes, J.; Gatchell, D.H.; Derby, N.E.; Alghamdi, R.; Leitner, Z.R.; Wick, A.; Acharya, U. Crop and Soil Responses to On-Farm Conservation Tillage Practices in the Upper Midwest. *Agric. Environ. Lett.* **2019**, *4*, 190012. [CrossRef]
7. Alghamdi, R.S.; Cihacek, L.; Daigh, A.L.M.; Rahman, S. Post-harvest crop residue contribution to soil N availability or unavailability in North Dakota. *Agrosyst. Geosci. Environ.* **2022**, *5*, e20233. [CrossRef]
8. Bakermans, W.A.P.; De Wit, C.T.G. Crop husbandry on naturally compacted soils. *Neth. J. Agric. Sci.* **1970**, *18*, 225–246. [CrossRef]
9. Bandel, V.A.; Dzienia, S.; Stanford, G.; Legg, G.J.O. N Behavior Under No-Till vs. Conventional Corn Culture. I. First-Year Results Using Unlabeled N Fertilizer 1. *Agron. J.* **1975**, *67*, 782–786. [CrossRef]
10. Bandel, V.A. Nitrogen fertilization of no-tillage corn. In Proceedings of the Abstr. Am. Soc. Agron. N.E. Branch Meeting. Rutgers Univ., New Brunswick, NJ, USA, 24–27 June 1979; pp. 15–20.
11. Alghamdi, R.S.; Cihacek, L.J. Do post-harvest crop residues in no-till systems provide for nitrogen needs of following crops? *Agron. J.* **2021**, *114*, 835–852. [CrossRef]
12. Miller, K.S.; Geisseler, D. Temperature sensitivity of nitrogen mineralization in agricultural soils. *Biol. Fertil. Soils* **2018**, *54*, 853–860. [CrossRef]
13. Schoenau, J.J.; Campbell, C.A. Impact of crop residues on nutrient availability in conservation tillage systems. *Can. J. Plant Sci.* **1996**, *76*, 621–626. [CrossRef]
14. Blevins, R.L.; Smith, M.S.; Thomas, G.W. Changes in soil properties under no-tillage. In *No-Tillage Agriculture Principles and Practices*; Phillips, R.E., Phillips, S.H., Eds.; Van Nostrand Reinhold Company Inc.: New York, NY, USA, 1984; pp. 190–230. ISBN 978-0442277314.
15. Green, C.J.; Blackmer, A.M. Residue Decomposition Effects on Nitrogen Availability to Corn following Corn or Soybean. *Soil Sci. Soc. Am. J.* **1995**, *59*, 1065–1070. [CrossRef]
16. Li, L.-J.; Han, X.-Z.; You, M.-Y.; Yuan, Y.-R.; Ding, X.-L.; Qiao, Y.-F. Carbon and nitrogen mineralization patterns of two contrasting crop residues in a Mollisol: Effects of residue type and placement in soils. *Eur. J. Soil Biol.* **2013**, *54*, 1–6. [CrossRef]
17. Satchell, J.E. Litter—Interface of animate/inanimate matter. In *Biology of Plant Litter Decomposition*; Dickinson, C.H., Pugh, G.J.F., Eds.; Academic Press: London, UK; Academic Press: New York, NY, USA, 1974; Volume 1, pp. xiv–xl. ISBN 9780323151061.
18. Stemmer, M.; Von Lützow, M.; Kandeler, E.; Pichlmayer, F.; Gerzabek, M.H. The effect of maize straw placement on mineralization of C and N in soil particle size fractions. *Eur. J. Soil Sci.* **1999**, *50*, 73–85. [CrossRef]
19. Vanhie, M.; Deen, W.; Lauzon, J.D.; Hooker, D.C. Effect of increasing levels of maize (*Zea mays* L.) residue on no-till soybean (*Glycine max* Merr.) in Northern production regions: A review. *Soil Tillage Res.* **2015**, *150*, 201–210. [CrossRef]
20. Coppens, F.; Garnier, P.; Findeling, A.; Merckx, R.; Recous, S. Decomposition of mulched versus incorporated crop residues: Modelling with PASTIS clarifies interactions between residue quality and location. *Soil Biol. Biochem.* **2007**, *39*, 2339–2350. [CrossRef]
21. Al-Kaisi, M.M.; Kwaw-Mensah, D.; Ci, E. Effect of Nitrogen Fertilizer Application on Corn Residue Decomposition in Iowa. *Agron. J.* **2017**, *109*, 2415–2427. [CrossRef]
22. Al-Kaisi, M.M.; Guzman, J.G. Effects of tillage and nitrogen rate on decomposition of transgenic Bt and near-isogenic non-Bt maize residue. *Soil Tillage Res.* **2013**, *129*, 32–39. [CrossRef]
23. Reis, E.M.; Baruffi, D.; Remor, L.; Zanatta, M. Decomposition of corn and soybean residues under field conditions and their role as inoculum source. *Summa Phytopathol.* **2011**, *37*, 65–67. [CrossRef]
24. Kirschbaum, M.U.F. The temperature dependence of soil organic matter decomposition, and the effect of global warming on soil organic C storage. *Soil Biol. Biochem.* **1995**, *27*, 753–760. [CrossRef]
25. Vigil, M.F.; Kissel, D.E. Rate of Nitrogen Mineralized from Incorporated Crop Residues as Influenced by Temperature. *Soil Sci. Soc. Am. J.* **1995**, *59*, 1636–1644. [CrossRef]
26. Aher, G.; Cihacek, L.J.; Cooper, K. An Evaluation of C and N of Fresh and Aged Crop Residue from Mixed Long-Term No-Till Cropping Systems. *J. Plant Nutr.* **2016**, *40*, 177–186. [CrossRef]
27. NOAA (National Oceanic and Atmospheric Administration). Climate of North Dakota. 2020. Available online: https://www.ncdc.noaa.gov/climatenormals/clim60/states/Clim_ND_01.pdf (accessed on 29 January 2020).
28. Carter, M.R. *Conservation Tillage in Temperate Agrosystems*; CRC Press: Boca Raton, FL, USA, 1994; ISBN 0-87371-571-3.

29. Shields, J.A.; Paul, E.A.; Lowe, W.E. Factors influencing the stability of labelled microbial materials in soils. *Soil Biol. Biochem.* **1974**, *6*, 31–37. [CrossRef]
30. Rovira, A.; Greacen, E. The effect of aggregate disruption on the activity of microorganisms in the soil. *Aust. J. Agric. Res.* **1957**, *8*, 659–673. [CrossRef]
31. Witkamp, M. Environmental effects on microbial turnover of some mineral elements: Part I—Abiotic factors. *Soil Biol. Biochem.* **1969**, *1*, 167–176. [CrossRef]
32. Gasser, J.K.R. Use of Deep-Freezing in the Preservation and Preparation of Fresh Soil Samples. *Nature* **1958**, *181*, 1334–1335. [CrossRef]
33. Soulides, D.A.; Allison, F.E. Effect of drying and freezing soils on carbon dioxide production, available mineral nutrients, aggregation, and bacterial population. *Soil Sci.* **1961**, *91*, 291–298. [CrossRef]
34. Ivarson, K.C.; Sowden, F.J. Effect of frost action and storage of soil at freezing temperatures on the free amino acids, free sugars and respiratory activity of soil. *Can. J. Soil Sci.* **1970**, *50*, 191–198. [CrossRef]
35. Ross, D.J. Effects of freezing and thawing of some grassland topsoils on oxygen uptakes and dehydrogenase activities. *Soil Biol. Biochem.* **1972**, *4*, 115–117. [CrossRef]
36. Schimel, J.P.; Clein, J.S. Microbial response to freeze-thaw cycles in tundra and taiga soils. *Soil Biol. Biochem.* **1996**, *28*, 1061–1066. [CrossRef]
37. Franzen, D.W.; Wick, A.F.; Bu, H.; Ressler, L.; Bell, J.; Berti, M.T.; Gasch, C. Nitrogen non-cycling from cover crops grown before corn and spring wheat—unexpected early project results. In Proceedings of the 48th North Central Extension-Industry Soil Fertility Conference, Des Moines, IA, USA, 14–15 November 2018; pp. 21–26.
38. Soil Survey Staff. Forman Soil Series. Natural Resources Conservation Service, United States Department of Agriculture. Official Soil Series Descriptions; 1998. Available online: https://soilseries.sc.egov.usda.gov/osdname.aspx (accessed on 30 March 2022).
39. Stanford, G.; Smith, S.J. Nitrogen Mineralization Potentials of Soils. *Soil Sci. Soc. Am. J.* **1972**, *36*, 465–472. [CrossRef]
40. SAS Institute Inc. *SAS/ACCESS®9.4 Interface to ADABAS: Reference*; SAS Institute Inc.: Cary, NC, USA, 2013.
41. Knops, J.M.H.; Bradley, K.L.; Wedin, D.A. Mechanisms of plant species impacts on ecosystem nitrogen cycling. *Ecol. Lett.* **2002**, *5*, 454–466. [CrossRef]

Article

Spatial Variability in Inorganic Soil Nitrogen Production in a Mixed-Vegetation Urban Landscape

Juma Bukomba [1,2] and Mary G. Lusk [2,*]

[1] Agricultural Sciences Program, Formally at EARTH University, Limon 70101, Costa Rica; jbukomba@ufl.edu
[2] Soil and Water Sciences Department, Gulf Coast Research and Education Center, University of Florida, Wimauma, FL 33598, USA
* Correspondence: mary.lusk@ufl.edu

Abstract: Urban landscapes are not homogeneous, and small-scale variations in plant community or management inputs can give rise to a large range of environmental conditions. In this paper, we investigated the small-scale variability of soil nitrogen (N) properties in a single urban landscape that has distinctly different patches or types of cover. We specifically measured soil net N mineralization, nitrification, and exchangeable forms of inorganic N for patches with traditional turfgrass versus patches with common turfgrass alternatives such as ornamental grasses, groundcovers, and mulches. All soil N properties were variable among landscape patches, showing that soil N processing can vary on scales of a few meters. Notably, both mineralization and nitrification were the highest in a patch covered with perennial peanut, but exchangeable nitrate (NO_3^-) was low for the same soil, indicating that soils under perennial peanut may be producing high levels of inorganic N but that the produced N does not stay in the soil, possibly leaching to underlying groundwater. We recommend future studies on the mechanisms that drive the variable N properties seen under distinct urban landscape patches, with special emphasis on potential patterns in N losses for mixed-vegetation landscapes.

Keywords: nitrogen mineralization; nitrification; turfgrass; residential landscapes; landscape patches; urban soils; perennial peanut; urban landscapes

Citation: Bukomba, J.; Lusk, M.G. Spatial Variability in Inorganic Soil Nitrogen Production in a Mixed-Vegetation Urban Landscape. *Nitrogen* **2022**, *3*, 118–127. https://doi.org/10.3390/nitrogen3010009

Academic Editor: Jacynthe Dessureault-Rompré

Received: 18 January 2022
Accepted: 10 March 2022
Published: 11 March 2022

Publisher's Note: MDPI stays neutral with regard to jurisdictional claims in published maps and institutional affiliations.

1. Introduction

A well-managed, aesthetically pleasing landscape is associated with wellbeing in many residential landscapes, and the area of turfgrass in the United States is larger than that for any irrigated crop [1]. For many urban homeowners in the United States, turfgrass is the predominant lawn cover. To maintain a healthy, attractive turfgrass cover, regular fertilization, irrigation, and pest control management plan is often established [2]. Besides the associated economic implications, the fate and potential loss of nutrients (nitrogen and phosphorus) from fertilized turfgrass have important implications for aquatic ecosystems since nutrients mobilized by leaching and/or runoff may impair receiving waterbodies through eutrophication and algal proliferation [3–5].

While numerous studies have indicated that nutrient losses via leaching and runoff are minimal from healthy, properly maintained turfgrass and that turfgrass lawns are sinks of nitrogen (N) in urban watersheds [6–8], turfgrass fertilizers are increasingly targeted by management practices and policies aimed at reducing anthropogenic nutrient inputs to aquatic ecosystems [9]. For example, in Florida, more than 50 counties and municipalities have enacted fertilizer ordinances that often prohibit any application of N- and phosphorus (P)-bearing fertilizers to urban lawns during Florida's summer rainy season (June to September) each year [10,11]. The premise behind these bans is that summer rains may lead to increased leaching and runoff losses of N and P applied as fertilizer to lawns, in turn leading to increased anthropogenic nutrient loading to nearby waterbodies. To date, the efficacy of the fertilizer ban ordinances has not been demonstrated, and they remain controversial strategies for urban nutrient management [12,13].

To address urban nutrient management needs, we need not only more studies on the mechanisms and extent of nutrient mobilization from urban landscapes but also a better understanding of soil nutrient dynamics in these landscapes [14,15]. Nutrients associated with soil organic matter pools have been identified as one source of N and P that may be mobilized from urban lawns to stormwater runoff [5,16,17]. We also need research focused on a wider variety of potential urban landscape covers, such as mulches or ornamental plants other than turfgrass. Only a few studies have investigated differences in soil nutrient cycling between turfgrass monoculture lawns and lawns with a mix of species. In one such study, Erickson et al. [8] compared N leaching and runoff from a St. Augustine turfgrass monoculture versus a mixed-species lawn in Florida. For both lawn types, N losses via surface runoff were minimal, but leaching losses were substantial, especially for the mixed-species lawns, which lost up to 48.3 kg N ha^{-1} via leaching. By comparison, the turfgrass lawns in the study lost 4.1 kg N ha^{-1} via leaching, indicating that the turfgrass was more effective than mixed-species vegetation for preventing N leaching from lawns. In another study, Amador et al. [18], studied pore water nitrate (NO_3^-) concentrations at 60 cm soil depth under turfgrass versus various landscape covers, including flowers, shrubs, and unplanted mulched beds. In their study, flowers, managed turfgrass, and ornamental deciduous and evergreen trees represented a lower risk of NO_3^- loss from the soil than unplanted mulched areas, which the authors recommended should be used sparingly in urban landscapes because of the potential for NO_3^- leaching to groundwater. Amador et al. [18] suggested that these unplanted mulched areas were more susceptible to NO_3^- leaching because there was no plant sink for N produced by mineralization of soil organic N. That study also noted that unplanted mulched landscape beds lost via leaching nearly twice the NO_3^- input to the landscape through atmospheric deposition, making the unplanted mulched areas net sources of NO_3^- to the underlying groundwater.

The sparse studies on soil N dynamics in mixed-species urban landscapes are in line with the framework for urban soil ecology presented by Byrne [19], in which the heterogeneity commonly found in urban landscapes gives rise to a specific "habitat structure," or a unique composition of physical matter with consequent unique effects on local ecological variables. Variation in habitat structure gives rise to differences in soil pH, moisture content, microbial populations, temperature, and vegetation cover, which in turn may all cause variations in the N cycling processes within an urban landscape [15,20–22]. This means that distinct patches may emerge, creating a landscape mosaic where ecological variables can vary at scales of just a few meters or less. In this study, we selected a mixed-species urban landscape with this type of small-scale patchiness in vegetative cover and investigated the spatial variability of inorganic N production in soils. We hypothesized that differences in landscape cover at a spatial scale of meters would result in varying levels of soil inorganic N production, as measured by nitrification and net N mineralization rates. We tested this hypothesis by evaluating N cycling processes in soils under traditional turfgrass and several common ornamental alternatives to turfgrass.

This work is important because it helps constrain and fill knowledge gaps related to N cycling processes in urban soils by focusing on small-scale differences that are common in urban landscapes. While previous studies mentioned above [8,18] have focused solely on N leaching from mixed-vegetation landscapes, this work adds the body of knowledge by focusing on N cycling processes, namely mineralization, and nitrification. Typically, urban lawns and other green spaces are broadly categorized as "lawn" or "turfgrass," when in reality, they are seldom turfgrass monocultures but instead a mosaic of various vegetative types and ground covers. As we grapple with water quality degradation associated with excess nutrients from urban landscapes, research models may be used to predict the transformations and movement of N in urban soils—and accounting for the expected mosaic of variable N processing in urban soils can lead to improvements in those modeling efforts. For example, in Florida, the Nitrogen Source Inventory and Loading Tool (NSILT) is a model used to predict the fate of fertilizer N applied to urban landscapes. The model is based on the best available data on N inputs and N transformations in urban soils but is informed only by data on N cycling in turfgrass soils, without consideration for other land cover types (e.g., flower beds or mulched areas) [23]. In this work, we show how N cycling processes in a single urban landscape can be highly variable, and we argue that efforts to constrain the fate of urban nutrient sources can be improved by greater attention to mixed-vegetation landscape scenarios as well as data at finer spatial scales.

2. Materials and Methods

2.1. Study Site

The research was carried out at the University of Florida's Gulf Coast Research and Education Center (UF-GCREC) in Hillsborough County, Florida (27°45′39.2″ N 82°13′47.4″ W), which is part of the Tampa metropolitan area. The research center was constructed in 2005 on abandoned citrus farmland. Its main building was surrounded by a variety of ornamental vegetation cover that is typical of urban residential and commercial land uses in the Tampa, Florida area, including turfgrass, landscape trees, and flowering shrubs in mulched beds. The annual average precipitation of the site was 1325 mm, and the annual temperature ranged from 11–32 °C. The soils at the research center were mostly sandy spodosols of the Zolfo series (sandy, siliceous, hyperthermic oxyaquic Alorthods) [24].

2.2. Selection of Landscape Patches and Soil Sampling

We selected 6 distinct urban landscape patches surrounding UF-GCREC and 1 remnant agricultural (Agr) patch for soil sampling and analysis of inorganic N production (Table 1). The non-agricultural patches were turfgrass (TfGr) and potential turfgrass alternatives common for residential landscapes in Florida, which included 1 patch of perennial peanut (PP), a low-growing alternative ground cover; 2 patches with ornamental grasses and shrubs (OMx and OGa); and 2 patches consisting mostly of mulch with little or no other vegetation (MulP and MulC). Table 1 describes the vegetation and other characteristic features of each landscape patch.

Table 1. Description of landscape patches.

Landscape Patch	Image	Description and Management Practice
TfGr		• Bermuda turfgrass (*Cynodon dactylon*) • Rainfall supplemented with sprinkler irrigation to achieve at least 0.5 inches of water per week • Mowed to maintain a height of about 3–4 inches
PP		• Leguminous perennial peanut (*Arachis pintoi*) • Rainfall not supplemented with irrigation
OMx		• Beds of mixed ornamental vegetation, including spider plant (*Chlorophytum comosum*), white ginger plant (*Hedychium coronarium*), giant spider lily (*Crinum asiaticum*), and creeping lantana (*Lantana montevidensis*) • Rainfall supplemented with sprinkler irrigation to achieve at least 0.5 inches of water per week
OGa		• Beds of ornamental gamma grass (*Tripsacum dactyloides*) • Rainfall supplemented with sprinkler irrigation to achieve at least 0.5 inches of water per week

Table 1. *Cont.*

Landscape Patch	Image	Description and Management Practice
MulP		• Mulched with pine needles; no vegetation • Rainfall not supplemented with irrigation
MulC		• Mulched with pine bark nuggets; no vegetation • Rainfall not supplemented with irrigation
Agr		• Fallow agricultural pasture used 6 months earlier for strawberry production • Selected to provide a contrast comparison of N between residential landscapes and a remnant agricultural field

Soil sampling took place on each landscape patch in September 2018. We used a hand auger to take 5 cm-diameter cores from 4 random locations within each patch to a depth of approximately 52 cm. All 4 cores for each patch were then each divided into 3 subsamples by depth (0–15 cm, 15–30 cm, and 30–52 cm) to provide 4 field replicates for each depth per landscape patch. Subsamples of the soils were then air dried and sieved through a standard #10 (2 mm) sieve and analyzed for soil pH, total soil N, and total soil C. Soil pH was measured with a soil:water 1:2 (w/v) slurry. Total soil N and C were measured on a Thermo Flash 1112 A NC Soil Analyzer in the University of Florida Wetland Biogeochemistry Laboratory. Field moist sieved soils were used to analyze for soil gravimetric water content, percent organic matter, and exchangeable inorganic N. Percent organic matter was determined via the loss on ignition method. Exchangeable inorganic N (NO_x and NH_4^+) was determined by extracting 10 g soil with 50 mL 2 M KCl for 1 h and analyzing for NO_3^- and NH_4^+ via a continuous segmented flow analyzer (AA3, SEAL Inc., Mequon, WI, USA) by EPA methods 353.2 and 350.1, respectively.

2.3. Inorganic Soil Nitrogen Production

We defined inorganic N production as the rates of production of NH_4-N (mineralization) and NO_3-N (nitrification). Mineralization is the microbial process whereby organic N is converted to mineral N, while nitrification is the process whereby ammonium N is converted to NO_3^-N by aerobic bacteria. Both soil net mineralization and nitrification rates were determined after methods described by Raciti et al. [15]. Briefly, 10 g field moist soil was incubated in Erlenmeyer flasks for 14 days with a 12:12 light/dark cycle and at lab ambient temperature (approximately 22 °C). Flasks were loosely covered with parafilm, and DI water (<2 mL) was periodically added to the soils as needed to maintain field capacity moisture conditions during the incubation. At the conclusion of the incubation period, soils were extracted with 50 mL 2 M KCl and analyzed for NO_3^- and NH_4^+ as above. A separate set of flasks and soils was likewise analyzed but without incubation and taken as the time zero, or initial, measurements. At the conclusion of the incubation period, net mineralization was calculated as the accumulation of total inorganic N ($NH_4^+ + NO_3^-$) and nitrification was calculated as the accumulation of NO_3^-. Results were expressed in mg N kg soil^{-1} day^{-1}, after using the initial gravimetric soil water content to base calculations on the equivalent dry mass of soil in each flask. Results were examined for each landscape patch by depth as well as on a whole-core basis [15].

2.4. Statistical Analysis

Descriptive statistics to relate mean, median, and standard errors for rates of inorganic N production were developed for each landscape patch with JMP v.15 software (SAS

Institute, Inc., Cary, NC, USA). Regression analysis was used to examine possible linear relationships between mineralization and nitrification rates for each landscape patch, also with JMP v.15 software.

3. Results and Discussion

3.1. Inorganic Soil Nitrogen Production

For all landscape patches, median whole core net mineralization and nitrification were highest in the Agr soils, with median rates of 1.1 mg and 0.7 mg N kg/d, respectively (Figure 1; Table S2, Supplementary Materials). Whole core net mineralization and nitrification were next highest in the PP soils (median 0.8 and 0.5 mg N kg/d, respectively). The relatively high inorganic N production rate results for the PP patch could be related to the ability of perennial peanut to fix atmospheric N in the soil in the presence of rhizobia bacteria [25]. In the process of N fixation, atmospheric dinitrogen is cleaved and enzymatically catalyzed to produce two molecules of ammonia, leading to subsidies of NH_4^+ to the soil. When soil NH_4^+ levels are initially low, net mineralization is slow because inorganic N is preferentially incorporated into microbial biomass [26]. On the other hand, increasing levels of soil NH_4^+ have been shown to increase N mineralization rates in soils [26]. In this way, it is likely that N fixation by the perennial peanut led to high initial levels of soil NH_4^+, which in turn promoted the high N mineralization rates seen in the PP soil. We are aware of no studies on the fate of fixed N when perennial peanut is used as a turfgrass alternative in urban landscapes. However, as discussed below, exchangeable NO_3^- was lower in the PP soils than all other landscape patches, indicating that even though the PP soils had high levels of NO_3^- production (nitrification), they were not maintaining high NO_3^- levels. Potential fates of the produced NO_3^- to explain its loss in the PP soils include denitrification, plant uptake, and leaching. The well-oxygenated and low organic matter conditions of our sandy soils make denitrification unlikely, since denitrification requires oxygen-limited environments and an organic carbon source [27]. We recommend future studies to investigate whether the inorganic N produced in soils under perennial peanut is readily lost via leaching, especially in sandy soils where NO_3^- mobility would be high. While perennial peanut offers water-saving advantages over traditional turfgrass [28], it may come with the tradeoff of greater inorganic N leaching potential, though this has not been studied yet for urban soils.

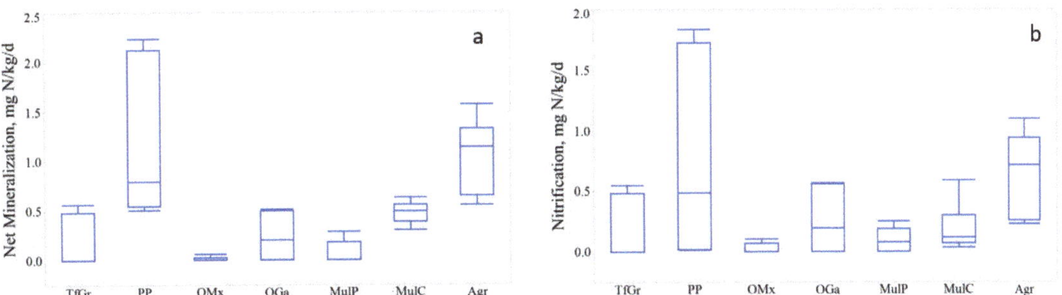

Figure 1. Net N mineralization (**a**) and nitrification (**b**) rates in landscape patches. Values based on whole-core basis. Agr; remnant agricultural field, TfGr; turfgrass, PP; perennial peanut, OMx; patches with a mixture of ornamental grasses, OGa; patch with gamma grass, MulP; patch with pine needle mulch, MulC; patch with pine bark nuggets.

After Agr and PP, the TfGr soils had the third highest median rates of net N mineralization and nitrification (Figure 1; Table S2). There were likely fertilizer subsidies of inorganic N in both the Agr and TfGr patches. As with the PP soils, relatively high initial levels of inorganic N for Agr and TfGr probably led to higher inorganic N production rates during the soil incubations when compared to non-perennial peanut turfgrass alternatives (OMx,

Oga, MulP, and MulC patches) [26]. For all landscape patches except the mulched areas (MulP and MulC), regression analysis revealed a significant linear relationship between nitrification and net mineralization on a whole-core basis (Figure 2). This is not surprising since nitrifying bacteria need ammonia as an energy source thus that nitrification generally increases with increasing ammonia levels in soils [27]. Identifying the underlying mechanism responsible for the lack of this relationship under the mulched landscape areas was beyond the scope of this paper, but factors that may inhibit nitrification include soil C:N ratios above ~22, acid soil pH conditions, and low oxygen availability [15]. We observed that the MulP landscape patch had the highest C:N among all landscape types (~22–23, Table S1), possibly indicating suppressed nitrification and a relatively less active N cycling microbial community in the MulP soils.

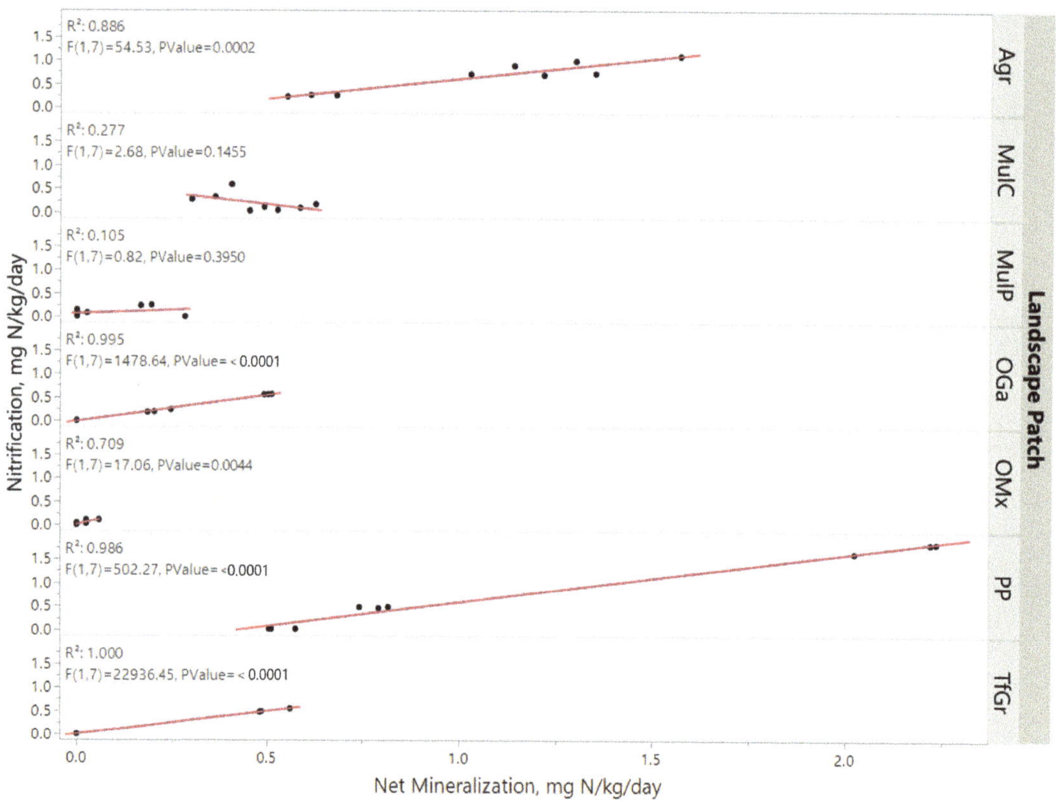

Figure 2. Regressions of nitrification against net N mineralization in the landscape patches, on a whole-core basis. Agr; remnant agricultural field, TfGr; turfgrass, PP; perennial peanut, OMx; patch with a mixture of ornamental grasses, OGa; patch with gamma grass, MulP; patch with pine needle mulch, MulC; patch with pine bark nuggets.

3.2. Exchangeable Inorganic Soil Nitrogen

Among the non-agricultural soils (all patches except Agr), the PP soils contained the highest median KCl-exchangeable NH_4^+ (1.4 mg/kg) but the lowest KCl-exchangeable NO_3^- (0.4 mg/kg) on a whole-core basis (Figure 3). As discussed above, this relatively low exchangeable NO_3^- pool for the PP soils indicates that N losses via leaching may be taking place. While the use of perennial peanut for urban landscapes has not been studied in terms of its potential for increased N leaching, several studies have looked at this for agricultural settings where perennial peanut is used for forage crops or as a cover crop.

Woodard et al. [29] observed that a forage system with perennial peanut leached more N than a comparable one with bermudagrass. In that study, the authors concluded that perennial peanut might pose threats to groundwater quality under agricultural systems because of its apparent role in subsidizing high NO_3^- levels in the soil.

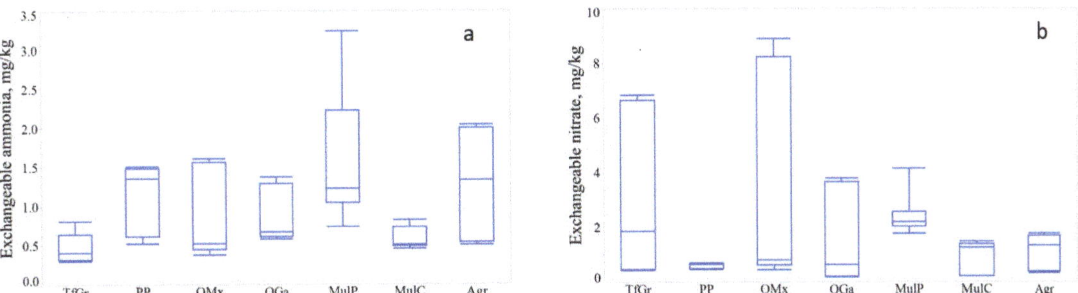

Figure 3. Exchangeable soil ammonium (**a**) and nitrate (**b**) in the landscape patches. Values are on a whole-core basis. Agr; remnant agricultural field, TfGr; turfgrass, PP; perennial peanut, OMx; patch with a mixture of ornamental grasses, OGa; patch with gamma grass, MulP; patches with pine needle mulch, MulC; patch with pine bark nuggets.

The mulched landscape patches (MulP and MulC), for which vegetation was absent, contained higher mean pools of exchangeable NH_4^+ and NO_3^- than the turfgrass and ornamental patches, though there was considerable spread in the data (Figure 3). It is likely that the mulched landscapes had higher mean exchangeable inorganic N due to the lack of a plant sink to remove inorganic N from the soils [18]. It is also likely that the mulches contained their own pools of N, which leached from the mulch and added to soil N pools [30]. In a study of several urban landscape alternatives, Loper et al. [30] observed that organic soil amendments such as composts could increase inorganic N leaching when compared to soils without amendments. In that study, the authors recommended that landowners did not apply nutrients to mulched areas and that landscape managers recognize that different patches of the same landscape may have different management needs.

In general, exchangeable NH_4^+ was highest in the surface 0–15 cm for all soils, while NO_3^- was generally higher with soil depth, perhaps indicating a tendency for NO_3^- to move downward in the soil profiles that are predominantly sand (Figure 4). At the 30–52 cm soil depth, exchangeable NO_3^- was higher in 4 out of 6 non-agricultural soils than it was in the Agr soil, indicating that the urban landscape patches may pose a greater threat of NO_3^- leaching than the agricultural soil (Figure 4). Turfgrass N fertilization has been found to contribute to the leaching of NO_3^- in Florida's sandy soils. For example, Shaddox and Surtain [31] found that applied N in Bermuda grass turf leached between 8% and 12% on a 10% slope. Notably, the NO_3^- concentrations in the 30–52 cm depth for the mixed ornamental (OMx) and the gamma grass patch (OGa) were in sharp contrast to each other, with much more NO_3^- at depth for the mixed ornamental patch than for the gamma grass patch. Gamma grass is one of the most deeply rooted ornamental grasses, with root depth up to 1.8 m [32], and its deep rooting may be contributing to N uptake deep in the soil profile.

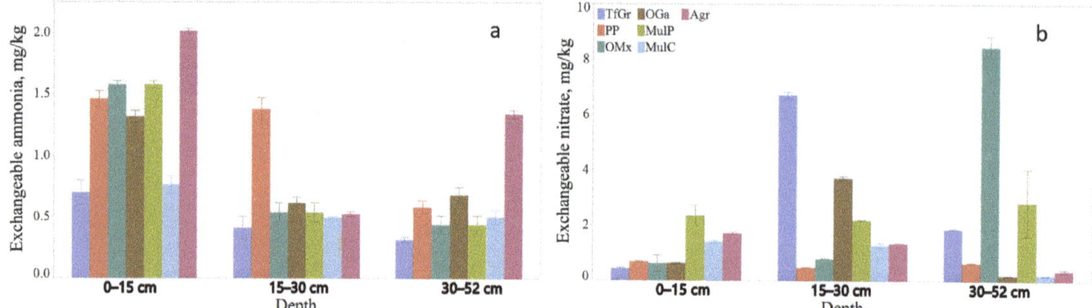

Figure 4. Exchangeable soil ammonium (**a**) and nitrate (**b**) in each landscape patch by soil depth. Agr; remnant agricultural field, TfGr; turfgrass, PP; perennial peanut, OMx; patch with a mixture of ornamental grasses, OGa; patches with gamma grass, MulP; patch with pine needle mulch, MulC; patch with pine bark nuggets.

4. Management Implications and Future Research Needs

It is beyond the scope of this paper to make recommendations about which landscape patch is preferable from a N availability or groundwater protection standpoint, as our main intent was to only investigate fine-scale variability in N production in a mixed-vegetation urban landscape. However, some key takeaways are manifest in our results. The first of these is that inorganic N availability and production were, in fact, highly variable among landscape patches at scales of a few meters, in support of our hypothesis. This finding supports the habitat structure framework of urban soil ecology presented by Bryne [19], who argued that the high spatial heterogeneity of human activities in urban landscapes leads to a large range of soil properties at often small scales. The presence of soil amendments such as mulches or plants such as perennial peanuts that can subsidize soil N pools is just two ways that humans influence urban soil properties. We have shown here that inorganic N production is variable in a single urban landscape and have presented some discussion as to why this may be the case for our specific landscape patch types. At a minimum, we argue that urban landscapes should not be broadly managed but that landscape managers should recognize that the nutrient needs of patches within a single landscape may be different and that the nutrient fates in soil under those patches may also be different. This has important implications for how urban soil properties are included in models that aim to predict nutrient transport from urban landscapes. Very likely, we need spatially finer-scale investigations of nutrient fate and transport in urban landscapes, especially those with mixed vegetation or heterogeneous human influences. Too often, urban landscapes are assumed to be covered with turfgrass only, and other possible land covers need to be more fully considered by those who aim to quantify the effects of urbanization on the environment.

A second key takeaway of this work is the implication related to potential N losses by leaching from certain landscape patches. The high nitrification potential of perennial peanut seen here suggests that this turfgrass alternative may produce NO_3^- that can be readily leached from sandy soils. However, our sample size was small, and similar work should be conducted in the future to expand the temporal and spatial scope of this work. In particular, we recommend studying N leaching under perennial peanut when it is used as an urban groundcover. We also recommend future work to better understand the mechanisms that drive the differences in potential N losses we observed among landscape patches. Some of the potential drivers include irrigation status, soil pH, moisture content, organic matter content, and variations in the soil microbial community. Finally, because urban landscapes are tied to human activities, we recognize that sociocultural variables may drive differences in soil properties as well [33,34]. These sociocultural variables may include the choice of plants that are used in ornamental beds. For example, we observed that the two ornamental patches had very different levels of inorganic N at the greatest soil depth, likely due to greater uptake of N at depth by the deeply rooted gamma grass. This human choice as to which plants to include in a landscape may therefore impact N losses by leaching through the soil profile.

Supplementary Materials: The following are available online at https://www.mdpi.com/article/10.3390/nitrogen3010009/s1, Table S1, Basic soil physical/chemical properties for landscape patches used in this study; Table S2, Raw data showing nitrification, mineralization, and exchangeable inorganic N for all landscape patches and soil depths.

Author Contributions: Conceptualization, M.G.L. and J.B.; methodology, M.G.L.; formal analysis, J.B. and M.G.L.; investigation, J.B.; resources, M.G.L.; data curation, J.B.; writing—original draft preparation, J.B.; writing—review and editing, M.G.L.; visualization, J.B.; supervision, M.G.L.; project administration, M.G.L.; funding acquisition, M.G.L. All authors have read and agreed to the published version of the manuscript.

Funding: This research was funded in part by the Center for Land Use Efficiency at the University of Florida.

Institutional Review Board Statement: Not applicable.

Informed Consent Statement: Not applicable.

Data Availability Statement: Data used in this manuscript is available in Table S2.

Conflicts of Interest: The authors declare no conflict of interest.

References

1. Milesi, C.; Running, S.W.; Elvidge, C.D.; Dietz, J.B.; Tuttle, B.T.; Nemani, R.R. Mapping and modeling the biogeochemical cycling of turf grasses in the United States. *Environ. Manag.* **2005**, *36*, 426–438. [CrossRef] [PubMed]
2. Stier, J.C.; Horgan, B.P.; Bonos, S.A. *Turfgrass: Biology, Use, and Management*; American Society of Agronomy Madison: Wisconsin, DC, USA, 2013.
3. Shuman, L. Phosphorus and nitrate nitrogen in runoff following fertilizer application to turfgrass. *J. Environ. Qual.* **2002**, *31*, 1710–1715. [CrossRef] [PubMed]
4. Bell, G.E.; Moss, J.Q.; Nett, M.; Carroll, M.; Horgan, B.; Martin Petrovic, A. (Eds.) Management practices that reduce runoff transport of nutrients and pesticides from turfgrass. In *The Fate of Nutrients and Pesticides in the Urban Environment*; ACS Symp Series; ACS Publications: Washington, DC, USA, 2008.
5. Krimsky, L.S.; Lusk, M.G.; Abeels, H.; Seals, L. Sources and concentrations of nutrients in surface runoff from waterfront homes with different landscape practices. *Sci. Total Environ.* **2021**, *750*, 142320. [CrossRef] [PubMed]
6. Easton, Z.M.; Petrovic, A.M. Determining nitrogen loading rates based on land use in an urban watershed. *Fate Nutr. Pestic. Urban Environ.* **2008**, *997*, 19–42.
7. Lusk, M.G.; Toor, G.S.; Inglett, P.W. Characterization of dissolved organic nitrogen in leachate from a newly established and fertilized turfgrass. *Water Res.* **2018**, *131*, 52–61. [CrossRef]
8. Erickson, J.; Cisar, J.; Volin, J.; Snyder, G. Comparing nitrogen runoff and leaching between newly established St. Augustinegrass turf and an alternative residential landscape. *Crop Sci.* **2001**, *41*, 1889–1895. [CrossRef]
9. Yang, Y.-Y.; Lusk, M.G. Nutrients in Urban Stormwater Runoff: Current State of the Science and Potential Mitigation Options. *Curr. Pollut. Rep.* **2018**, *4*, 112–127. [CrossRef]
10. Hartman, R. The Spread of Fertilizer Ordinances in Florida. *Sea Grant Law Policy J.* **2008**, *1*, 98.
11. Dukes, M.D.; Krimsky, L.; Lusk, M.G.; Trenholm, L.; Unruh, B.; Atkinson, M.; Mylavarapu, R. Urban Fertilizer Ordinances in the Context of Environmental Horticulture and Water Quality Extension Programs: Frequently Asked Questions. *EDIS* **2020**, *2020*. [CrossRef]
12. Hochmuth, G.; Nell, T.; Sartain, J.; Unruh, J.B.; Martinez, C.; Trenholm, L.; Cisar, J. Urban Water Quality and Fertilizer Ordinances: Avoiding Unintended Consequences: A Review of the Scientific Literature. *EDIS* **2009**, *2009*. [CrossRef]
13. Ryan, C.D.; Unruh, J.B.; Kenworthy, K.E.; Lamm, A.J.; Erickson, J.E.; Trenholm, L.E. Culture, Science, and Activism in Florida Lawn and Landscape Fertilizer Policy. *HortTechnology* **2019**, *29*, 854–865. [CrossRef]
14. Raciti, S.M.; Groffman, P.M.; Fahey, T.J. Nitrogen Retention in Urban Lawns and Forests. *Ecol. Appl.* **2008**, *18*, 1615–1626. [CrossRef] [PubMed]
15. Raciti, S.M.; Groffman, P.M.; Jenkins, J.C.; Pouyat, R.V.; Fahey, T.J.; Pickett, S.T.; Cadenasso, M.L. Nitrate production and availability in residential soils. *Ecol. Appl.* **2011**, *21*, 2357–2366. [CrossRef] [PubMed]
16. Yang, Y.-Y.; Toor, G.S. Sources and mechanisms of nitrate and orthophosphate transport in urban stormwater runoff from residential catchments. *Water Res.* **2017**, *112*, 176–184. [CrossRef]

17. Jani, J.; Yang, Y.-Y.; Lusk, M.G.; Toor, G.S. Composition of nitrogen in urban residential stormwater runoff: Concentrations, loads, and source characterization of nitrate and organic nitrogen. *PLoS ONE* **2020**, *15*, e0229715. [CrossRef]

18. Amador, J.A.; Hull, R.J.; Patenaude, E.L.; Bushoven, J.T.; Görres, J.H. Potential nitrate leaching under common landscaping plants. *Water Air Soil Pollut.* **2007**, *185*, 323–333. [CrossRef]

19. Byrne, L.B. Habitat structure: A fundamental concept and framework for urban soil ecology. *Urban Ecosyst.* **2007**, *10*, 255–274. [CrossRef]

20. Hope, D.; Zhu, W.; Gries, C.; Oleson, J.; Kaye, J.; Grimm, N.B.; Baker, L.A. Spatial variation in soil inorganic nitrogen across an arid urban ecosystem. *Urban Ecosyst.* **2005**, *8*, 251–273. [CrossRef]

21. Kaye, J.P.; McCulley, R.; Burke, I. Carbon fluxes, nitrogen cycling, and soil microbial communities in adjacent urban, native and agricultural ecosystems. *Glob. Chang. Biol.* **2005**, *11*, 575–587. [CrossRef]

22. Scharenbroch, B.C.; Lloyd, J.E.; Johnson-Maynard, J.L. Distinguishing urban soils with physical, chemical, and biological properties. *Pedobiologia* **2005**, *49*, 283–296. [CrossRef]

23. Katz, B.G.; Eller, K. The Nitrogen Source Inventory and Loading Tool (NSILT) and restoration of water-quality impaired springs. *Fla. Sci.* **2016**, *79*, 299–310.

24. Web Soil Survey. Natural Resources Conservation Service, US Department of Agriculture. Available online: https://websoilsurvey.sc.egov.usda.gov/App/HomePage.htm (accessed on 1 November 2018).

25. Miavitiz, E. (Ed.) Rhizomal perennial peanut in the urban landscape. In Proceedings of the Florida State Horticultural Society, Marco Island, FL, USA, 2–4 June 2002; Volume 115, pp. 136–138.

26. White, J.; Reddy, K. Influence of selected inorganic electron acceptors on organic nitrogen mineralization in Everglades soils. *Soil Sci. Soc. Am. J.* **2001**, *65*, 941–948. [CrossRef]

27. Reddy, K.R.; DeLaune, R.D. *Biogeochemistry of Wetlands: Science and Applications*; CRC Press: Boca Raton, FL, USA, 2008.

28. Rouse, R.E.; Miavitz, E.M.; Roka, F.M. Guide to using rhizomal perennial peanut in the urban landscape. *EDIS* **2004**, *2004*. [CrossRef]

29. Woodard, K.R.; French, E.C.; Sweat, L.A.; Graetz, D.A.; Sollenberger, L.E.; Macoon, B.; Portier, K.M.; Rymph, S.J.; Wade, B.L.; Prine, G.M. Nitrogen removal and nitrate leaching for two perennial, sod-based forage systems receiving dairy effluent. *J. Environ. Qual.* **2003**, *32*, 996–1007. [CrossRef]

30. Loper, S.J.; Shober, A.L.; Wiese, C.; Denny, G.C.; Stanley, C.D. Nutrient leaching during establishment of simulated residential landscapes. *J. Environ. Qual.* **2013**, *42*, 260–270. [CrossRef]

31. Shaddox, T.; Sartain, J. (Eds.) *Fate of Nitrogen During Grow-In of a Golf Course Fairway under Different Nitrogen Management Practices*; Soil and Crop Science Society of Florida: Tallahassee, FL, USA, 2001.

32. Brown, R.N.; Percivalle, C.; Narkiewicz, S.; DeCuollo, S. Relative rooting depths of native grasses and amenity grasses with potential for use on roadsides in New England. *Hortscience* **2010**, *45*, 393–400. [CrossRef]

33. Grimm, N.B.; Grove, J.G.; Pickett, S.T.; Redman, C.L. Integrated approaches to long-term studies of urban ecological systems: Urban ecological systems present multiple challenges to ecologists—Pervasive human impact and extreme heterogeneity of cities, and the need to integrate social and ecological approaches, concepts, and theory. *BioScience* **2000**, *50*, 571–584.

34. Law, N.; Band, L.; Grove, M. Nitrogen input from residential lawn care practices in suburban watersheds in Baltimore County, MD. *J. Environ. Plan. Manag.* **2004**, *47*, 737–755. [CrossRef]

 *Nitrogen*

Article

Oilseed Rape Cultivars Show Diversity of Root Morphologies with the Potential for Better Capture of Nitrogen

László Kupcsik [1], Claudia Chiodi [1], Taraka Ramji Moturu [1], Hugues De Gernier [1,†], Loïc Haelterman [1], Julien Louvieaux [1,2], Pascal Tillard [3], Craig J. Sturrock [4], Malcolm Bennett [4], Philippe Nacry [3] and Christian Hermans [1,*]

[1] Crop Production and Biostimulation Laboratory, Interfaculty School of Bioengineers, Université libre de Bruxelles, Campus Plaine CP 245, Bd du Triomphe, 1050 Brussels, Belgium; koopac@gmail.com (L.K.); claudia.chiodi@ulb.be (C.C.); taraka.ramji.moturu@ulb.be (T.R.M.); hugues.degernier@psb.vib-ugent.be (H.D.G.); loic.haelterman@ulb.be (L.H.); julien.louvieaux@ulb.be (J.L.)

[2] Laboratory of Applied Plant Ecophysiology, Haute Ecole Provinciale de Hainaut Condorcet, Centre pour l'Agronomie et l'Agro-industrie de la Province de Hainaut, 11 rue Paul Pastur, 7800 Ath, Belgium

[3] Biochemistry and Plant Molecular Physiology, UMR 5004 CNRS/INRAE/SupAgro-M/UM2, Integrative Biology Institute for Plants, Place Viala, CEDEX 1, 34060 Montpellier, France; tillardp@orange.fr (P.T.); philippe.nacry@inrae.fr (P.N.)

[4] Sutton Bonington Campus, School of Biosciences, University of Nottingham, Leicestershire LE12 5RD, UK; craig.sturrock@nottingham.ac.uk (C.J.S.); malcolm.bennett@nottingham.ac.uk (M.B.)

* Correspondence: christian.hermans@ulb.be; Tel.: +32-2-650-5416

† Present address: Department of Plant Biotechnology and Bioinformatics, Ghent University, 71 Technologiepark, 9052 Ghent, Belgium.

Citation: Kupcsik, L.; Chiodi, C.; Moturu, T.R.; De Gernier, H.; Haelterman, L.; Louvieaux, J.; Tillard, P.; Sturrock, C.J.; Bennett, M.; Nacry, P.; et al. Oilseed Rape Cultivars Show Diversity of Root Morphologies with the Potential for Better Capture of Nitrogen. *Nitrogen* **2021**, 2, 491–505. https://doi.org/10.3390/nitrogen2040033

Academic Editors: Jacynthe Dessureault-Rompré and Marouane Baslam

Received: 1 October 2021
Accepted: 8 December 2021
Published: 14 December 2021

Publisher's Note: MDPI stays neutral with regard to jurisdictional claims in published maps and institutional affiliations.

Abstract: The worldwide demand for vegetable oils is rising. Oilseed rape (*Brassica napus*) diversifies cereal dominated crop rotations but requires important nitrogen input. Yet, the root organ is offering an untapped opportunity to improve the nitrogen capture in soil. This study evaluates three culture systems in controlled environment, to observe root morphology and to identify root attributes for superior biomass production and nitrogen use. The phenotypic diversity in a panel of 55 modern winter oilseed rape cultivars was screened in response to two divergent nitrate supplies. Upon in vitro and hydroponic cultures, a large variability for root morphologies was observed. Root biomass and morphological traits positively correlated with shoot biomass or leaf area. The activities of high-affinity nitrate transport systems correlated negatively with the leaf area, while the combined high- and low-affinity systems positively with the total root length. The X-ray computed tomography permitted to visualize the root system in pipes filled with soil. The in vitro root phenotype at germination stage was indicative of lateral root deployment in soil-grown plants. This study highlights great genetic potential in oilseed rape, which could be manipulated to optimize crop root characteristics and nitrogen capture with substantial implications for agricultural production.

Keywords: *Brassica napus*; natural variation; nitrogen nutrition; root system architecture

1. Introduction

Agriculture is facing the challenge of producing more food while reducing the negative environmental impact of nitrogen (N) fertilization. Synthetic N fertilizers come with elevated prices. Such costs are not only economical but also environmental, as fertilizers cause groundwater pollution by nitrate leaching [1,2] and air pollution by nitrous oxide emission [3,4]. Hence, there is a need to reduce the environmental footprint by lowering the fertilizer inputs. Breeding crops with greater Nitrogen Use Efficiency (NUE) is one solution to achieve that goal [5]. Agronomically, it is calculated as the ratio of crop yield to N units available. The NUE is a combination of the Nitrogen Uptake Efficiency (NUpE)—the ability of the crop to take up N from the soil and the Nitrogen Utilization Efficiency (NUtE)—the ability of the crop to utilize the absorbed N for producing yield [6,7]. The root system architecture defines the spatial distribution of roots in the soil, which translates the ability of the plant to acquire soil resources [8]. At present, root-based approaches are being developed to improve NUE in various crops [9,10]. Still, methods for phenotyping crop root system architecture can be challenging.

Oilseed rape (*Brassica napus*) has poor NUE, with a low ratio of seeds produced per N unit applied [11]. Thus, improving NUE is essential to ensure the environmental and economic sustainability of that crop production [12,13]. Breeding programs are largely focusing on NUtE in above-ground organs [14–16] and research on soil and below-ground processes are limited [17]. Yet, the root organ holds potential for NUpE improvement [18,19]. Oilseed rape offers a large genetic diversity of root morphologies [20–23]. Nonetheless, literature on the influence of N nutrition on root growth and development in that crop is still scarce compared to parented *Arabidopsis thaliana*. In the model species, a dual effect of nitrate on lateral root development is described: (i) a systemic inhibition of uniformly elevated nitrate concentrations occurs on lateral root elongation at the post-emergence developmental stage and (ii) a localized stimulation of nitrate-rich patches triggers lateral root growth in a low N environment, known as the foraging capacity [24–27]. The exploitation of Arabidopsis has unveiled some molecular networks shaping root morphology in response to nutritional factors [28–32].

Several study cases showed that root morphology optimization results in greater plant productivity and increased nutrient stress tolerance [33,34]. A rapid root proliferation may be a determining factor for accessing N from deep layers of quickly leaching or drying soils [18]. The NUpE can be improved by redesigning a more branched root system that explores a larger soil volume to prevent N leaching [18,35]. There are indications that at low N supply, high-yielding oilseed rape cultivars are characterized by expansive root growth during the vegetative stage or after the flower stem extension [36,37] and by great N uptake [38]. These observations support the hypothesis that large root phenotypes may enhance N acquisition. Nevertheless, conflicting views may be expressed regarding a possible trade-off between profuse root system, contributing to the capacity of N uptake, and metabolic costs associated with the growth and maintenance of the root organ [39].

This study explores the natural variation of root morphology in response to the nitrate supply among a diversity panel of 55 modern winter oilseed rape cultivars. The technical objective was to set up and assess the relevance of three culture systems to observe root morphology at different growth stages. Eventually, we sought to strengthen the premise that (i) root biomass production and morphological traits could be positive indicators of above-ground biomass production and (ii) larger root system size would confer greater N uptake.

2. Materials and Methods

2.1. Plant Material

Seeds of 55 winter oilseed rape cultivars were obtained from the Walloon Agricultural Research Centre (CRA-W), Gembloux, Belgium. The description of these lines is given in Table S1.

2.2. In Vitro Culture and 2D Root Morphology Analysis

The in vitro culture was conducted in a growth chamber at a constant temperature of 21 °C and a photoperiod of 16 h light (40 µmol photons m^{-2} s^{-1})/8 h darkness. Seeds of 55 cultivars were sterilized in 2 mL microtubes after sequentially adding 70% (*v:v*) ethanol for 10 min, 20% (*v:v*) hypochlorite solution for 5 min and rinsing twice with sterile water. Then, seeds were plated on square (12 cm × 12 cm) dishes filled with a modified Murashige-Skoog (MS) medium containing 0.010 mM KNO_3 + 9.090 mM KCl (hereby referred as low nitrate treatment, N−) or 10 mM KNO_3 (high nitrate, N+) and 2% (*w:v*) plant agar (Duschefa, Haarlem, the Netherlands) [40,41]. Seeds were stratified during two days at 4 °C. After three days, seedlings were transferred to a fresh growth medium of identical composition. The top portion (~3 cm) of the agar medium was cut and four seedlings were inserted between the gel matrix and the bottom of the plate (Figure 1a). Three plates containing four seedlings per cultivar and per N condition were prepared. Seedlings grew during four days prior to harvest, corresponding to seven days after germination. Root and shoot organs were separated and dried at 60 °C. The pooled dry weights of four organs from one plate were measured. For root morphology observation, the dishes were scanned at a resolution of 400 dpi. The scans were annotated using the RootNav image analysis software [42].

Figure 1. Culture systems used to grow oilseed rape cultivars. (**a**) Vertically placed agar plates (size: 12 cm × 12 cm); (**b**) Hydroponic containers (capacity: 8 L); (**c**) Column filled with soil (d: 5 cm, h: 30 cm).

2.3. Hydroponic Culture, Physiological Characterization, and Nitrate Uptake Assays

The hydroponic culture was performed in a growth chamber at a constant temperature of 20 °C, a photoperiod of 8 h light (150 µmol photons m^{-2} s^{-1})/16 h darkness and a relative humidity of 60%. First, seeds of 12 selected cultivars were sown in peat-based soil (DCM Maison et Jardin, EAN:5413448070033) and stratified during two days at 4 °C. After five days, the roots were carefully rinsed with water and seedlings transferred for seven days into hydroponic containers, with dimensions 35 cm × 25 cm × 13.5 cm and filled with 8 L of nutrient solution (Figure 1b). Plants were first fed during nine days with a nutrient solution containing 1 mM $Ca(NO_3)_2$ [43,44], then during seven days with solutions containing 0.1 mM $Ca(NO_3)_2$ + 1.9 mM $CaCl_2$ (N−) or 1 mM $Ca(NO_3)_2$ (N+). Five plants per cultivar and per N condition were cultivated. Plants were harvested after nine days of treatment, corresponding to 21 days after germination. Roots and leaves were separated and scanned at a resolution of 400 dpi. Total root lengths were measured with the image analysis software Optimas 6.0 (Meyer Instruments, Inc., Houston, TX, USA). Organs were dried at 60 °C and dry weight was measured. Between 10 and 50 mg of crushed organ samples were analyzed with a Vario MAX Cube (Elementar, Langenselbold, Germany) for simultaneous carbon and nitrogen determination at the Centre pour l'Agronomie et l'Agro-Industrie de la province de Hainaut (CARAH), 11 rue Paul Pastur, 7800 Ath, Belgium.

The ^{15}N labelling procedure for the nitrate uptake assay was conducted as described in [45]. Plants were fed during 21 days with a nutrient solution containing 1 mM $Ca(NO_3)_2$ [43,44]. Then, they were incubated for 90 min in a solution containing 0.1 mM or 5 mM KNO_3. Roots were sequentially bathing for 1 min in 10 mL of $CaSO_4$ solution (0.1 mM), for 5 min in 10 mL of a solution containing 0.1 mM or 5 mM $K^{15}NO_3$ (Sigma-Aldrich, Saint-Quentin-Fallavier, France) and for 1 min in 10 mL $CaSO_4$ solution (0.1 mM). Root and shoot organs were immediately separated and dried at 60 °C, prior to dry weight measurement. Samples were analyzed with an integrated system for continuous flow isotope ratio mass spectrometry (Euro-EA elementar Analyzer, EuroVector IRMS Isoprime Elementar) at the

Stable Isotope Analytical platform of the Biochemistry and Plant Molecular Physiology, INRAE Montpellier, Place Viala, CEDEX 1, 34060 Montpellier, France.

2.4. Soil Culture and Micro-Scale Computed Tomography Imaging

The soil culture was conducted between 20 April and 18 May 2015, in a greenhouse located on the Sutton Bonington campus, Nottingham University (UK), with the control on light and temperature levels. Seeds of four selected cultivars were germinated in soil-filled pipes (d = 5 cm, h = 30 cm) for the purpose of X-ray computed tomography (Figure 1c). The soil was collected on a site ($50°36'47''$ N, $3°46'12''$ E) without recorded agricultural activity during the last decades. The loamy soil characteristics were total C = 1.05%, organic N = 1.13‰, $N\text{-}NH_4^+$ = 0.73 mg kg^{-1}, $N\text{-}NO_3^-$ = 10.48 mg kg^{-1}, Ca = 224 mg 100 g^{-1}, K = 31.76 mg 100 g^{-1}, Mg = 10.89 mg 100 g^{-1} and p = 16.08 mg 100 g^{-1}. The columns were watered from the top with 500 mL of a nutrient solution without nitrate (N−) or with 50 mM KNO_3 (N+) [43,44]. Watering was stopped two days prior to scanning. All images were acquired using a Phoenix v | tome | x m scanner (GE Measurement and Control Solutions, Billerica, MA, USA). The accelerating voltage was 150 kV and the current 160 μA. The X-rays were filtered through a 0.1 mm copper plate. For each sample, there were 4×2160 projections taken with an exposure time of 200 ms, and signal averaging of two and one skip per projection. These were then assembled into a single 3D image with a 58 μm spatial resolution. The root images were segmented and measured manually using VGStudio MAX.

2.5. Statistical Treatment

Analysis of variance for the in vitro and hydroponic culture dataset was performed using a linear model implemented with the lmer function in the {lme4} package in R statistical language. The model consisted of fixed effects for cultivar and nitrate treatment and their interactions. Pearson correlations were calculated for all combinations of phenotypic traits in every nitrate condition, using R and XLSTAT. The principal component analyses of the in vitro and hydroponic culture dataset were conducted with XLSTAT. The Tukey's Honesty Significant Difference (HSD) test for the nitrate uptake assay was performed using R.

3. Results

Biomass production and root morphology were examined in a set of 55 winter oilseed rape cultivars (Table S1). Three culture systems were employed to examine the root system in response to the nitrate supply. In a pilot screen, the full set grew in vitro (Figure 1a). Eventually, a dozen of cultivars with contrasting root morphologies were further characterized in hydroponics (Figure 1b), and four of them upon soil culture (Figure 1c).

3.1. Variation for Biomass Production and Root Morphologies in Seedlings Cultivated In Vitro

Seedlings grew on vertical dishes filled with agar medium, as illustrated in Figure 1a. The germination characteristics of seedlings grown with 0.01 mM (N−) or 10 mM (N+) nitrate are presented in Figure 2. In this case, 12 selected cultivars were ranked from the poorest to the greatest total root length at N−: Cardiff (CAR), SY Carlo (CLO), Exocet (EXO), NK Aviator (AVI), DK Exquisite (EXQ), Recordie (REC), Limone (LIM), SY Saveo (SAV), Hertz (HER), Troy (TRO), Battz (BAT) and ES Jason (JAS). On average for the diversity panel cultivated in vitro, the root biomass (R, +28%) and the root-to-shoot biomass ratio (R:S, +39%,) were superior, while the shoot biomass (S, −18%) and the total biomass (R + S, −8%) were inferior at N− compared to N+. The length of primary root (L_{PR}, +18%), the lengths of primary root zone 2 and zone 4 (L_{Z2} and L_{Z4}, +24%), the number of lateral roots (N_{LR}, +38%), the sum of lateral root lengths (ΣL_{LR}, +63%), the total root length (TLR, +41%), the density of lateral roots in zone 1 ($D_{LR}\text{-}Z1$, +22%) and the specific root length (SRL, +15%) increased, while the length of primary root zone 3 (L_{Z3}, −351%) and the density of lateral roots in zone 2 (D_{LR}-Z2, -8%) decreased at N$-$ compared to N+ treatment. The cultivars showed considerable variation for biomass production and root morphology in response to N supply (Figure 3a,b). Variations for N_{LR} and $\sum L_{LR}$ were generally the largest ones, and they were amplified at N+ compared to N$-$. For instance, the percentage differences for $\sum L_{LR}$ between extreme cultivars were more than ten-fold (CAR vs. JAS at N$-$ and CLO vs. LIM at N+). The responsiveness of one cultivar to N depletion (i.e., increase/decrease of one trait value in response to N$-$ compared to N+) was evaluated (Figure 3c). A large variation in lateral root phenotypic plasticity was observed, with cultivars being poorly (CAR showing $\sum L_{LR}$ increase less than 10%) or greatly (BAT showing $\sum L_{LR}$ increase more than 500%) responsive to N$-$. The genotype (cultivar) effect, the environment (N treatment) effect and the interaction (cultivar $\times$ N) were significant ($p < 0.05$) for all traits, except the (cultivar $\times$ N) for S biomass (ns) (Table S2). The in vitro observations highlight an important variability among cultivars in terms of root morphological traits, and this leaves space for genetic selection targets in breeding programs.

A principal component analysis was used to compress and classify the biomass and root morphology data (Figure 4a,b). The two first principal components (PCs) explained together 71% of the total phenotypic variation. The PC1 (53.1%) was influenced the most by root morphological traits, while PC2 (17.8%) by the shoot biomass production (Figure 4a). The distribution of the cultivars across PC1 and PC2 permitted to clearly distinguish between the two N treatments (Figure 4b). Cultivars with low PC1 scores exhibited long primary and lateral roots, while those with elevated PC2 scores produced important aerial biomass.

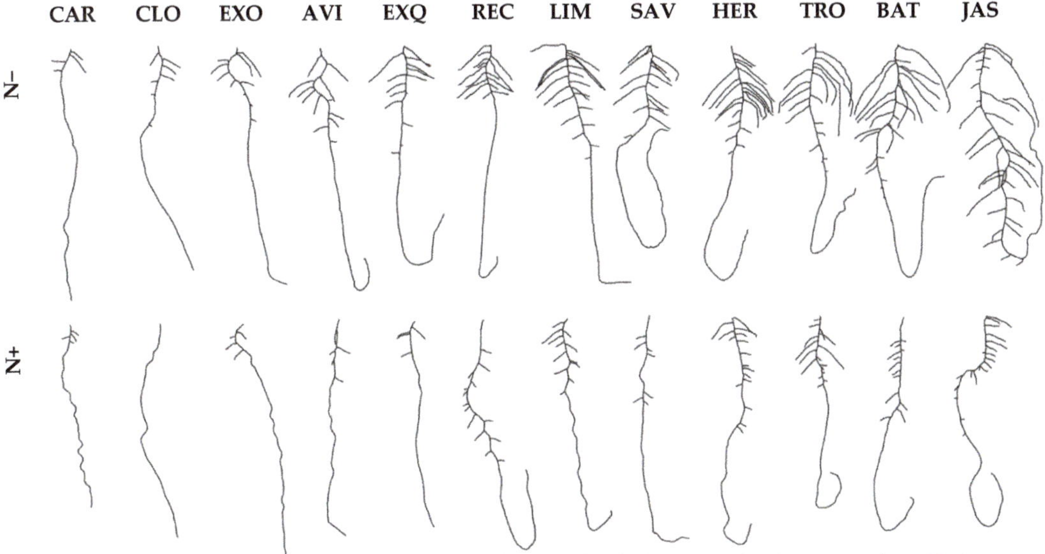

Figure 2. Root morphologies of oilseed rape cultivars grown in vitro with two divergent nitrate supplies. Pictures of representative root organs of 12 contrasting oilseed rape accessions grown with 0.01 mM (upper row, N$-$) or 10 mM (lower row, N+) nitrate supplies. Accessions are ranked from the left to the right by increasing total root length (TLR) measured at N$-$. The TRL values were ranging between 12.1 ± 0.9 cm (CAR) and 41.5 ± 2.9 cm (JAS). Cultivar full names are listed in Table S1. Scale bar: 2 cm.

Spearman correlation coefficients were calculated between the traits measured in vitro (Figure S1). Some correlations were found between derivative traits and their components (Table 1) but also between unrelated traits. The strongest and most significant correlation was found between L_{PR} and $\sum L_{LR}$ ($\varrho^2 = 0.57$; $p < 0.001$) at N−. Furthermore, S biomass showed weak but significant positive correlations with N_{LR} and $\sum L_{LR}$ ($0.15 < \varrho^2 < 0.26$, $p < 0.001$) during both N treatments.

Table 1. Abbreviations and definitions of the measured traits.

Biomass Production	
R	Root biomass
S	Shoot biomass
R + S	Total biomass
R:S	Root to shoot biomass ratio
Root Morphology	
L_{PR}	Length of primary root; $L_{PR} = L_{Z2} + L_{Z3} + L_{Z4}$
L_{Z2}	Length of primary root zone 2, between the first and last lateral roots; $= 0$ if $N_{LR} = 0$ or $N_{LR} = 1$
L_{Z3}	Length of primary root zone 3, between the hypocotyl junction and the first lateral root; $= L_{PR}$ if $N_{LR} \leq 1$
L_{Z4}	Length of primary root zone 4, between the last lateral root and the primary root tip; $= 0$ if $N_{LR} \leq 1$
N_{LR}	Number of lateral roots > 1 mm
$\sum L_{LR}$	Sum of lateral root lengths
TRL	Total root length; $= L_{PR} + \sum L_{LR}$
D_{LR}-Z1	Density of lateral roots in zone 1; $= N_{LR}/L_{PR}$
D_{LR}-Z2	Density of lateral roots in zone 2; $= (N_{LR}\text{-}1)/L_{Z2}$, not defined if $N_{LR} \leq 1$
SRL	Specific root length; $= (L_{PR} + \sum L_{LR})/R$
Shoot Morphology	
LA	Leaf area
SLA	Specific leaf area; $= LA/S$
Carbon and Nitrogen Analyses	
C_R	Carbon concentration in root tissues
C_S	Carbon concentration in shoot tissues
N_R	Nitrogen concentration in root tissues
N_S	Nitrogen concentration in shoot tissues
NUI	Nitrogen utilization index; $= S/((S \times N_S) + (R \times N_R))$
Nitrate Uptake Assay	
$^{15}NQ_R$	^{15}N amount in root tissues
$^{15}NQ_S$	^{15}N amount in shoot tissues
HATS/LATS plan^{-1}	
	HATS/LATS mediated ^{15}N influx rate per hour and per plant; $= (^{15}NQ_R + {}^{15}NQ_S)/5 \times 60$
HATS/LATS root biomass^{-1}	
	HATS/LATS mediated ^{15}N influx rate per hour and per root biomass; $= (^{15}NQ_R + {}^{15}NQ_S)/5 \times 60 \times R$

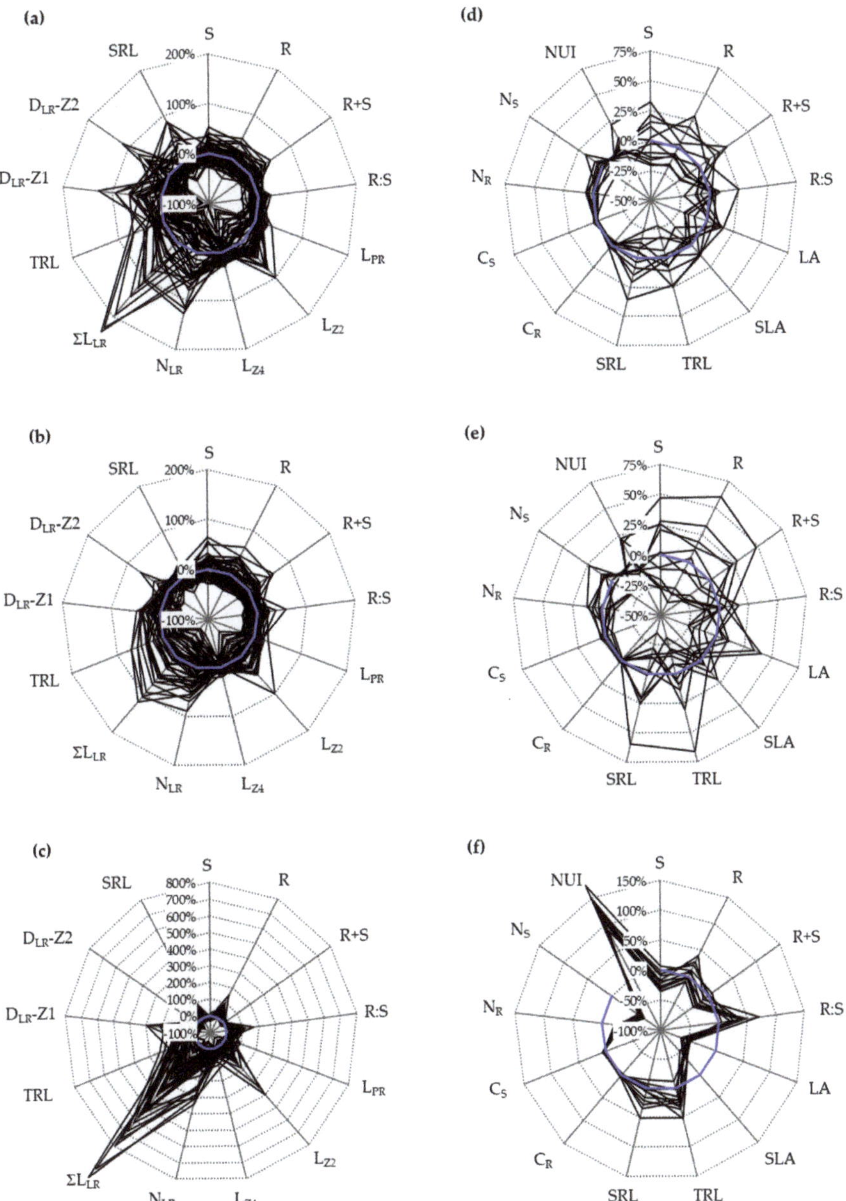

Figure 3. Relative variation of phenotypic traits measured in 55 oilseed rape cultivars cultivated in vitro (**a–c**) and in 12 cultivars in hydroponics (**d–e**). The spider plots (**a,b,d,e**) show the percentage variation of a given trait for every cultivar, normalized by the mean value of the panel, measured during N+ (**a,d**) or N− (**b,e**) conditions. Zero percent (blue circle) indicates no difference compared to the mean value of the panel in one condition. The spider plots (**c,f**) show the percentage variation of a given trait for every cultivar grown during N−, normalized by the value observed during N+ conditions. This defines the responsiveness of one trait to nitrate depletion. Zero percent (blue circle) indicates no difference compared to elevated nitrate conditions. n = 15 plants per genotype and per N condition cultivated in vitro, and 5 plants in hydroponics. Traits are defined in Table 1.

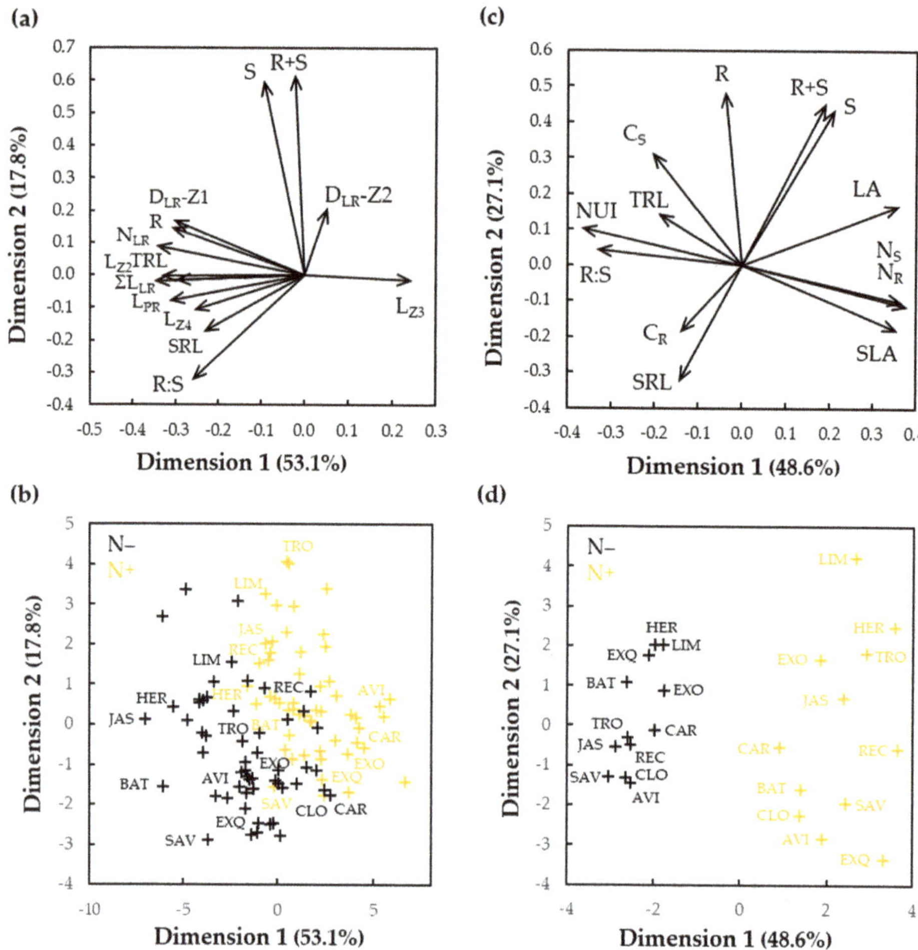

Figure 4. Principal component analysis of 14 phenotypic traits measured in 55 oilseed rape cultivars cultivated in vitro (**a,b**) and of 13 phenotypic traits in 12 cultivars in hydroponics (**c,d**). (**a,c**) Principal component (PC) biplot showing the compositions of the first two PCs, with cumulative variance; (**b,d**) Representation of the cultivars. Symbols indicate position of the cultivars as determined by their trait values in the two first PCs. Black symbols refer to low nitrate (N−) and yellow symbols to high nitrate (N+) conditions. Traits are defined in Table 1 and cultivars listed in Table S1.

3.2. Characterization of Selected Cultivars in Hydroponics

The vertical plate culture system delivered a rapid and extensive examination of 2D root morphology at seed germination. Next, we examined how these early observations were relating with later developmental stages. Based on the total lateral root length observed in vitro (see above), 12 contrasting cultivars were retained for further physiological analysis upon hydroponic culture (Figure 1b). Plants were fed with a nutrient solution containing 0.2 mM (N−) or 2 mM (N+) nitrate. On average for the core set cultivated in hydroponics, R (+7%) and R:S (+26%) were superior, whereas S (−24%) and R + S (−20%) were inferior at N− compared to N+ (Table S3). The leaf area (LA) (−117%) and the specific leaf area (SLA) (−66%) decreased, while TRL (+18%) and SRL (+10%) increased in the same conditions. The nitrogen concentration in root tissues (N_R) (−36%) and in shoot tissues (N_S) (−154%) decreased, while the carbon concentration in root tissues (C_R) (+1%) and in shoot tissues (C_S) (+3%) increased. The nitrogen utilization index (NUI), used as a surrogate for the nitrogen use efficiency at a vegetative stage, increased by 56%. They were large biometric variations between cultivars (Figure 3d,e). For instance, the percentage differences between the two most extreme cultivars were in the range of 34% (AVI and REC vs. LIM) and 65% (AVI vs. HER) for LA, and of 52% (CAR vs. BAT) and 82% (EXQ vs. BAT) for TRL, respectively, at N− and N+. The responsiveness of cultivars to N depletion was also assessed (Figure 3f). A large variation in phenotypic plasticity was observed, with cultivars poorly or greatly responsive to N treatment for LA (e.g., SAV and TRO vs. RC) and TRL (e.g., CAR vs. BAT). The cultivar effect for all traits, the N treatment effect for all traits except R and NS, the interaction (cultivar × N) for all traits except R, S, SLA and TRL were significant ($p < 0.05$) (Table S3).

The two first PCs explained together three quarters of the total phenotypic variation. The PC1 (48.6%) was influenced the most by root morphological traits, while PC2 (27.1%) by the shoot biomass production (Figure 4c). The distribution of the cultivars across PC1 and PC2 permitted to clearly distinguish between the two N treatments (Figure 4d). Cultivars with low PC1 scores exhibited large leaves with elevated N concentration in tissues, while those with elevated PC2 scores produced important root biomass. We noted that the data collected from seedlings grown on vertical plates (Figure 4a,b) and those from plants grown in hydroponics (Figure 4c,d) showed the same pattern of response to the nitrate supply. The notable effect of N starvation was mainly reflected on R:S biomass ratio, TRL and SRL.

Spearman correlation coefficients were calculated between the traits measured in hydroponics (Figure S2). The strongest correlation was found between R and S biomasses ($\varrho^2 = 0.63$ at N−, $\varrho^2 = 0.95$ at N+). The SLA correlated negatively with C_S ($\varrho^2 = -0.59$ at N−, $\varrho^2 = -0.87$ at N+) and NUI ($\varrho^2 = -0.79$ at N−, $\varrho^2 = -0.76$ at N+), and positively with N_S ($\varrho^2 = 0.72$ at N−, $\varrho^2 = 0.77$ at N+). At N+, the LA correlated positively with R ($\varrho^2 = 0.88$) and S ($\varrho^2 = 0.77$). At N−, C_R correlated negatively with C_S ($\varrho^2 = -0.66$), and SRL positively with N_R ($\varrho^2 = 0.61$). All reported correlations are significant ($p < 0.001$).

A nitrate influx assay with ^{15}N tracer was performed to assess the relationship between uptake capacity of roots and plant morphological traits. The activities of the nitrate High-(HATS) and Low-(LATS) Affinity Transport Systems were measured in 10 cultivars grown in hydroponics at N+ (Figure 5).

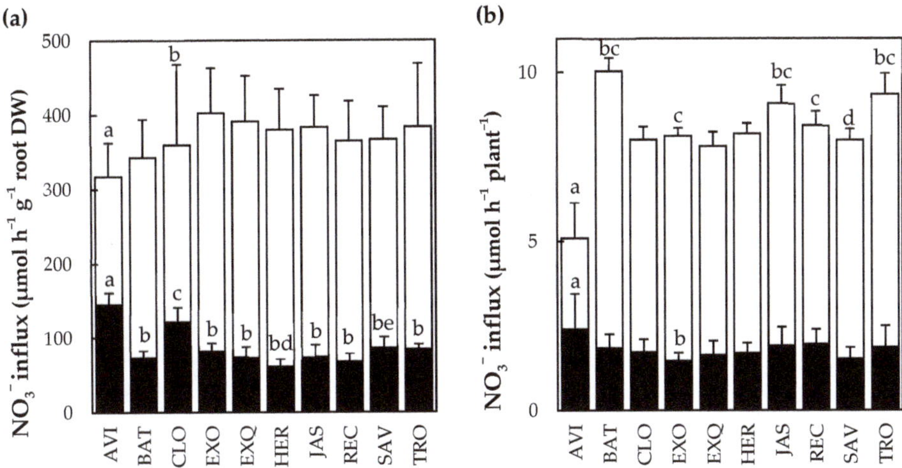

Figure 5. Variability of HATS and combined HATS and LATS activities between oilseed rape cultivars. (**a**) Nitrate influx expressed per hour and per root biomass. (**b**) Nitrate influx expressed per hour and per plant. Black bars: HATS activity, white bars: HATS + LATS activity. N = 6–12 plants ± std. Bars with different letters are significantly different ($p < 0.05$) according to Tukey's HSD test.

The HATS expressed per root biomass were four-fold less important than the HATS and LATS combined values. Variations between cultivars were in the range of 200% for HATS and 25% for HATS + LATS. The AVI and CLO cultivars had significant ($p < 0.05$) greater HATS values expressed per root biomass (Figure 5a). The AVI and BAT showed, respectively, the lowest and greatest HATS + LATS values expressed per plant (Figure 5b). The HATS expressed per root biomass was negatively correlated with LA ($\varrho^2 = -0.74$; $p < 0.01$), while the HATS + LATS expressed per plant was positively correlated with TRL ($\varrho^2 = 0.59$; $p < 0.05$) (Figure S3).

3.3. In-Soil Root Phenotyping

The X-ray computed tomography was used to examine the root system architecture of four cultivars (BAT, CLO, HER and SAV) in pipes filled with soil (Figure 1c). The profile of lateral roots along the taproot was assessed after watering with a solution without (N−) or with 50 mM KNO$_3$ (N+) (Figure 6). The CLO cultivar was showing poor soil exploration along with poor shoot development (data not shown), and it was weakly responsive to the N treatment. It may indicate that the culture conditions were particularly unfavorable for that cultivar. On the contrary, the lateral root deployment of BAT, HER and SAV was more profuse and concentrated close to the soil surface. The root system of the BAT cultivar was repressed in the upper soil portion, while the one of SAV was clearly stimulated by N− condition. The HER genotype showed subtle profile change in response to N treatment. There was no obvious difference close to the surface but less lateral root branching down to the pipe, during N− in that cultivar.

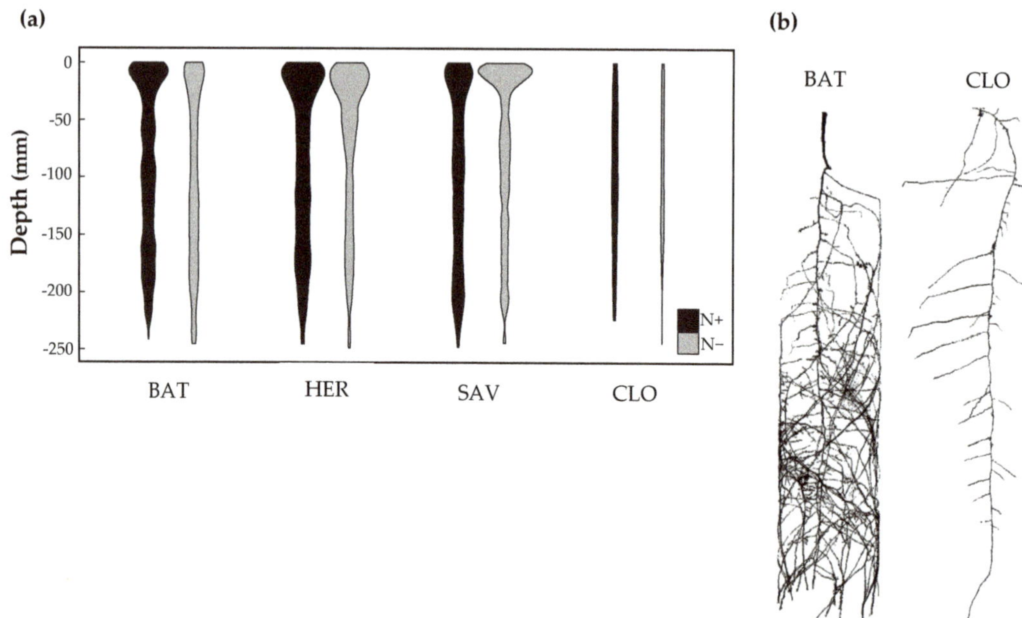

Figure 6. Characterization of the root system architecture in soil-filled column. (**a**) The violin plots show the distribution of lateral root density, as a function of the column depth. The estimates are based on five replicates of four oilseed rape cultivars cultivated with low (N−) or high (N+) nitrate conditions; (**b**) Reconstructed images of two contrasting cultivars at N−. Scale bar = 5 cm.

4. Discussion

Increasing NUE is essential to ensure the environmental and economic sustainability of oilseed rape production. Nonetheless, NUE is notoriously difficult to work with and this is due to the many peripherical processes being involved. The root system is still considered as a 'black box' and very few studies have focused on oilseed rape root response to N supply. Indeed, the below-ground organ has been poorly considered in breeding programs. Today, it is recognized as a lever to improve soil resources uptake and to reduce the negative environmental impact of mineral fertilization [35,46–48]. A diversity panel of modern winter oilseed rape cultivars was challenged with divergent nitrate supplies to identify contrasting root morphologies. Such a screening strategy in laboratory conditions may speed up for breeders, the delivery of genotypes with desired root morphological features. However, given the complexity of the agricultural environment, these data obtained in laboratory conditions may not ultimately reflect the field situation [18].

In the quest for root phenotyping methods, we employed three different procedures in controlled environment and evaluated their practical values and limitations.

(i) The in vitro culture system on vertical agar plate in sterile conditions, is a reliable approach for germinating seeds and observing two-dimensional root morphology. We have adapted the system, which is conveniently used for Arabidopsis model [41,49], to the more vigorous growth of oilseed rape (Figure 1a). That method is low cost but handling agar plates and analyzing scanned images are two laborious tasks. Despite, some algorithms could be used to speed up the extraction of root morphological traits from image data [50]. Finally, agar plate for culturing sprouts constitutes one first step towards further characterization of later developmental stages.

(ii) The hydroponic culture in containers filled with nutrient solution (Figure 1b), is suitable for mineral nutrition studies. That culture system without substrate gives easy access to clean root organs, for physiological and molecular characterization data [51,52]. While plant biometrics (e.g., biomass and total root surface) can easily be monitored over time, no such detailed analysis of root morphological traits can be provided.

(iii) The technique based on X-ray computed tomography allows to visualize the three-dimensional structure of the root system [48,53,54]. We recovered the segmentation of the main root and first-order lateral roots of four oilseed rape cultivars in a column filled with soil (Figure 1c). That experimental setting is low throughput, as image capture and analysis are time-consuming. The technique relies on discriminating differences in X-ray attenuation between roots and soil matrix. In this case, distinguishing between water-filled pore space and fine roots quickly became a challenge. Hence, the technique only provides a qualitative assessment of the root system in a small size pipe. Nonetheless, this may constitute a final laboratory phase prior to field phenomics [48]. For a small number of cultivars, the root phenotype at germination during in vitro culture (Figure 2) was indicative of the root system deployment in the soil (Figure 6). However, these observations should be treated with caution, as studies in other crops (e.g., wheat) are reporting a lack of correlation between the root systems at a young developmental stage in the laboratory environment and at a reproductive stage in the field [55].

Overall, the three complementary culture systems permitted to observe wide phenotypic variation among the diversity panel and to discriminate between cultivars with contrasting behaviors. For example, the cultivar CLO was poorly performing across all experimental settings and N conditions, contrary to HER consistently demonstrating superior characteristics (Figures 2, 4 and 6).

Arabidopsis is a prime model organism for studying root biology. Fundamental knowledge can be translated to *Brassica* crops, which are genetically related [56,57]. The model species may serve as a resource base for studying root morphogenesis in response to N availability and N acquisition. We observed some similar root morphological responses between the two species, despite different root growth rates. Nitrate depletion stimulated the lateral root outgrowth of Arabidopsis [41] and oilseed rape (Figures 2 and 3) seedlings during in vitro culture on agar plates. However, the number of lateral roots varied less but the length of primary root (mainly L_{Z4}) was more influenced by nitrate depletion in crop compared to Arabidopsis.

Learning about mechanisms of lateral root growth stimulation or repression by nitrate availability will help drawing strategies to optimize root system architecture. We aimed at observing the elaboration of root traits in oilseed rape cultivars and how this relates to biomass production and N capture. A premise is that root biomass production and morphological traits could be positive indicators of above-ground biomass production (and presumably yield). A rationale shared by several authors is that a branched root system that explores an important soil volume would limit nitrate leaching [18,58–61]. This study shows no negative correlation between root biomass production or total root length and shoot biomass production during in vitro and hydroponic cultures (Figures S1 and S2). Large root system deployed by BAT, JA and TRO, can support great shoot biomass production. Hence, selection for root and shoot biomass production under laboratory settings may be equally effective for improving yield.

Identifying oilseed rape cultivars with greater N capture is a pressing issue for reducing N loss in soil. This study demonstrates some potential for increased sustainability of oilseed rape production by targeting traits related to N uptake. An important variability degree for the root system size (Figure 3) and the N uptake capacity (Figure 5) was uncovered. This likely reflects no direct selection for root traits in breeding history of oilseed rape. A positive relationship was found between the total root length and the nitrate influx of HATS + LATS expressed per plant (Figure S3). This indicates that the N uptake capacity was to some extent, root-morphology dependent. Interestingly, the HATS expressed per root biomass correlated negatively with the leaf area (Figure S3). It is known that plants are modulating root N acquisition in response to shoot N demand, through long-distance mobile proteins and peptides [62,63]. That negative correlation suggests a feedback regulation of N products from the shoot on the root N uptake.

5. Conclusions

We illustrated that gel and liquid cultures rapidly deliver quantitative plant biometrics, and tomography a more qualitative outcome. The study revealed a large diversity of root morphologies and variability of N uptake capacities. These traits can be introgressed into one performant NUpE genotype possessing large root surface (e.g, HER cultivar), great N uptake rate per root mass (e.g., AVI, CLO) and reducing the adverse impact of N products on root uptake.

Supplementary Materials: The following are available online at https://www.mdpi.com/article/10.3390/nitrogen2040033/s1, Figure S1: Inter-trait phenotypic correlations in 55 winter oilseed rape cultivars grown in vitro, Figure S2: Inter-trait phenotypic correlations in 12 winter oilseed rape cultivars grown in hydroponics, Figure S3: Correlation between nitrate influxes and leaf area and total root length in 10 oilseed rape cultivars grown in hydroponics, Table S1: List of oilseed rape cultivars used in the study, Table S2: Statistical treatment for the data set measured in vitro, Table S3: Statistical treatment for the data set measured in hydroponics.

Author Contributions: Conceptualization and original draft preparation: C.H., L.K.; Investigation, methodology: C.H., C.J.S., H.D.G., J.L., L.K., M.B., P.N., P.T., T.R.M.; editing, C.C., L.H. All authors have read and agreed to the published version of the manuscript.

Funding: This research was funded by Fonds de la Recherche Scientifique (F.R.S.-FNRS) (MIS and PDR T.0116.19), and by European Plant Phenotyping Network (EPPN) (Grant Agreement No. 284443).

Institutional Review Board Statement: Not applicable.

Informed Consent Statement: Not applicable.

Data Availability Statement: The data presented in this study are available on request from the corresponding author.

Acknowledgments: L.H. and C.H. are, respectively, PhD fellow and research associate from F.R.S.-FNRS.

Conflicts of Interest: The authors declare no conflict of interest.

References

1. Huang, T.; Ju, X.; Yang, H. Nitrate leaching in a winter wheat-summer maize rotation on a calcareous soil as affected by nitrogen and straw management. *Sci. Rep.* **2017**, *8*, 42247. [CrossRef] [PubMed]
2. Mateo-Marìn, N.; Quìlez, D.; Isla, R. Utility of stabilized nitrogen fertilizers to reduce nitrate leaching under optimal management practices. *J. Plant Nutr. Soil Sci.* **2020**, *183*, 567–578. [CrossRef]
3. Velthof, G.L.; Lesschen, J.P.; Webb, J.; Pietrzak, S.; Miatkowski, Z.; Pinto, M.; Kros, J.; Oenema, O. The impact of the Nitrates Directive on nitrogen emissions from agriculture in the EU-27 during 2000–2008. *Sci. Total. Environ.* **2014**, *468–469*, 1225–1233. [CrossRef]
4. He, T.; Yuan, J.; Luo, J.; Wang, W.; Fan, J.; Liu, D.; Ding, W. Organic fertilizers have divergent effects on soil N_2O emissions. *Biol. Fert. Soils* **2019**, *55*, 685–699. [CrossRef]
5. Sharma, L.K.; Bali, S.K. A review of methods to improve nitrogen use efficiency in agriculture. *Sustainability* **2018**, *10*, 51. [CrossRef]
6. Moll, R.H.; Kamprath, E.J.; Jackson, W.A. Analysis and interpretation of factors which contribute to efficiency to nitrogen utilization. *Agron. J.* **1982**, *74*, 562–564. [CrossRef]
7. Bouchet, A.-S.; Laperche, A.; Bissuel-Belaygue, C.; Baron, C.; Morice, J.; Rousseau-Gueutin, M.; Dheu, J.-E.; George, P.; Pinochet, X.; Foubert, T.; et al. Genetic basis of nitrogen use efficiency and yield stability across environments in winter rapeseed. *BMC Genet.* **2016**, *17*, 131. [CrossRef]
8. Smith, S.; De Smet, I. Root system architecture: Insights from Arabidopsis and cereal crops. *Phil. Trans. R. Soc. B* **2012**, *367*, 1441–1452. [CrossRef] [PubMed]
9. Garnett, T.; Conn, V.; Kaiser, B.N. Root based approaches to improve nitrogen use efficiency in plants. *Plant Cell Environ.* **2009**, *32*, 1272–1283. [CrossRef]
10. Rosolem, C.A.; Ritz, K.; Cantarella, H.; Galdos, M.V.; Hawkesford, M.J.; Whalley, W.R.; Mooney, S.J. Enhanced plant rooting and crop system management for improving N Use Efficiency. *Adv. Agron.* **2017**, *146*, 205–239.
11. Sylvester-Bradley, R.; Kindred, D.R. Analysing nitrogen responses of cereals to prioritize routes to the improvement of nitrogen use efficiency. *J. Exp. Bot.* **2009**, *60*, 1939–1951. [CrossRef]
12. Han, M.; Okamoto, M.; Beatty, P.H.; Rothstein, S.J.; Good, A.G. The genetics of nitrogen use efficiency in crop plants. *Annu. Rev. Genet.* **2015**, *49*, 269–289. [CrossRef] [PubMed]
13. Wang, J.; Dun, X.; Shi, J.; Wang, X.; Liu, G.; Wang, H. Genetic dissection of root morphological traits related to nitrogen use efficiency in *Brassica napus* L. under two contrasting nitrogen conditions. *Front. Plant Sci.* **2017**, *8*, 1709. [CrossRef]
14. Girondé, A.; Etienne, P.; Trouverie, J.; Bouchereau, A.; Le Cahérec, F.; Leport, L.; Orsel, M.; Niogret, M.-F.; Nesi, N.; Carole, D.; et al. The contrasting N management of two oilseed rape genotypes reveals the mechanisms of proteolysis associated with leaf N remobilization and the respective contributions of leaves and stems to N storage and remobilization during seed filling. *BMC Plant Biol.* **2015**, *15*, 59. [CrossRef]
15. Koelin-Findeklee, F.; Becker, M.A.; van der Graaff, E.; Roitsch, T.; Horst, W.J. Differences between winter oilseed rape (*Brassica napus* L.) cultivars in nitrogen starvation-induced leaf senescence are governed by leaf-inherent rather than root-derived signals. *J. Exp. Bot.* **2015**, *66*, 3669–3681. [CrossRef]
16. Han, Y.; Liao, J.; Yu, Y.; Song, H.; Rong, N.; Guan, C.; Lepo, J.E.; Ismail, A.M.; Zhang, Z. Exogenous abscisic acid promotes the nitrogen use efficiency of *Brassica napus* by increasing nitrogen remobilization in the leaves. *J. Plant Nutr.* **2017**, *40*, 18. [CrossRef]
17. Williams, S.T.; Vail, S.; Arcand, M.M. Nitrogen Use Efficiency in parent vs. hybrid canola under varying nitrogen availabilities. *Plants* **2021**, *10*, 2364. [CrossRef] [PubMed]
18. Louvieaux, J.; Leclercq, A.; Haelterman, L.; Hermans, C. In-field observation of root growth and nitrogen uptake efficiency of winter oilseed rape. *Agronomy* **2020**, *10*, 105. [CrossRef]
19. Vazquez-Carrasquer, V.; Laperche, A.; Bissuel-Bélaygue, C.; Chelle, M.; Richard-Molard, C. Nitrogen Uptake Efficiency mediated by fine root growth early determines temporal and genotypic variations in Nitrogen Use Efficiency of Winter Oilseed Rape. *Front. Plant Sci.* **2021**, *12*, 641459. [CrossRef] [PubMed]
20. Fletcher, R.S.; Mullen, J.L.; Heiliger, A.; McKay, J.K. QTL analysis of root morphology, flowering time, and yield reveals trade-offs in response to drought in *Brassica napus*. *J. Exp. Bot.* **2015**, *66*, 245–256. [CrossRef] [PubMed]
21. Thomas, C.L.; Alcock, T.D.; Graham, N.S.; Hayden, R.; Matterson, S.; Wilson, L.; Young, S.D.; Dupuy, L.X.; White, P.J.; Hammond, J.P.; et al. Root morphology and seed and leaf ionomic traits in a *Brassica napus* L. diversity panel show wide phenotypic variation and are characteristic of crop habit. *BMC Plant Biol.* **2016**, *16*, 214. [CrossRef] [PubMed]
22. Thomas, C.L.; Graham, N.S.; Hayden, R.; Meacham, M.C.; Neugebauer, K.; Nightingale, M.; Dupuy, L.X.; Hammond, J.P.; White, P.J.; Broadley, M.R. High-throughput phenotyping (HTP) identifies seedling root traits linked to variation in seed yield and nutrient capture in field-grown oilseed rape (*Brassica napus* L.). *Ann. Bot.* **2016**, *118*, 655–665. [CrossRef]
23. Zhang, Y.; Thomas, C.L.; Xiang, J.; Long, Y.; Wang, X.; Zou, J.; Luo, Z.; Ding, G.; Cai, H.; Graham, N.S.; et al. QTL meta-analysis of root traits in *Brassica napus* under contrasting phosphorus supply in two growth systems. *Sci. Rep.* **2016**, *6*, 33113. [CrossRef] [PubMed]

24. Sun, C.; Yu, J.; Hu, D. Nitrate: A crucial signal during lateral roots development. *Front. Plant Sci.* **2017**, *8*, 485. [CrossRef] [PubMed]

25. Muller, B.; Guédon, Y.; Passot, S.; Lobet, G.; Nacry, P.; Pagès, L.; Wissuwa, M.; Draye, X. Lateral roots: Random diversity in adversity. *Trends Plant Sci.* **2019**, *24*, 810–825. [CrossRef]

26. Jia, Z.; von Wirén, N. Signaling pathways underlying nitrogen-dependent changes in root system architecture: From model to crop species. *J. Exp. Bot.* **2020**, *71*, 4393–4404. [CrossRef]

27. Meier, M.; Liu, Y.; Lay-Pruitt, K.S.; Takahashi, H.; von Wirén, N. Auxin-mediated root branching is determined by the form of available nitrogen. *Nat. Plants* **2020**, *6*, 1136–1145. [CrossRef]

28. Cai, Q.; Ji, C.; Yan, Z.; Jiang, X.; Fang, J. Anatomical responses of leaf and stem of *Arabidopsis thaliana* to nitrogen and phosphorus addition. *J. Plant Res.* **2017**, *130*, 1035–1045. [CrossRef]

29. Li, J.; Song, X.; Kong, X.; Wang, J.; Sun, W.; Zuo, K. Natural variation of *Arabidopsis thaliana* root architecture in response to nitrate availability. *J. Plant Nutr.* **2019**, *42*, 723–736. [CrossRef]

30. Meyer, R.C.; Gryczka, C.; Neitsch, C.; Müller, M.; Bräutigam, A.; Schlereth, A.; Schön, H.; Weigelt-Fischer, K.; Altmann, T. Genetic diversity for nitrogen use efficiency in *Arabidopsis thaliana*. *Planta* **2019**, *250*, 41–57. [CrossRef]

31. Yan, Z.; Eziz, A.; Tian, D.; Li, X.; Hou, X.; Peng, H.; Han, W.; Guo, Y.; Fang, J. Biomass allocation in response to nitrogen and phosphorus availability: Insight from experimental manipulations of *Arabidopsis thaliana*. *Front. Plant Sci.* **2019**, *10*, 598. [CrossRef] [PubMed]

32. Yue, K.; Fornara, D.A.; Li, W.; Ni, X.; Peng, Y.; Liao, S.; Tan, S.; Wang, D.; Wu, F.; Yang, Y. Nitrogen addition affects plant biomass allocation but not allometric relationships among different organs across the globe. *J. Plant Ecol.* **2021**, *14*, 361–371. [CrossRef]

33. McGrail, R.K.; Van Sanford, D.A.; McNear, D.H., Jr. Trait-based root phenotyping as a necessary tool for crop selection and improvement. *Agronomy* **2020**, *10*, 1328. [CrossRef]

34. Khan, M.A.E.; Gemenet, D.C.; Villordon, A. Root system architecture and abiotic stress tolerance: Current knowledge in root and tuber crops. *Front. Plant Sci.* **2016**, *7*, 1584. [CrossRef]

35. Louvieaux, J.; De Gernier, H.; Hermans, C. Exploiting genetic variability of root morphology as a lever to improve nitrogen use efficiency in oilseed rape. In *Engineering Nitrogen Utilization in Crop Plants*; Springer: Berlin/Heidelberg, Germany, 2018; pp. 185–201.

36. Ulas, A.; Erley, G.S.A.; Kamh, M.; Wiesler, F.; Horst, W.J. Root-growth characteristics contributing to genotypic variation in nitrogen efficiency of oilseed rape. *J. Plant Nutr. Soil Sci.* **2012**, *175*, 489–498. [CrossRef]

37. Guo, X.; Ma, B.; McLaughlin, N.B.; Wu, X.; Chen, B.; Gao, Y. Nitrogen utilisation-efficient oilseed rape (*Brassica napus*) genotypes exhibit stronger growth attributes from flowering stage onwards. *Funct. Plant Biol.* **2021**, *48*, 755–765. [CrossRef] [PubMed]

38. Berry, P.M.; Spink, J.; Foulkes, M.J.; White, P.J. The physiological basis of genotypic differences in nitrogen use efficiency in oilseed rape (*Brassica napus* L.). *Field Crops Res.* **2010**, *119*, 365–373. [CrossRef]

39. Lynch, P.J. Root phenes that reduce the metabolic costs of soil exploration: Opportunities for 21st century agriculture. *Plant Cell Environ.* **2015**, *38*, 1775–1784. [CrossRef] [PubMed]

40. Hermans, C.; Porco, S.; Verbruggen, N.; Bush, D. Chitinase-like protein CTL1 plays a role in the root system plasticity in response to multiple environmental signals. *Plant Physiol.* **2010**, *152*, 904–917. [CrossRef]

41. De Pessemier, J.; Chardon, F.; Juraniec, M.; Delaplace, P.; Hermans, C. Natural variation of the root morphological response to nitrate supply in *Arabidopsis thaliana*. *Mech. Develop.* **2013**, *130*, 45–53. [CrossRef]

42. Pound, M.P.; French, A.P.; Atkinson, J.A.; Wells, D.M.; Bennett, M.J.; Pridmore, T. RootNav: Navigating images of complex root architectures. *Plant Physiol.* **2013**, *162*, 1802–1814. [CrossRef] [PubMed]

43. Hermans, C.; Vuylsteke, M.; Coppens, F.; Craciun, A.; Inzé, D.; Verbruggen, N. The early transcriptomic changes induced by magnesium deficiency in *Arabidopsis thaliana* reveal the alteration of circadian clock genes expression in roots and the triggering of ABA-responsive genes. *New Phytol.* **2010**, *187*, 119–131. [CrossRef]

44. Hermans, C.; Vuylsteke, M.; Coppens, F.; Cristescu, S.; Harren, F.J.M.; Inzé, D.; Verbruggen, N. System analysis of the responses to long term magnesium deficiency and restoration in *Arabidopsis thaliana*. *New Phytol.* **2010**, *187*, 132–144. [CrossRef]

45. Malagoli, P.; Laineé, P.; Deunff, E.; Rossato, L.; Ney, B.; Ourry, A. Modeling nitrogen uptake in oilseed rape cv Capitol during a growth cycle using influx kinetics of root nitrate transport systems and field experimental data. *Plant Physiol.* **2004**, *134*, 388–400. [CrossRef] [PubMed]

46. Wasson, A.P.; Richards, R.A.; Chatrath, R.; Misra, S.C.; Prasad, S.V.; Rebetzke, G.J.; Kirkegaard, J.A.; Christopher, J.; Watt, M. Traits and selection strategies to improve root systems and water uptake in water-limited wheat crops. *J. Exp. Bot.* **2012**, *63*, 3485–3498. [CrossRef]

47. Voss-Fels, K.P.; Qian, L.; Parra-Londono, S.; Uptmoor, R.; Frisch, M.; Keeble-Gagnère, G.; Appels, R.; Snowdon, R.J. Linkage drag constrains the roots of modern wheat. *Plant Cell Environ.* **2017**, *40*, 717–725. [CrossRef]

48. Louvieaux, J.; Spanoghe, M.; Hermans, C. Root morphological traits of seedlings are predictors of seed yield and quality in winter oilseed rape hybrid cultivars. *Front. Plant. Sci.* **2021**, *11*, 568009. [CrossRef]

49. Hu, Y.; Omary, M.; Hu, Y.; Doron, O.; Hoermayer, L.; Chen, Q.; Megides, O.; Chekli, O.; Ding, Z.; Friml, J.; et al. Cell kinetics of auxin transport and activity in Arabidopsis root growth and skewing. *Nat. Commun.* **2021**, *12*, 1657. [CrossRef]

50. Dupuy, L.X.; Wright, G.; Thompson, J.A.; Taylor, A.; Dekeyser, S.; White, C.P.; Thomas, W.T.B.; Nightingale, M.; Hammond, J.P.; Graham, N.S.; et al. Accelerating root system phenotyping of seedlings through a computer-assisted processing pipeline. *Plant Methods* **2017**, *13*, 57. [CrossRef] [PubMed]

51. Xu, Z.; Ma, J.; Lei, P.; Wang, Q.; Feng, X.; Xu, H. Poly-γ-glutamic acid induces system tolerance to drought stress by promoting abscisic acid accumulation in *Brassica napus* L. *Sci. Rep.* **2020**, *14*, 252. [CrossRef] [PubMed]

52. Murad, M.A.; Razi, K.; Benjamin, L.K.; Lee, J.H.; Kim, T.H.; Muneer, S. Ethylene regulates sulfur acquisition by regulating the expression of sulfate transporter genes in oilseed rape. *Physiol. Plant.* **2021**, *171*, 533–545. [CrossRef] [PubMed]

53. Mairhofer, S.; Pridmore, T.; Johnson, J.; Wells, D.M.; Bennett, M.J.; Mooney, S.J.; Sturrock, C.J. X-ray Computed Tomography of crop plant root systems grown in soil. *Curr. Protoc. Plant Biol.* **2017**, *2*, 270–286. [CrossRef] [PubMed]

54. Pfeifer, J.; Kirchgessner, N.; Colombi, T.; Walter, A. Rapid phenotyping of crop root systems in undisturbed field soils using X-ray computed tomography. *Plant Methods* **2015**, *11*, 41. [CrossRef] [PubMed]

55. Watt, M.; Moosavi, S.; Cunningham, S.C.; Kirkegaard, J.A.; Rebetzke, G.J.; Richards, R.A. A rapid, controlled-environment seedling root screen for wheat correlates well with rooting depths at vegetative, but not reproductive, stages at two field sites. *Ann. Bot.* **2013**, *112*, 447–455. [CrossRef] [PubMed]

56. Leijten, W.; Koes, R.; Roobeek, I.; Frugis, G. Translating flowering time from *Arabidopsis thaliana* to Brassicaceae and Asteraceae crop species. *Plants* **2018**, *7*, 111. [CrossRef]

57. Stephenson, P.; Stacey, N.; Brüser, M.; Pullen, N.; Ilyas, M.; O'Neill, C.; Wells, R.; Østergaard, L. The power of model-to-crop translation illustrated by reducing seed loss from pod shatter in oilseed rape. *Plant Reprod.* **2019**, *32*, 331–340. [CrossRef]

58. Li, X.; Zeng, R.; Liao, H. Improving crop nutrient efficiency through root architecture modifications. *J. Integr. Plant Biol.* **2016**, *58*, 193–202. [CrossRef] [PubMed]

59. Thorup-Kristensen, K.; Kirkegaard, J. Root system-based limits to agricultural productivity and efficiency: The farming systems context. *Ann. Bot.* **2016**, *118*, 573–592. [CrossRef]

60. Pierret, A.; Maeght, J.; Clément, C.; Montoroi, J.; Hartmann, C.; Gonkhamdee, S. Understanding deep roots and their functions in ecosystems: An advocacy for more unconventional research. *Ann. Bot.* **2016**, *118*, 621–635. [CrossRef]

61. Van der Bom, F.J.T.; Williams, A.; Bell, M.J. Root architecture for improved resource capture: Trade-offs in complex environments. *J. Exp. Bot.* **2020**, *71*, 5752–5763. [CrossRef]

62. Gu, M.G.; Hu, X.; Wang, T.; Xu, G. Modulation of plant root traits by nitrogen and phosphate: Transporters, long-distance signaling proteins and peptides, and potential artificial traps. *Breed Sci.* **2021**, *71*, 62–75. [CrossRef] [PubMed]

63. Ohkubo, Y.H.; Tanaka, M.; Tabata, R.; Ogawa-Ohnishi, M.; Matsubayashi, Y. Shoot-to-root mobile polypeptides involved in systemic regulation of nitrogen acquisition. *Nat. Plants* **2017**, *3*, 17029. [CrossRef] [PubMed]

Nitrogen

Article

Surface Properties and Adherence of *Bradyrhizobium diazoefficiens* to *Glycine max* Roots Are Altered When Grown in Soil Extracted Nutrients

Armaan Kaur Sandhu [1], Senthil Subramanian [1,2] and Volker S. Brözel [1,3,*]

1   Department of Biology and Microbiology, South Dakota State University, Brookings, SD 57006, USA;
    Armaankaur.sandhu@sdstate.edu (A.K.S.); Senthil.subramanian@sdstate.edu (S.S.)
2   Department of Agronomy, Horticulture and Plant Science, South Dakota State University,
    Brookings, SD 57006, USA
3   Department of Biochemistry, Genetics and Microbiology, University of Pretoria, Pretoria 0004, South Africa
*   Correspondence: Volker.brozel@sdstate.edu; Tel.: +1-605-688-6144

Citation: Sandhu, A.K.;
Subramanian, S.; Brözel, V.S. Surface
Properties and Adherence of
Bradyrhizobium diazoefficiens to *Glycine
max* Roots Are Altered When Grown
in Soil Extracted Nutrients. *Nitrogen*
2021, 2, 461–473. https://doi.org/
10.3390/nitrogen2040031

Academic Editor:
Jacynthe Dessureault-Rompré

Received: 14 October 2021
Accepted: 11 November 2021
Published: 15 November 2021

Abstract: Soybean roots are colonized and nodulated by multiple strains of compatible nitrogen-fixing rhizobia primarily belonging to the Genus *Bradyrhizobium*. Motility towards the root and attachment to root hairs are key determinants of competitive colonization and subsequent nodulation. Bacterial surface properties and motility are known to vary with chemical composition of the culture medium, and root adhesion and nodulation occur in a soil environment rather than laboratory medium. We asked whether the nodulation-promoting factors motility, surface hydrophobicity and surface adhesion of *Bradyrhizobium* are affected by growth in a soil nutrient environment. *B. diazoefficiens* USDA 110, 126, 3384, and *B. elkanii* USDA 26 were grown in mineral salt medium with peptone, yeast extract and arabinose (PSY), and in a soil extracted soluble organic matter (SESOM) medium. Surface hydrophobicity was determined by partitioning into hydrocarbon, motility by transition through soft agar, and surface-exposed saccharides by lectin profiling, followed by biofilm formation and soybean root adhesion capacity of populations. SESOM-grown populations were generally less motile and more hydrophobic. They bound fewer lectins than PSY-grown populations, indicating a simpler surface saccharide profile. SESOM populations of USDA 110 did not form detectable biofilm, but showed increased binding to soy roots. Our results indicate that growth in a soil environment impacts surface properties, motility, and subsequent soy root adhesion propensity. Hence, evaluation of *Bradyrhizobium* for nodulation efficiency should be performed using soil from the specific field where the soybeans are to be planted, rather than laboratory culture media.

Keywords: *Bradyrhizobium*; attachment; root; biofilm; lectin; soybean; soil; hydrophobicity

1. Introduction

Rhizobium bacteria contribute combined nitrogen to many leguminous plants while occurring inside specialized root structures called nodules [1]. The productivity of soybeans (*Glycine max*) is enhanced significantly through nodule occupancy by prolific fixers such as *B. diazoefficiens* USDA 110 [2]. Early during plant growth, root hairs signal soil-borne rhizobia by releasing isoflavonoids [3]. Rhizobia respond to isoflavonoids such as genistein by synthesizing Nod Factors, which in turn initiate curling of soy root hairs, followed by infection thread and nodule formation [4,5]. The nitrogen contributing efficacy of rhizobia varies widely, necessitating screening and evaluation of candidate bacterial strains for application in fields. Successful colonization of the infection thread and nodules require temporal adhesion and colonization of rhizobia at the root hair surface. Adhesion to the root surface is dependent on physical proximity, facilitated by bacterial motility in the soil. Adherence and colonization by proximal bacteria depend on the physiochemical compatibility of both the root and rhizobial surfaces.

Surface properties of bacterial cells have a considerable influence on attachment to the root surface, so compatible surface properties contribute to increased adhesion. More hydrophobic mutants of *Bradyrhizobium japonicum* have been associated with more competitive nodule formation [6]. Bacterial lipopolysaccharides (LPS) have direct relation to hydrophobicity and more hydrophobic LPS is positively correlated to surface hydrophobicity [7,8]. *Bradyrhizobium* producing altered or limited extracellular polysaccharides (EPS) were defective in biofilm formation and binding to soy lectin and resulted in pseudo nodules [9,10]. Along with surface properties, there is evidence that motility can provide the cell with some significant advantages to adhere to the root surface [11,12]. On the other hand, not much is known about soy root surface properties. Soy roots produce surface-exposed lectins which sequester cognate sugar moieties at the surface of bacteria [13,14], especially in acidic soil. Lectins are proteinaceous, specific carbohydrate-binding proteins. They interact with carbohydrates in a highly specific but non-covalent manner [15]. In more alkaline soil, root hairs produce the protein rhicadhesin which promotes the bacterial attachment [16]. The surface properties of rhizobia therefore play an important role in the initial adhesion to soy root surface.

Bacteria respond to specific chemical and physical cues by condition-specific gene expression and altered phenotype. This is also the case with *Bradyrhizobium*, as phenotypic variations have been reported for populations cultured in different sugar sources. While *B. diazoefficiens* swims by sub-polar flagella, L-arabinose induces the production of lateral flagella, which lead to swarming on moist surfaces [17]. The EPS composition of *B. japonicum* varies by the available sugar sources [9]. *Bradyrhizobium* are generally cultured using a mineral salt medium with peptone, yeast extract and either arabinose or mannitol as a carbon source. We asked whether *Bradyrhizobium* adapts its adhesion-specific phenotype when growing in soil. Arguing that the primary drivers of phenotypic change would be water-diffusible substances able to enter the cell, we cultured *Bradyrhizobium* in soybean field aqueous soil extract and characterized adhesion-pertaining phenotypes.

2. Materials and Methods

2.1. Bacterial Strains and Culture Media

B. diazoefficiens USDA 110, 126, 3384 and *B. elkanii* USDA 26 were obtained from the NRRL Culture Collection of the Agricultural Research Service, United State Department of Agriculture. Cultures were grown in liquid or on solid PSY and SESOM, a filter-sterilized liquid soil extract [18]. PSY was prepared as mentioned by Mesa et al. [19], containing per liter: KH_2PO_4, 300 mg; Na_2HPO_4, 300 mg; $CaCl_2 \cdot 2H_2O$, 5 mg; $MgSO_4 \cdot 7H_2O$, 100 mg; peptone, 3 g; yeast extract, 1 g; H_3BO_3, 10 mg; $ZnSO_4 \cdot 7H_2O$, 1 mg; $CuSO_4 \cdot 5H_2O$, 0.5 mg; $Na_2MoO_4 \cdot 2H_2O$, 0.1 mg; $MnCl_2 \cdot 4H_2O$, 0.1 mg; $FeCl_3$, 0.19 mg; thiamine-HCl, 1 mg; biotin, 1 mg; Na-panthothenate, 1 mg; L-arabinose.

Soil for SESOM was obtained from a field under soybean cultivation directly after harvest, was dried at 55 °C and stored at 4 °C. SESOM was prepared as described previously [18] by adding 200 g of dry soil to 1 l of prewarmed (60 °C) sterile MOPS buffer (10 mM, pH 7.0) in a 2 L Erlenmeyer and shaken for 4 h at 150 rpm. To remove soil particles, the suspension was filtered sequentially through filter paper and cellulose acetate filters of pore sizes 8, 4, 1.2, 0.8, 0.4 and 0.2 μM. The clear extract was filtered to sterility through a 0.2 μM bottle top filter. Each extract was checked for sterility by plating 20 μL on R2A agar and incubating at 30 °C for 72 h. The sterile SESOM was supplemented with 0.1 g/L Bacto peptone and 0.1 g/L L(+)-Arabinose.

2.2. Motility

The effect of culture medium on swimming motility was determined by inoculating the center of low percentage agar plates (0.35%). Precultures were prepared by inoculating grown in PSY and SESOM and incubating at 30 °C for 24 h while shaking at 250 rpm. Cultures were diluted to A_{600} of 0.100 and 20 μL was drop inoculated at the center of PSY or SESOM low agar plates. Plates were incubated at 30 °C for 10 d when the colony radius was measured.

2.3. Microbial Adhesion to Hydrocarbons

Microbial adhesion to hydrocarbons (MATH) was assessed to analyze surface hydrophobicity of cultures. Exponential phase cultures (50 mL) prepared as outlined above were harvested by centrifugation for 10 min at $10,000 \times g$ and resuspended in 15 mL sterile Phosphate Urea Magnesium Sulfate Buffer (PUM) containing 22.2 g/L $K_2HPO_4 \cdot 3H_2O$, 7.26 g/L KH_2PO_4, 1.8g/L Urea and 0.02 g/L $MgSO_4 \cdot 7H_2O$ with pH adjusted to 7.1 [20,21]. Following a second round of harvesting, cells were suspended in 15 mL PUM by vortexing. To three acid washed glass tubes, we added 4 mL cell suspension, and supplemented with 1 mL n-hexadecane, retaining the final 3 mL suspension as blank. Cell suspended in PUM were exposed to hexadecane by vortexing for 1 min and allowing the mixture to separate at room temperature for 1 h. After separation, 1 mL of the aqueous phase was withdrawn very carefully using acid washed, sterile glass pipettes, and transferred to a quartz cuvette. The absorbance was measured at 600 nm, using hexadecane-supplemented PUM without bacteria added as blank, termed Absorbance Math (A_M). The absorbance of untreated cell suspensions was measured at 600 nm, using PUM buffer as blank, and termed Absorbance original (A_O). The fraction of cells that partitioned to the hydrocarbon phase was calculated as:

$$Fp = 1 - (A_M / A_O)$$

2.4. Lectin Binding Assay

Surface-exposed sugar moieties were characterized using a collection of 24 lectins (Vector Laboratories, Burlingame, CA, USA) (Table 1). Nine of these were fluorescein-conjugated, and the remaining fifteen were biotinylated. Exponential phase 24 h old cultures were prepared as described above, harvested, and washed with HSB buffer (2.383 g/L HEPES, 8.766 g/L NaCl, 0.011 g/L $CaCl_2$ and 0.8 g/L sodium azide) (Vector Laboratories), and resuspended to absorbance of 0.100 at 600 nm. Lectins were diluted to 20 µg/ mL in HSB, and 100 µL added to 50 µL of cell suspension. After mixing by brief vortexing, lectins were allowed to bind for 10 min at room temperature, followed by vortexing and a further 10 min rest period. The cells were harvested by centrifugation at $16,200 \times g$ for 7 min. Unbound lectin was removed by three wash cycles using HSB buffer. For fluorescein-conjugated lectins, the resulting pellet was resuspended in 20 µL of HSB and the entire sample transferred to a clean microscopic slide. The drop of cell suspension was covered with a glass coverslip, pressed slightly to reduce cell movement, and the sides sealed using clear nail varnish to prevent evaporation. For biotinylated lectins, lectin-exposed washed cells were supplemented with 50 µL of 20 µg/mL streptavidin-FITC and left to bind for 30 min. Unbound streptavidin-FITC was removed by washing cells twice. The pellet was resuspended in 20 µL HSB and transferred to a microscope slide.

Samples were viewed by fluorescence microscopy using an Olympus BX53 Upright Compound Microscope with 466/40 nm excitation and a 525/50 nm emission filter, captured using an Olympus DP70 digital camera. Sample images were captured using 100, 200, 400, 600, 800, 1000 and 1200 ms exposure time. Binding intensity was scored by the shortest exposure time yielding visible fluorescence, with 100 ms scoring level 4, and <1200 ms scoring level 0. The proportion of cells displaying fluorescence was determined by Image J by preparing the binary image under Process. The resulting binary image was further processed for analysis by using the following programs sequentially > Mask > Watershed > Fill holes. For the analyses, the analyze particles program was selected under the Analyze icon.

Table 1. List of lectins used.

Lectin Name	Lectin Abbreviation	Sugar Specificity *	Conjugated with
Concanavalin A	Con A	Branched and terminal mannose [High-Man, Man α-1,6(Manα-1,3)]	FITC
Dolichos biflorus agglutinin	DBA	GlcNAc β-1,4 GlcNAc oligomers and LacNAc (Gal β 1,4 GlcNAc)	FITC
Peanut agglutinin	PNA	Terminal Gal (β-OR)	FITC
Ricinus communis agglutinin I	RCA I	Gal	FITC
Soybean agglutinin	SBA	α- or βLinked terminal GalNAc, GalNAc α-1,3 Gal	FITC
Ulex europaeus agglutinin I	UEA I	α-Fucose	FITC
Wheat germ agglutinin	WGA	β-GlcNAc, sialic acid, GalNAc	FITC
Datura Stramonium Lectin	DSL	GlcNAc β-1,4 GlcNAc oligomers and LacNAc (Gal β 1,4 GlcNAc)	FITC
Galanthus Nivalis Lectin	GNL	Terminal α-1, 3 mannose	FITC
Lens culinaris agglutinin	LCA	Complex (man/GlcNAc core with α-1,6 Fuc)	Biotin
Pisum sativum agglutinin	PSA	Man, (Fuc a-1,6 GlcNAc, α-D-Glc, α-D-Man)	Biotin
Griffonia simplicifolia lectin I	GSL-I	α-Galactose, also binds some GalNAc	Biotin
Sophora japonica agglutinin	SJA	βGalNAc	Biotin
Vicia villosa agglutinin	VVA	GalNAc	Biotin
Lycopersicon esculentum (tomato) lectin	LEL	β-1,4 GlcNAc oligomers	Biotin
Solanum tuberosum (potato) lectin	STL	GlcNAc oligomers, LacNAc	Biotin
Griffonia simplicifolia lectin II	GSL-II	Terminal GlcNAc	Biotin
Succinylated Wheat germ agglutinin	sWGA	GalcNAc	Biotin
Erythrina cristagalli lectin	ECL	Gal β-1,4 GalNAc	Biotin
Artocarpus integrifolia (Jacalin)	J	Gal β-1,3 GalNAc	Biotin
Phaseolus vulgaris erythroagglutinin	PHA-E	Complex-type N-glycans with outer Gal and bisecting GlcNAc	Biotin
Phaseolus vulgaris leucoagglutinin	PHA-L	β-1,6 Brandched tri mannosyl core N-linked glycans	Biotin
Maackia amurensis lectin I	MAL-I	Galactosyl (β-1,4) N-acetylglucos amine, (α-2,3) sialic acid	Biotin
Maackia amurensis lectin II	MAL-II	α-2,3 sialic acid-LacNAc structure	Biotin

* Vector Laboratories.

2.5. Biofilm Formation

The effect of culture medium on biofilm formation was determined by the widely used polystyrene crystal violet assay [22,23], with some modifications. Exponential phase cultures were prepared as described above, harvested by centrifugation, suspended in sterile water to remove residual medium, harvested, and resuspended in fresh PSY or SESOM to absorbance of 0.100 at 600 nm. Nunclon Delta Surface Polystyrene 96-well plates (Thermo Scientific, Waltham, MA, USA; catalog 167008) were loaded with 150 μL of cells in 7 replicates. The sterile culture media were loaded into additional wells to serve as blank. After 24 h incubation at 30 °C, the planktonic cells were removed by pipetting. Each inoculated well of the plate was washed twice by pipetting using sterile water to facilitate removal of all planktonic cells. Biofilms were stained by adding 200 μL of 0.1% aqueous crystal violet solution, shaking at room temperature for 20 min. The crystal violet solution was removed from the wells by pipette, followed by two cycles of careful washing with water by pipetting. The plate was left open for 30 min to dry and 200 µL of 95% ethanol was added to dissolve the crystal violet. The plates were incubated for 10 min at room temperature with shaking and absorbance determined at 570 nm.

2.6. Root Adherence

To determine bacterial adherence to young roots, exponential phase cultures were exposed to germinating soy seeds. Seeds were sterilized as described previously [24]. Briefly, seeds were sterilized by washing in 30% of concentrated bleach for 5 min while shaking gently, followed by three consecutive washes with sterilized deionized water under shaking. Seeds were suspended in 70% ethanol while being shaken gently for 20 min, followed by seven consecutive washes with sterilize deionized water under shaking. Following a final 20 min soak in sterile deionized water, seven seeds were placed onto each R2A agar plate (Difco) and left in the dark at 30 °C for 9 d. R2A comprised of yeast extract 0.5 g/L, proteose peptone No. 3 0.5 g/L, casamino acids 0.5 g/L, dextrose 0.5 g/L, soluble starch 0.5 g/L, sodium pyruvate 0.3 g/L, dipotassium phosphate 0.3 g/L, magnesium sulfate 0.05 g/L and agar 15 g/L. The plates were inspected every 2 d, and plates with seedlings showing microbial outgrowth on the agar were discarded. By day 9, roots varied from 8 to 12 cm in length.

The exponential phase cultures were prepared as described above and diluted in fresh medium to absorbance of 0.100 at 600 nm. Aliquots were taken, and the culturable count was determined using the droplet plate count technique (Lindsay and VonHoly, 1999). Briefly, samples were serially diluted and 20 µL volumes spotted onto R2A plates. After incubation at 30 °C for 4 d, colonies were counted. Five seedlings were added to each 250 mL flask containing 50 mL of exponential phase cultures (50 mL) prepared in either PSY or SESOM as outlined above, left for 60 min while shaking gently. Seeds were removed aseptically using sterile tweezers, washed twice with sterilize deionized water and transferred to 50 mL conical tubes containing 50 mL PBS with 0.02% Tween 20 [24]. Tubes were immersed in a sonicator bath (Fisher FS20) and exposed to ultrasound for 30 s, and then transferred to ice water for 30 s to prevent undue heating. This sonication and cooling cycle was repeated 4 times. One mL of treated sample was drawn to perform serial dilutions and triplicate 20 µL volumes were plated on R2A plates. After incubation at 30 °C for 4 d, colonies were counted.

3. Results

3.1. Growth Medium Can Significantly Influence Cell Surface Hydrophobicity and Motility of Bradyrhizobium Strains

Surface properties of bacterial cells affect interaction with surfaces. We determined surface hydrophobicity of populations cultured in PSY and SESOM using the MATH partitioning assay. Hydrophobicity of the four strains varied, when grown in PSY medium with arabinose. USDA 110 was most hydrophilic while USDA 26 was the most hydrophobic (with USDA126 and USDA3384 displaying intermediate hydrophobicity), indicating strain-specific surface phenotype (Figure 1). When grown in the soil extract medium, SESOM, two of the strains underwent a significant increase in surface hydrophobicity. For USDA 3384, the fraction partitioned to the hydrocarbon phase increased by 700% and for USDA 110 by 300%. For USDA 110, some absorbance values of aqueous phase after partitioning were negative, suggesting minor emulsification due to hydrocarbon droplets into the aqueous phase. However, no significant change was observed for USDA 26 and 126 (Figure 1). These results indicate that surface hydrophobicity of *Bradyrhizobium* has some inherent strain-specific differences, and it is affected by the culture medium, with increased hydrophobicity predicted for cells growing in soil.

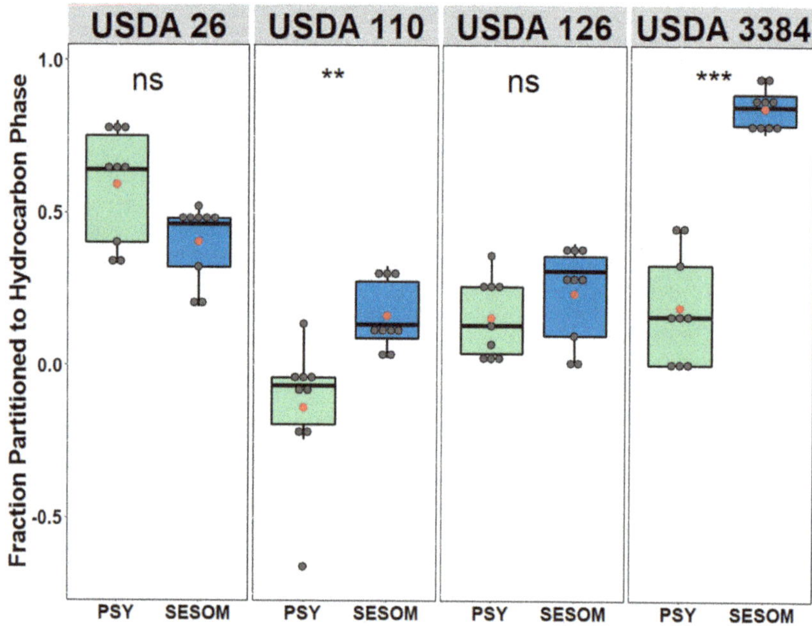

Figure 1. Surface hydrophobicity of *Bradyrhizobium* strains cultured in either PSY-Arabinose or SESOM to mid exponential phase and partitioned into hexadecane as an apolar solvent. Data were analyzed by Student *t*-test, and *p* values indicate statistical significance (ns not significant, ** $p < 0.01$, *** $p < 0.001$).

To determine whether growth in a soil environment affects motility, strains were inoculated into classical motility low agar plates (0.35%) in either PSY or SESOM. The strains varied in degree of motility in either medium, with USDA 110 and 126 the most motile, USDA 3384 less motile and 26 the least motile (Figure 2). USDA 26, 110 and 3384 showed a significant decrease in motility when grown in SESOM, while 126 remained unchanged (Figure 2). Strains 110 and 3384 showed the largest change. Collectively, USDA 110 and USDA 3384 became more hydrophobic and less motile when cultured in SESOM.

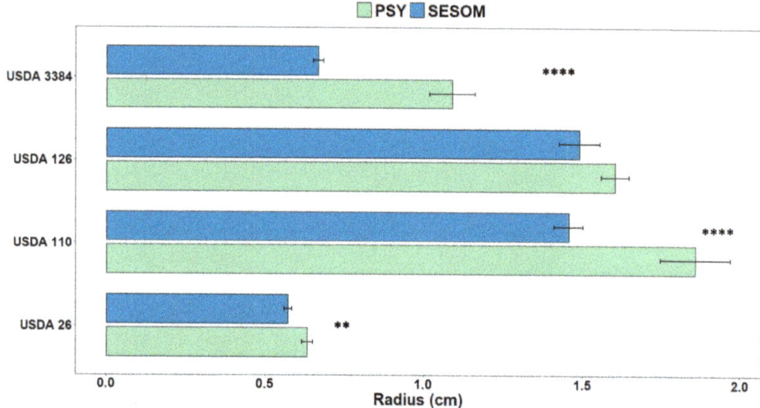

Figure 2. Motility radius of *Bradyrhizobium* strains spot-inoculated into either PSY-Arabinose or SESOM solidified with 0.35% agar and incubated for 10 d. Data were analyzed by Student *t*-test, and *p* values indicate statistical significance (ns not significant, ** $p < 0.01$, **** $p < 0.0001$).

3.2. Lectin Binding Profiles of Bradyrhizobium Strains Change When Cultured in SESOM

The presence of surface-exposed sugars was characterized semi-quantitatively using 24 lectins. Lectin binding intensity was scored by visible fluorescence after different exposure times, yielding a lectin fingerprint for each strain (Figure 3). USDA 110 bound to the highest number of lectins, followed by strains 26, 126 and then 3384. Each strain bound not one but multiple lectins, but also displayed binding to a unique combination of lectins, indicating different combinations of surface-exposed sugars. The fluorescence intensity varied among lectins and strains, indicating differences in the quantity of bound lectin. Importantly, SESOM-grown populations all bound fewer lectins than those grown in PSY. This indicates a decrease in the diversity of surface-exposed sugars and therefore shifts in extracellularly expressed polysaccharides. SESOM-grown USDA 110 bound only four of the 13 lectins bound ex PSY. Intriguingly, no new lectin binding was observed ex SESOM, except for USDA 126 which bound sWGA and STL. Collectively, this indicates a decrease in the variety of surface-exposed sugars in *Bradyrhizobium* when growing in a soil environment.

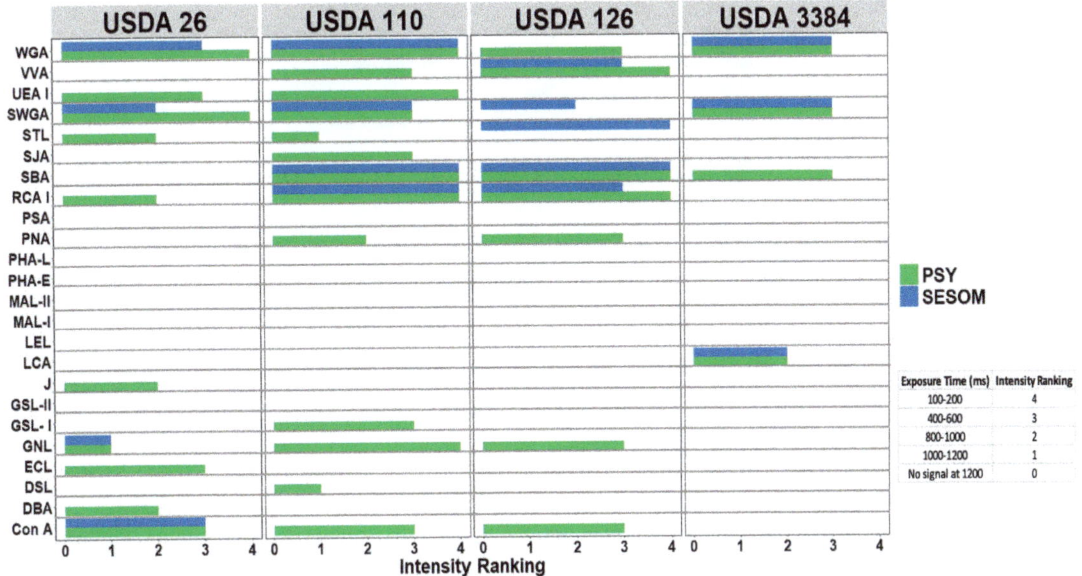

Figure 3. Lectin binding profiles of *Bradyrhizobium* strains cultured in either PSY-Arabinose or SESOM to mid-exponential phase. Binding intensity was scored by the shortest exposure time yielding visible fluorescence, with 100 ms scoring level 4, and <1200 ms scoring level 0.

Only a small proportion of cells displayed fluorescence, so cells appeared to either bind or not bind a specific lectin (Figure 4). To ascertain whether this was due to an insufficient quantity of lectin in the binding reaction, cultures were exposed to increased lectin concentrations, up to 300 µg/mL. This did not increase the proportion of lectin binding cells (data not shown), but instead indicated the phenotypic variability among exponential phase cells in a liquid culture. As the slow-growing *Bradyrhizobium* are more prone to contamination during culture, experimentation included regular sub-culturing on agar to ensure cultures were not contaminated. SESOM-cultured populations generally displayed lower proportions of lectin binding cells than ex PSY, with some exceptions such as USDA 110 binding to WGA and USDA 126 binding to RCA I. These results indicate a differentiation in surface properties among cells in *Bradyrhizobium* populations cultured under homogenous conditions (liquid shake culture).

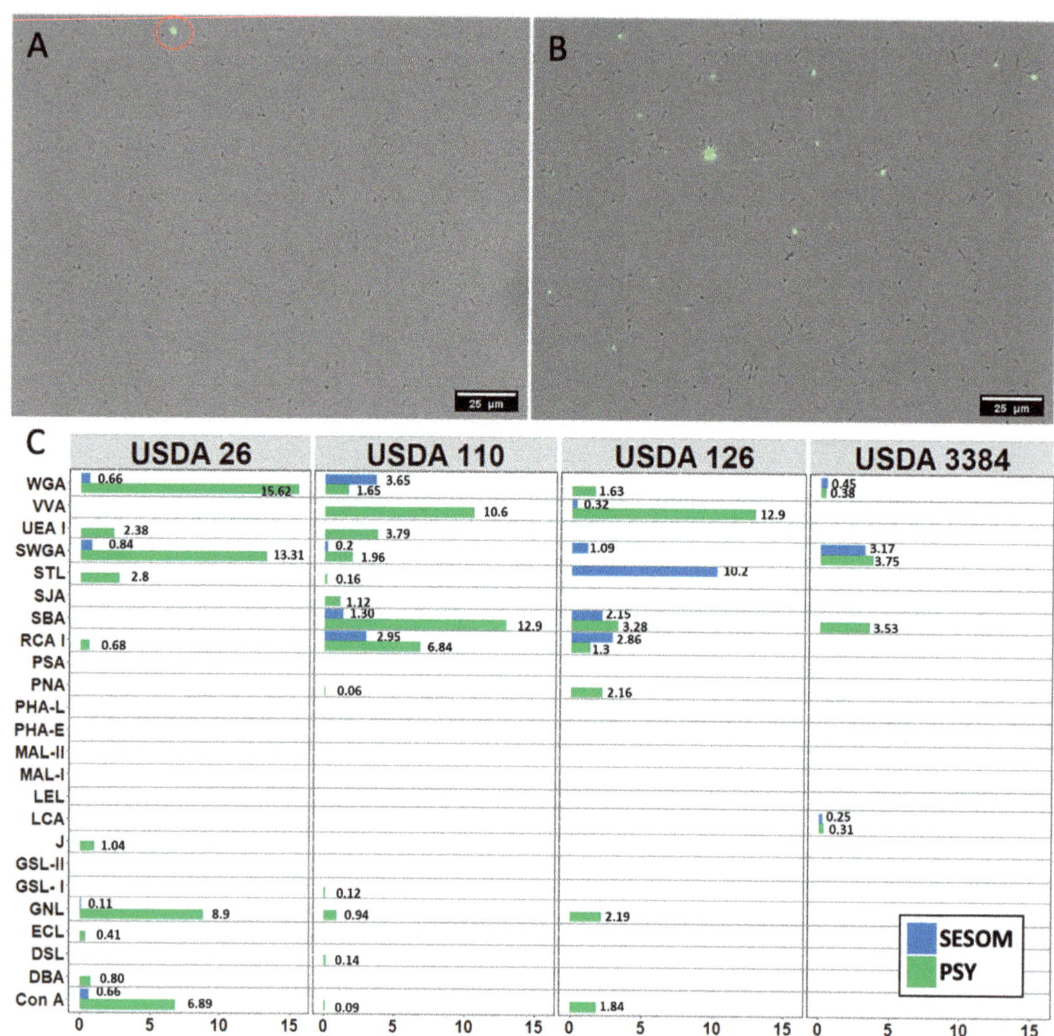

Figure 4. Lectin binds only to a subset of cells in the population. Binding of sWGA to *B diazoefficiens* USDA 26 grown in SESOM (**A**) and PSY (**B**), and the percentage of cells binding to the 25 lectins after culturing in PSY and SESOM (**C**).

3.3. SESOM-Grown Populations Form No Biofilm but Display Increased Adherence to Soy Roots

Bradyrhizobium growing in SESOM displayed altered attachment-associated traits, so we quantified in vitro biofilm formation using the standard microtitre plate biofilm assay. PSY-grown USDA 110 and 26 formed biofilms on polystyrene within 24 h (Figure 5A), while USDA 126 and 3384 did not, even when incubated for 96 h (data not shown). Intriguingly, no strain formed biofilm in SESOM, with cells remaining in the planktonic phase. These results indicate that *Bradyrhizobium* growing in soil would not attach and form biofilms on more hydrophilic surfaces such as the Nunclon Delta Surface Polystyrene used in these experiments.

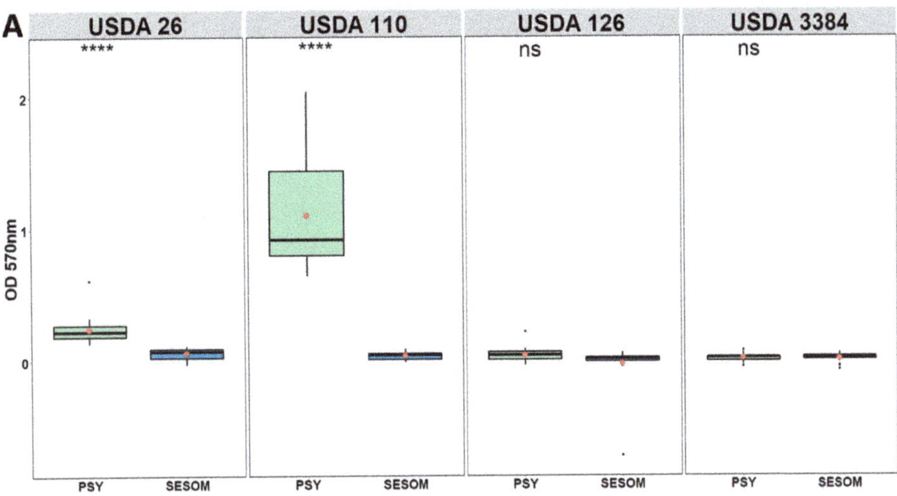

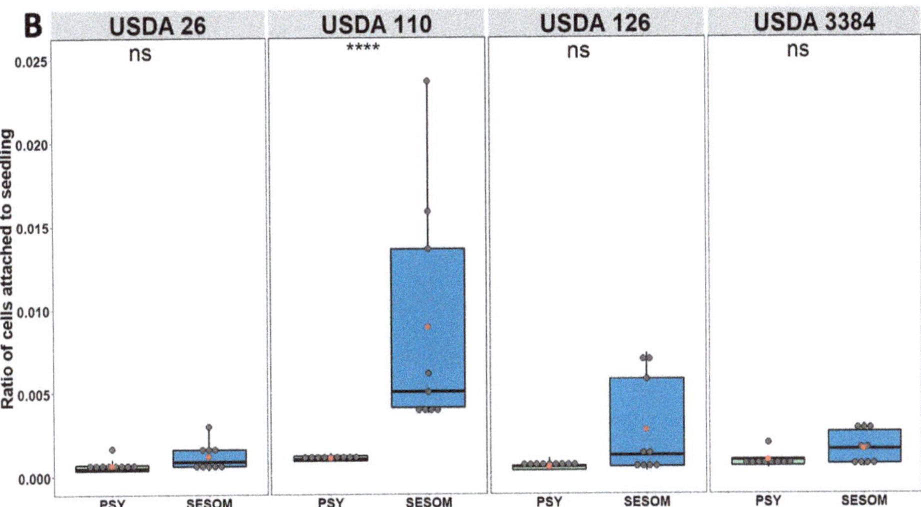

Figure 5. Biofilm (**A**) and soy root adherence (**B**) of *Bradyrhizobium* strains. Exponential phase *Bradyrhizobium* cultured in either PSY-Arabinose or SESOM were inoculated into 96-well polystyrene plates and biofilm formed after 24 h quantified by crystal violet staining (**A**), or exposed to 9 d old roots for 60 min (**B**). Data were analyzed by Student *t*-test, and *p* values indicate statistical significance (ns not significant, **** $p < 0.0001$).

To gain insight into association with young soy roots, *Bradyrhizobium* was grown in PSY or SESOM, suspensions of exponentially growing cells were exposed to 9-day-old gnotobiotic seedlings for 60 min, and the adhering cells were counted. Adherence of all four PSY-cultured populations was scant (Figure 5B). In contrast, SESOM-grown USDA 110 displayed significantly more root adhesion. The adhesion data obtained for SESOM-grown cultures were more varied than for ex PSY. While root adhesion of strains 26, 126 and 3384 was not found statistically significant, it was higher than in PSY in multiple cases (Figure 5B). This greater variability in observed phenotype among individual experiments may be due to either variability among young roots, or due to variation in *Bradyrhizobium* population phenotype, as is also observed in lectin binding experiments.

4. Discussion

Our data reveal that *Bradyrhizobium* adopts adhesion-related surface properties when grown in the liquid soil extract SESOM. We observed the effect of soil nutrient environment on the surface hydrophobicity, motility, lectin binding spectrum and adherence capacity of the cells to soy roots. SESOM-grown populations of strains 110 and 3384 were more hydrophobic in nature, whereas there was no significant difference with strains 26 and 126. An effect of growth media on the surface hydrophobicity of bacterial cells has been reported for other bacterial species [25–27]. The hydrophobic properties of both the root surface and the bacteria involved in symbiosis contributes to one of the strongest forces in the initial root association. Bacterial cell surface hydrophobicity has been correlated directly to root attachment and competitive nodule forming ability [6,28]. Structural changes brought in LPS increase the cell surface hydrophobicity, which likely is significantly involved in the endocytotic infection process [7]. It is likely that the soil nutrients make 110 and 3384 more hydrophobic, and hence increase their chance of attachment by supporting them in their competitiveness to other bacteria. This was observed in the root adherence assay with strain 110 but not 3384.

The motility of rhizobia in response to chemical signaling plays a critical role in the initial contact, attachment and colonization of the root hair [29]. Mutants of *Rhizobium meliloti LS-30* with defective motility took 5–20 times more time to adsorb to the root cell, and had delayed nodule formation, making them less competitive than other bacteria [11]. In this study, SESOM-grown populations were observed to be less motile. This could be because *B. japonicum* grown in L-arabinose as sugar source expresses lateral flagella along with subpolar flagella, making them move faster than those grown in other sugars [17]. Our lectin data indicated reduced surface sugar diversity when growing in SESOM. Alterations of the surface sugar moieties and their effect on symbiotic capability have been observed in *Bradyrhizobium* [9,30–32]. Surface sugar diversity has been reported to vary in composition when grown on different carbon sources [33,34]. These data suggest that the carbon sources in SESOM lead *Bradyrhizobium* to produce a different combination of extracellular polysaccharides and other sugar-containing polymers. Specifically, SESOM-grown populations appear to produce a smaller number of such extracellular polymers.

The contrast between biofilm formation on polystyrene and adherence to soy roots was unexpected, especially in the case of strain 110. While SESOM-grown populations failed to form biofilm on polystyrene, they were more prone to adhere to the soy roots. Arabinose as carbon source yielded opposite outcome. Medium characteristics such as osmolarity, nutrients, and factors derived from the biotic environment, may also affect attachment and/or biofilm formation. Little is known about the distribution of surface properties on young soybean roots, but the SESOM-grown, more hydrophobic strain 110 attached to the roots. In contrast it did not form detectable biofilm on the hydrophilic-modified Nunclon Delta Surface plates used. As confirmation, the hydrophilic ex PSY populations attached to the hydrophilic polystyrene, but not the soy roots. Biofilm formation of *Rhizobium* is affected by growth medium and incubation conditions [35]. Soybean agglutinin (SBA) lectin is thought to contribute to root-surface adherence [36]. While strains 110 and 126 displayed SBA binding under both culture condition (Figure 3), root adherence was significantly different. This suggest that root adherence may be driven largely by other factors.

The binding of specific lectins to only some cells in the population was surprising. Selective binding was not due to a shortage of lectin concentration as confirmed by experiments with increased concentration. The purity of the culture had also been verified regularly, so selective lectin binding pointed to differentiation in surface-exposed sugar polymers. This apparent differentiation of surface properties of cells in liquid shake cultures was unexpected. Our methodology included the removal of residual sugars from the culture medium by washing cells, and a wash step to remove unbound lectins after binding. These wash steps could have contributed to a loss of EPS into the liquid phase, leading to less or no bound lectin retained on cells. A similar observation of differential polysaccharide positioning has also been reported in *Agrobacterium tumefaciens* and

Rhizobium leguminosarum [37,38]. While phenotypic population differentiation has been studied extensively in *Bacillus subtilis* [39,40], it is better known from environments allowing niche-specific physicochemical conditions such as in biofilms. The ecophysiological role of phenotypic differentiation of *Bradyrhizobia* presents itself as a new field of study.

From the results, it is clear that growth in a soil nutrient environment does affect the surface properties and adherence capacity of *Bradyrhizobium*. This is true especially for USDA 110, which is an agriculturally important rhizobial bacteria for soybean nitrogen fixation, where SESOM increased hydrophobicity and root adherence capability significantly [41]. The observed differences in the surface and phenotypic properties of the soil nutrient grown population leads to the conclusion that *Bradyrhizobium* adjust with the chemical cues that are present around them. The results from this study emphasize the need to further explore the effect of (secreted) root exudates on the surface and phenotypic behavior of *Bradyrhizobium* cells and resulting root adherence.

Author Contributions: Conceptualization, S.S. and V.S.B.; methodology, A.K.S. and V.S.B.; formal analysis, A.K.S.; investigation, A.K.S.; resources, S.S.; data curation, A.K.S.; writing—original draft preparation, A.K.S.; writing—review and editing, A.K.S., S.S. and V.S.B.; visualization, A.K.S.; supervision, V.S.B.; project administration, S.S; funding acquisition, S.S. All authors have read and agreed to the published version of the manuscript.

Funding: This material is based upon work supported by the National Science Foundation/EPSCoR RII Track-1: Building on The 2020 Vision: Expanding Research, Education and Innovation in South Dakota, Award OIA-1849206 and by the South Dakota Board of Regents.

Institutional Review Board Statement: Not applicable.

Informed Consent Statement: Not applicable.

Acknowledgments: We thank the NRRL Culture Collection of the Agricultural Research Service, United State Department of Agriculture for providing the bacterial cultures used in this study. We acknowledge use of the SDSU-FGCF supported in part by NSF/EPSCoR Grant No. 0091948 and by the State of South Dakota. We thank K M Taufiqur Rahman for his help with image analysis.

Conflicts of Interest: The authors declare no conflict of interest. The funders had no role in the design of the study; in the collection, analyses, or interpretation of data; in the writing of the manuscript, or in the decision to publish the results.

Disclaimer: Any opinions: findings, and conclusions or recommendations expressed in this material are those of the author(s) and do not necessarily reflect the views of the National Science Foundation.

References

1. Peix, A.; Ramirez-Bahena, M.H.; Velazquez, E.; Bedmar, E.J. Bacterial Associations with Legumes. *Crit. Rev. Plant Sci.* **2015**, *34*, 17–42. [CrossRef]
2. Kaneko, T.; Nakamura, Y.; Sato, S.; Minamisawa, K.; Uchiumi, T.; Sasamoto, S.; Watanabe, A.; Idesawa, K.; Iriguchi, M.; Kawashima, K.; et al. Complete genomic sequence of nitrogen-fixing symbiotic bacterium *Bradyrhizobium japonicum* USDA110. *DNA Res.* **2002**, *9*, 189–197. [CrossRef]
3. Wang, Q.; Liu, J.E.; Zhu, H.Y. Genetic and Molecular Mechanisms Underlying Symbiotic Specificity in Legume-*Rhizobium* Interactions. *Front. Plant Sci.* **2018**, *9*, 313. [CrossRef]
4. Liu, C.W.; Murray, J.D. The Role of Flavonoids in Nodulation Host-Range Specificity: An Update. *Plants* **2016**, *5*, 33. [CrossRef] [PubMed]
5. Broghammer, A.; Krusell, L.; Blaise, M.; Sauer, J.; Sullivan, J.T.; Maolanon, N.; Vinther, M.; Lorentzen, A.; Madsen, E.B.; Jensen, K.J.; et al. Legume receptors perceive the rhizobial lipochitin oligosaccharide signal molecules by direct binding. *Proc. Natl. Acad. Sci. USA* **2012**, *109*, 13859–13864. [CrossRef] [PubMed]
6. Ozawa, T.; Ogata, H.; Doi, R.; Komai, Y. Isolation of transposon Tn5-induced hydrophobic mutants of a *Bradyrhizohium japonicum* strain with improved competitive nodulation abilities. *Soil Sci. Plant Nutr.* **1992**, *38*, 545–552. [CrossRef]
7. Kannenberg, E.L.; Carlson, R.W. Lipid A and O-chain modifications cause *Rhizobium* lipopolysaccharides to become hydrophobic during bacteroid development. *Mol. Microbiol.* **2001**, *39*, 379–391. [CrossRef]
8. Park, K.M.; So, J.S. Altered cell surface hydrophobicity of lipopolysaccharide-deficient mutant of *Bradyrhizobium japonicum*. *J. Microbiol. Meth.* **2000**, *41*, 219–226. [CrossRef]
9. Karr, D.B.; Liang, R.T.; Reuhs, B.L.; Emerich, D.W. Altered exopolysaccharides of *Bradyrhizobium japonicum* mutants correlate with impaired soybean lectin binding, but not with effective nodule formation. *Planta* **2000**, *211*, 218–226. [CrossRef]

10. Quelas, J.I.; Mongiardini, E.J.; Casabuono, A.; Lopez-Garcia, S.L.; Althabegoiti, M.J.; Covelli, J.M.; Perez-Gimenez, J.; Couto, A.; Lodeiro, A.R. Lack of Galactose or Galacturonic Acid in *Bradyrhizobium japonicum* USDA 110 Exopolysaccharide Leads to Different Symbiotic Responses in Soybean. *Mol. Plant Microbe Interact.* **2010**, *23*, 1592–1604. [CrossRef]

11. Caetanoanolles, G.; Wall, L.G.; Demicheli, A.T.; Macchi, E.M.; Bauer, W.D.; Favelukes, G. Role of Motility and Chemotaxis in Efficiency of Nodulation by *Rhizobium meliloti*. *Plant Physiol.* **1988**, *86*, 1228–1235. [CrossRef]

12. Mitchell, J.G.; Kogure, K. Bacterial motility: Links to the environment and a driving force for microbial physics. *FEMS Microbiol. Ecol.* **2006**, *55*, 3–16. [CrossRef]

13. Loderio, A.R.; Lopez-Garcia, S.L.; Vazquez, T.E.E.; Favelukes, G. Stimulation of adhesiveness, infectivity, and competitiveness for nodulation of *Bradyrhizobium japonicum* by its pretreatment with soybean seed lectin. *FEMS Microbiol. Lett.* **2000**, *191*, 157–158.

14. De Hoff, P.L.; Brill, L.M.; Hirsch, A.M. Plant lectins: The ties that bind in root symbiosis and plant defense. *Mol. Genet. Genom.* **2009**, *282*, 1–15. [CrossRef] [PubMed]

15. Ambrosi, M.; Cameron, N.R.; Davis, B.G. Lectins: Tools for the molecular understanding of the glycocode. *Org. Biomol. Chem.* **2005**, *3*, 1593–1608. [CrossRef] [PubMed]

16. Smit, G.; Tubbing, D.M.J.; Kijne, J.W.; Lugtenberg, B.J.J. Role of Ca^{2+} in the Activity of Rhicadhesin from *Rhizobium-Leguminosarum* Biovar Viciae, Which Mediates the 1st Step in Attachment of Rhizobiaceae Cells to Plant-Root Hair Tips. *Arch. Microbiol.* **1991**, *155*, 278–283. [CrossRef]

17. Covelli, J.M.; Althabegoiti, M.J.; Lopez, M.F.; Lodeiro, A.R. Swarming motility in *Bradyrhizobium japonicum*. *Res. Microbiol.* **2013**, *164*, 136–144. [CrossRef]

18. Vilain, S.; Luo, Y.; Hildreth, M.B.; Brozel, V.S. Analysis of the life cycle of the soil saprophyte *Bacillus cereus* in liquid soil extract and in soil. *Appl. Environ. Microbiol.* **2006**, *72*, 4970–4977. [CrossRef]

19. Mesa, S.; Hauser, F.; Friberg, M.; Malaguti, E.; Fischer, H.M.; Hennecke, H. Comprehensive assessment of the Regulons controlled by the FixLJ-FixK(2)-FixK(1) cascade in *Bradyrhizobium japonicum*. *J. Bacteriol.* **2008**, *190*, 6568–6579.

20. Rosenberg, M. Microbial adhesion to hydrocarbons: Twenty-five years of doing MATH. *FEMS Microbiol. Lett.* **2006**, *262*, 129–134. [CrossRef]

21. Zoueki, C.W.; Tufenkji, N.; Ghoshal, S. A modified microbial adhesion to hydrocarbons assay to account for the presence of hydrocarbon droplets. *J. Colloid Interface Sci.* **2010**, *344*, 492–496. [CrossRef]

22. O'Toole, G.A. Microtiter Dish Biofilm Formation Assay. *J. Vis. Exp. JoVE* **2011**, *47*, 2473. [CrossRef]

23. O'Toole, G.A.; Kolter, R. Initiation of biofilm formation in *Pseudomonas fluorescens* WCS365 proceeds via multiple, convergent signalling pathways: A genetic analysis. *Mol. Microbiol.* **1998**, *28*, 449–461. [CrossRef] [PubMed]

24. White, L.J.; Brozel, V.S.; Subramanian, S. Isolation of Rhizosphere Bacterial Communities from Soil. *Bio-protocol* **2015**, *5*, e1569. [CrossRef]

25. Ngwai, Y.B.; Sabiya, G. Cultivation in different growth media affects the expression of the cell surface hydrophobicity of bacteria. *Cameroon J. Exp. Biol.* **2007**, *3*, 26–29.

26. Nwanyanwu, C.E.; Abu, G. Influence of growth media on hydrophobicity of phenol-utilizing bacteria found in petroleum refinery effluent. *Int. Res. J. Biol. Sci.* **2013**, *2*, 6–11.

27. Deptula, A.; Mikucka, A.; Gospodarek, E. Effect of growth conditions on cell surface hydrophobicity of multiresistant *Pseudomonas aeruginosa* strains. *Med. Dosw. I Mikrobiol.* **2004**, *56*, 359–364.

28. Ozawa, T.; Tokuda, S.; Komai, Y. Correlation between Competitive Nodulation Ability and Cell Surface Hydrophobicity of *Bradyrhizobium japonicum*. *Bull. Jpn. Soc. Microb. Ecol.* **1991**, *6*, 19–23. [CrossRef]

29. Tambalo, D.D.; Yost, C.K.; Hynes, M.F. Motility and chemotaxis in the rhizobia. *Biol. Nitrogen Fixat.* **2015**, *1*, 337–348.

30. Noh, J.G.; Jeon, H.E.; So, J.S.; Chang, W.S. Effects of the *Bradyrhizobium japonicum waaL* (*rfaL*) Gene on Hydrophobicity, Motility, Stress Tolerance, and Symbiotic Relationship with Soybeans. *Int. J. Mol. Sci.* **2015**, *16*, 16778–16791. [CrossRef]

31. Lee, H.I.; In, Y.H.; Jeong, S.Y.; Jeon, J.M.; Noh, J.; So, J.S.; Chang, W.S. Inactivation of the *lpcC* gene alters surface-related properties and symbiotic capability of *Bradyrhizobium japonicum*. *Lett. Appl. Microbiol.* **2014**, *59*, 9–16. [CrossRef]

32. Chang, W.S.; Park, K.M.; Koh, S.C.; So, J.S. Characterization of the *Bradyrhizobium japonicum galE* gene: Its impact on lipopolysaccharide profile and nodulation of soybean. *FEMS Microbiol. Lett.* **2008**, *280*, 242–249. [CrossRef]

33. Staudt, A.K.; Wolfe, L.G.; Shrout, J.D. Variations in exopolysaccharide production by *Rhizobium tropici*. *Arch. Microbiol.* **2012**, *194*, 197–206. [CrossRef] [PubMed]

34. Janczarek, M. Environmental Signals and Regulatory Pathways That Influence Exopolysaccharide Production in Rhizobia. *Int. J. Mol. Sci.* **2011**, *12*, 7898–7933. [CrossRef] [PubMed]

35. Rinaudi, L.; Fujishige, N.A.; Hirsch, A.M.; Banchio, E.; Zorreguieta, A.; Giordano, W. Effects of nutritional and environmental conditions on *Sinorhizobium meliloti* biofilm formation. *Res. Microbiol.* **2006**, *157*, 867–875. [CrossRef] [PubMed]

36. Pérez-Giménez, J.; Mongiardini, E.J.; Althabegoiti, M.J.; Covelli, J.; Quelas, J.I.; López-García, S.L.; Lodeiro, A.R. Soybean Lectin Enhances Biofilm Formation by *Bradyrhizobium japonicum* in the Absence of Plants. *Int. J. Microbiol.* **2009**, *2009*, 8. [CrossRef]

37. Xu, J.; Kim, J.; Koestler, B.J.; Choi, J.H.; Waters, C.M.; Fuqua, C. Genetic analysis of *Agrobacterium tumefaciens* unipolar polysaccharide production reveals complex integrated control of the motile-to-sessile switch. *Mol. Microbiol.* **2013**, *89*, 929–948. [CrossRef]

38. Laus, M.C.; Logman, T.J.; Lamers, G.E.; Van Brussel, A.A.N.; Carlson, R.W.; Kijne, J.W. A novel polar surface polysaccharide from *Rhizobium leguminosarum* binds host plant lectin. *Mol. Microbiol.* **2006**, *59*, 1704–1713. [CrossRef]

39. Cao, H.J.; Kuipers, O.P. Influence of global gene regulatory networks on single cell heterogeneity of green fluorescent protein production in *Bacillus subtilis*. *Microb. Cell Factories* **2018**, *17*, 134. [CrossRef]
40. Gallegos-Monterrosa, R.; Christensen, M.N.; Barchewitz, T.; Koppenhofer, S.; Priyadarshini, B.; Balint, B.; Maroti, G.; Kempen, P.J.; Dragos, A.; Kovacs, A.T. Impact of Rap-Phr system abundance on adaptation of *Bacillus subtilis*. *Commun. Biol.* **2021**, *4*, 468. [CrossRef]
41. Zhang, L.; Liu, J.Y.; Gu, H.; Du, Y.F.; Zuo, J.F.; Zhang, Z.B.; Zhang, M.L.; Li, P.; Dunwell, J.M.; Cao, Y.R.; et al. *Bradyrhizobium diazoefficiens* USDA 110-Glycine max Interactome Provides Candidate Proteins Associated with Symbiosis. *J. Proteome Res.* **2018**, *17*, 3061–3074. [CrossRef] [PubMed]

MDPI
St. Alban-Anlage 66
4052 Basel
Switzerland
Tel. +41 61 683 77 34
Fax +41 61 302 89 18
www.mdpi.com

Nitrogen Editorial Office
E-mail: nitrogen@mdpi.com
www.mdpi.com/journal/nitrogen